W9-CRB-594

FUTURE GENERATION PHOTOVOLTAIC TECHNOLOGIES

FUTURE GENERATION PHOTOVOLTAIC TECHNOLOGIES

First NREL Conference

Denver, CO March 1997

EDITOR
Robert D. McConnell
National Renewable Energy Laboratory

AIP CONFERENCE
PROCEEDINGS 404

American Institute of Physics

Woodbury, New York

SEP/ru
PHYS

Authorization to photocopy items for internal or personal use, beyond the free copying permitted under the 1978 U.S. Copyright Law (see statement below), is granted by the American Institute of Physics for users registered with the Copyright Clearance Center (CCC) Transactional Reporting Service, provided that the base fee of $10.00 per copy is paid directly to CCC, 222 Rosewood Drive, Danvers, MA 01923. For those organizations that have been granted a photocopy license by CCC, a separate system of payment has been arranged. The fee code for users of the Transactional Reporting Service is: 1-56396-704-9/ 97 /$10.00.

© 1997 American Institute of Physics

Individual readers of this volume and nonprofit libraries, acting for them, are permitted to make fair use of the material in it, such as copying an article for use in teaching or research. Permission is granted to quote from this volume in scientific work with the customary acknowledgment of the source. To reprint a figure, table, or other excerpt requires the consent of one of the original authors and notification to AIP. Republication or systematic or multiple reproduction of any material in this volume is permitted only under license from AIP. Address inquiries to Office of Rights and Permissions, 500 Sunnyside Boulevard, Woodbury, NY 11797-2999; phone: 516-576-2268; fax: 516-576-2499; e-mail: rights@aip.org.

L.C. Catalog Card No. 97-74386
ISBN 1-56396-704-9
ISSN 0094-243X
DOE CONF- 970396

Printed in the United States of America

CONTENTS

TK2960
F84
1994
PHYS

CRITERIA FOR NEW PV CONCEPTS

ROLE OF GOVERNMENT, UNIVERSITIES, AND INDUSTRIES
IN DEVELOPING AND FUNDING NEW IDEAS

DYE-SENSITIZED PHOTOCHEMICAL SOLAR CELLS

NEW MATERIALS AND DEVICE ARCHITECTURES FOR PV

POSTER SESSION

PROGRAM COMMITTEE

Robert D. McConnell, Chairman, National Renewable Energy Laboratory
Harry Atwater, California Institute of Technology
David Carlson, Solarex
George Collins, Colorado State University
Lionel ("Kim") Kimerling, Massachusetts Institute of Technology
Michael Mauk, AstroPower, Inc.
Arthur Nozik, National Renewable Energy Laboratory
Peter Searson, Johns Hopkins University
P. Craig Taylor, University of Utah
David Ginley, Awards Chairman, National Renewable Energy Laboratory

CONFERENCE SPONSORS

U. S. Department of Energy
National Renewable Energy Laboratory

CONFERENCE COORDINATOR

Joan Ross, National Renewable Energy Laboratory

PROCEEDINGS

Robert D. McConnell, National Renewable Energy Laboratory

PREFACE

The National Renewable Energy Laboratory hosted 120 researchers from nine countries at the First Conference on Future Generation Photovoltaic Technologies, held March 24–26, 1997, in Denver, Colorado. The emphasis was on technical discussion and assessment of recent achievements of photovoltaic (PV) technologies that have a long-term potential to simultaneously achieve high performance and very low cost. For this conference, "future generation photovoltaic technologies" were those not in production and not expected to enter production soon because additional long-term research, development, and innovation are needed.

The theme for this conference was a leaping frog, which represented a technology that might make dramatic progress even in the short term. Consistent with this theme were "Calaveras Awards for Leapfrog Technologies" given for the most innovative presentation, the technology most likely to succeed, the best oral presentation, the best poster presentation, and the most fun presentation. The winners' certificates follow their manuscripts in these proceedings.

The 38 oral presentations in six sessions covered criteria for new PV concepts; the roles of government, industry, and universities in developing new concepts; dye-sensitized photochemical cells; single-crystal-like films on low-cost substrates; other innovative concepts; and new materials and device architectures. In addition, there were 12 poster presentations. Only a handful of these presentations are not represented in these proceedings.

One highlight of the conference was the report of over 10% conversion efficiency and the discovery of a new dye for dye-sensitized photochemical cells. Other interesting discussions included the fabrication of low-cost substrates that might be suitable for epitaxial growth of crystalline silicon or gallium arsenide, new thermophotovoltaic concepts, innovative concentrator approaches, solar cells containing nanocrystalline particles, organic semiconductor solar cells, direct absorption of the electromagnetic visible light waves followed by rectification, flywheel energy storage, new PV materials, and a fresh look at some old PV concepts. Indeed, many of these presentations referenced a long history for their ideas. What has changed, perhaps, is the availability of new technologies and discoveries that might make some of these concepts realistic. A business reality check for all of these concepts was provided by Kyle Lefkoff of Boulder Ventures, who gave an excellent banquet presentation on investment issues for future-generation PV technologies.

A fundamental principle guiding the U.S. Department of Energy's National Photovoltaics Program has always been to provide some support for high-risk, high-payoff PV technologies requiring long-term research, development, and innovation. The Program Committee thanks the National Renewable Energy Laboratory and the U.S. Department of Energy for their support of this conference. Joan Ross from the National Renewable Energy Laboratory skillfully coordinated preparations for the conference. She and the members of her group ensured a smooth flow of the scheduled activities and resolved individual attendees' problems. My personal thanks go to my wife, Suzie Star McConnell, for suggesting and hosting the conference casino party, a no-risk gaming event with lots of frog prizes.

I also wish to thank the program committee who identified so many excellent researchers to participate. Finally, I thank the many contributors who took the time and effort to prepare high-quality presentations for the conference and for these proceedings.

Robert D. McConnell
Program Committee Chairman

CRITERIA FOR NEW PV CONCEPTS

Materials Availability for Thin Film Solar Cells

Yunosuke Makita

Electrotechnical Laboratory
Tsukuba, Ibaraki 305, Japan

Abstract. Materials availability is one of the most important factors when we consider the mass-production of next generation photovoltaic devices. "In (indium)" is a vital element to produce high efficient thin film solar cells such as InP and $CuIn(Ga)Se_2$ but its lifetime as a natural resource is suggested to be of order of $10 \sim 15$ years. The lifetime of a specific natural resource as an element to produce useful device substances is directly related with its abundance in the earth's crust, consumption rate and recycling rate (if recycling is economically meaningful). The chemical elements having long lifetime as a natural resource are those existing in the atmosphere such as N (nitrogen) and O (oxygen); the rich elements in the earth's crust such as Si, Ca, Sr and Ba; the mass-used metals such as Fe (iron), Al (aluminum) and Cu (copper) that reached the stage of large-scale recycling. We here propose a new paradigm of semiconductor material-science for the future generation thin film solar cells in which only abundant chemical elements are used. It is important to remark that these abundant chemical elements are normally not toxic and are fairly friendly to the environment. β-$FeSi_2$ is composed of two most abundant and nontoxic chemical elements. This material is one of the most promising device materials for future generation energy devices (solar cells and thermoelectric device that is most efficient at temperature range of 700 - 900 ℃). One should remind of the versatility of β-$FeSi_2$ that this material can be used not only as energy devices but also as photodetector, light emitting diode and/or laser diode at the wavelength of 1.5μm that can be monolithically integrated on Si substrates due to the relatively small lattice mismatch.

INTRODUCTION

In order to realize high photovoltaic effect better than poly-crystalline- and crystalline- Si, a variety of chemical elements such as In, Ga, Al, As, P, Te, Se and Cd are used to fabricate III-V (InP, InGaP, GaAs and AlGaAs), II-VI (CdTe) and chalcopyrite ($CuInSe_2$) compound-semiconductor solar cells. We can readily notice that the majority of these chemical elements hold an extremely short lifetime as a natural resource to match the demand of the mass-production

CP404, *Future Generation Photovoltaic Technologies: First NREL Conference*, edited by McConnell
© 1997 The American Institute of Physics 1-56396-704-9/97/$10.00

as low-cost solar cells. It is therefore a matter of course that materials availability is one of the key factors when we consider paractically the future generation photovoltaic devices. Materials availability is directly related with the lifetime of the chemical element as a natural resource which is governed by its remaining amount as a reserve and its consumption rate. For some specific chemical elements, one should also take into consideration the recycling rate if the recycling has economically (and sometimes environmentally) enough meaning. The abundance of a chemical element in the earth's crust is rationally related with its abundance in the space outside the earth. He (helium) atom is known to be a fundamental substance to produce various atoms in the space. Since the atomic mass of He is four, an atom holding a mass with multiple of four is abundant in the space than the remaining atoms. The probability to find these atoms (O $(16=4x4)$, Ne $(20=4x5)$, ---, Ca $(40=4x10)$, ----, Fe$(56=4x14)$, ---) is therefore specifically high in the space (and nearly necessarily in the earth's crust).

The existence ratio of chemical elements in the earth's crust is demonstrated in Fig.1-a) and -b) for the chemical elements lighter and heavier than Fe (iron), respectively [1]. After the creation of the earth with the elapse of time, chemical elements with high specific gravity concentrated gradually into the magma and accordingly chemical elements with relatively low specific gravity gathered at the surface of the earth, the earth's crust. The majority of chemical elements in the form of gas that do not react with other metal elements return into the space beyond the earth's gravity or stay above the topmost surface of the earth eventually to shape the atmosphere. These physical and chemical processes brought about a modified existence ratio of chemical elements in the earth's crust that is fairly different from that in the space. These situations are faithfully reflected in Fig.1 in which one can easily recognize the abundance of O, Si, Al, Fe, Ca and other alkaline and alkaline-earth elements. The elements having long lifetime as natural resources are accordingly i) those existing in the atmosphere such as N (nitrogen) and O (oxygen), ii) the rich substances in the earth's crust such as Si, groups IA alkaline- (Na and K) and IIA alkaline-earth (Sr and Ba) elements, iii) C (carbon) existing as CO_2 in the atmosphere and as an essential constituent element in fossil fuel and as a principal element of organic substances, iv) Fe (iron), Al (aluminum) and Cu (copper) that reach the evolution-stage of large-scale recycling.

In order to obtain the practical lifetime of the chemical elements necessary for the fabrication of currently-fabricated high efficient solar cells, the reserves and the refinery production of individual element are shown in Fig. 2 [1]. One notes that the reserves of the frequently used group II, III and V elements such as Cd, In and As are extremely small compared with Si, Al, Fe and Ca.

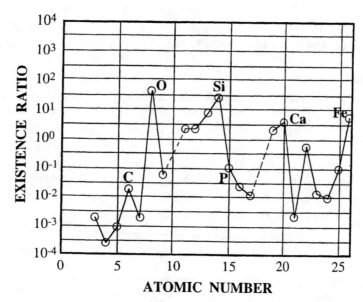

FIGURE 1-a) The existence ratio of chemical elements lighter than Fe (iron) in the earth's crust

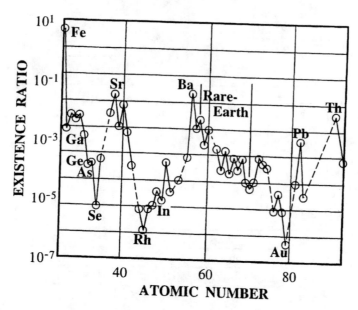

FIGURE 1-b) The existence ratio of chemical elements heavier than Fe (iron) in the earth's crust

From the figure we can readily remark that the estimated lifetime of the two important substances, In (indium) and As (arsenic) for high efficient compound-semiconductor solar cells (InP, CuInSe$_2$ and GaAs), is fairly short of the order of 15 and 50 years, respectively. Recently, the consumption rate of In has been significantly grown to produce transparent electrodes (InSnO$_x$) for LCR (liquid-crystal) panel display and solar-cell windows. This trend is at present steeply accelerated by the fast expanding production of personal computers.

The lifetime of group II and IV chemical elements are also quite short which are in a great extent used for the fabrication of II-VI semiconductor solar cells or window layers for III-V or CuInSe$_2$ solar cells. Particularly the lifetime of Cd (cadmium) and S (sulfur) that are quite common host materials for high efficient CdTe solar cell and window (CdS) layers, is short and is estimated to be of the order of 20 and 30 years, respectively.

TABLE 1   Resources' lifetime and possible energy generation

Resource	Estimated reserves (R)	Production rate per year(P)	Lifetime(R/P) in Year
In	870	62.2	14
Ga	110,000	16.2	6,780
As	708,000	24,000	47
Cd	331,400	17,200	19
Te	20,500	110	185
Se	64,000	1,440	45
Ge	3,300	85	38

We here propose a new paradigm of semiconductor materials-science for the future generation thin film solar cells by using only the abundant elements as natural resources. These elements should be those which are not appearing in Fig.2 as poor chemical elements in natural resources or those which reaches the stage of large-scale and high cost-performance recycling. The abundant chemical elements which at least for 100 - 200 years will not face the shortage of supply are Si, Fe, Ca (calcium) and Ba (barium) together with other ubiquitously existing chemical elements to compose animals and plants texture such as H (hydrogen), C, N and O. Cu, Fe and Al already reached the stage of high efficient recycling, while other metals such as In are recycled only by 5 % annually.

Since the wide occurrence of specific chemical elements in the earth's crust automatically means that the abundant materials have long physically contacted with biological organs, these materials generally give physiologically no harmful effects to biological substances and are normally nontoxic and friendly to the environment.

Using the above abundant chemical elements, for the moment two semiconductor materials are considered to be useful as solar cell materials, β-$FeSi_2$ and Ca_2Si. Fe-Si system has been long investigated not as a highly sophisticated semiconductor material but as a simply sintered material for the fabrication of thermoelectric device. As for Ca_2Si, little is known for its physical properties and optical band-gap is reported to be 1.9eV [3].

β-$FeSi_2$ AS AN ENERGY-DEVICE MATERIAL

Fe and Si form seven compounds and cubic ε-FeSi (an indirect band-gap of 0.1 eV) and orthorhombic β-$FeSi_2$ present semiconductor features [4,5]. β-$FeSi_2$ is stable below 937°C. It was theoretically predicted and experimentally demonstrated that β-$FeSi_2$ has a direct band-gap of E_g=0.84 - 0.88 eV [6]. It is empirically well known that a compound semiconductor consisting of atom(s) having d-electron(s) in most outer-orbit normally exhibits high optical absorption coefficient. One of the examples is $CuInSe_2$ in which Cu holds ten 3d electrons in the most outer orbit. Since β-$FeSi_2$ is composed of Fe, holding d-electrons, β-$FeSi_2$ ought to indicate a high optical absorption coefficient. Theoretical calculation yielded a conversion efficiency of 23 % [2]. We here show that β-$FeSi_2$ has practically a high optical absorption coefficient ($\sim$1x10^5 cm^{-1}) [7, 8]. These features are particularly advantageous to produce low-cost β-$FeSi_2$ solar cells since high optical absorption coefficient leads to the reduced film thickness and accordingly to the minimized required energy to fabricate films. β-$FeSi_2$ has therefore the high potential for use in solar cells [2, 9] and infrared detectors [10].

The above optical features of β-FeSi$_2$ and its relatively small lattice-mismatch (~ 2%) to Si (100)-oriented substrate are preferable conditions for the fabrication of laser or light-emitting diode that can be integrated on Si. β-FeSi$_2$ has also high potential as an thermoelectric power device (figure of merit : ~10^{-4} K^{-1}) and it presents the highest Seebeck coefficient in the temperature range of 700-900℃ [10]. Additional important features pertinent to β-FeSi$_2$ are its high resistance to oxidation and chemical stability owing to the high decomposing temperature compared with ordinary II-VI and III-V compound-semiconductor solar cells.

β-FeSi$_2$ thin films and bulk crystals have been fabricated by many groups using various techniques. Here we present the fabrication of β-FeSi$_2$ bulk crystals by horizontal gradient freeze method (HGF) [12] and β-FeSi$_2$ films by ion beam synthesis (IBS) method using high-energy ion-implantation [13-16].

Samples with Fe:Si ratio of 1:2 were synthesized and were melted between 1300℃ and 1500℃. After cooling, the samples were kept at 800℃ and 900℃ during 66 to 100 hours, leading to transformation from α to β phase. Neutron diffraction measurements were carried out for annealed FeSi$_2$ crystals using high resolution powder diffractometer. The diffraction data and the differential signals obtained by comparing experimental data and theoretical

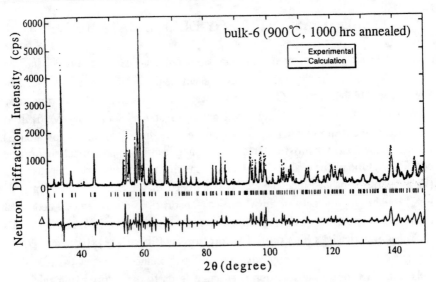

FIGURE 2   Neutron diffraction pattern of bulk β-FeSi$_2$
crystal prepared by HGF method

simulation (RITAN) are shown in Fig.2. The results apparently indicate that our FeSi$_2$ bulk crystal grown by HGF method is a high-quality sample and is predominantly consisting of semiconducting β-phase.

The synthesis of β-FeSi$_2$ films was achieved by high-dose Fe ion implantation into Si (100) substrate at 350°C. XRD analysis revealed that polycrystalline β-FeSi$_2$ layers were formed after annealing at 875°C~915°C. The thickness of β-FeSi$_2$ was estimated to be ~75 nm from Rutherford backscattering measurements. Using room temperature optical absorption measurements, the nature of the optical band-gap (E_g) of β-FeSi$_2$ layers was confirmed to be indirect for the sample annealed at T_a=875°C for annealing duration time of t_a=1 ~ 120 min. E_g was obtained to be 0.493 ~ 0.506 eV. For the samples with T_a= 915°C and t_a=1 ~ 120 min, they showed a direct band-gap of E_g=0.853 ~ 0.878 eV. The samples with T_a=900°C and t_a=1~120 min, however, revealed the existence of a minimum indirect band-gap of E_g =0.694 ~ 0.731 eV and a slightly larger direct band-gap of E_g =0.842 ~ 0.856 eV. In all cases, high optical absorption coefficient extending to 10^5 cm^{-1} was observed above 1.0 eV. These features are illustrated in Fig.3 together with the data of other solar-cell materials.

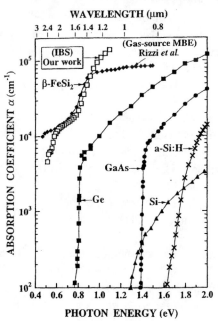

FIGURE 3   Optical absorption spectra of β-FeSi$_2$ at band-edge

The above results suggest that the nature of the band-gap of β-FeSi$_2$ changes from an indirect to a direct with increasing T_a and does not depend on t_a and could be significantly influenced by a subtle strain related to the crystallinity of the Si matrix [6]. We can therefore vary the band-gap nature and its energy value for β-FeSi$_2$ by precisely controlling the fluctuations of strain in this compound.

ACKNOWLEDGMENTS

The author is grateful to Dr. H. Katsumata, Dr. T. Iida and Mr. H. Kakemoto for the preparation of the manuscript. Special thanks are due to Prof. H. Atwater and Dr. R. McConnell for the arrangement of the participation in the meeting and Prof. K.Takeda for the permission of the reproduction of Fig.1.

REFERENCES

1. Takeda, K., *Boundary (in Japanese)* , **12**, No.6, 2-8 (1996); Nishiyama,T., *Boundary (in Japanese)* , **12**, No.1, p4-10 (1996); Maeda,M., *Boundary (in Japanese)* , **12**, No.1, p12-19 (1996).

2. Powalla, M. and Herz, K. , *Appl. Surf. Sci.* **65/66**, 482-488 (1993) ; Libezny, M., Poortmans, J., Vermeulen, T., Nijs, J., Amesz, P.H., Herz, K., and Powalla, M., *Book of Abstracts of 13th European Photovoltaic Solar Energy Conference and Exhibition*, **PO8B**., 49 (1995).

3. Derrien, J., Chevrier, J., Le Tranh, V., and Mahan, J. E., *Appl. Surface Sci.* **56-58**, 382-393 (1992).

4. Massalski, T. B., Murray,J.L., Bennett, L.H., and Baker, H., *Binary Alloy Phase Diagrams* (American Society for Metals), Vol. **2**, p. 1108 (1986).

5. Fu, C., Krijin, M., and Doniach, S., *Phys. Rev.* **B49**, 2219 (1994).

6. Dimitrdis, C. A., Werner, J. H., Logothetidis, S., Stutzmann,M., Weber, J., and Nesper, R., *J. Appl. Phys.* **68**, 1726 (1990).

7. Katsumata, H., Makita, Y., Kobayashi, N., Shibata, H., Hasegawa, M., and Uekusa, S., Aksenov, I., Kimura, S., Obara, A., and Uekusa, S., J. Appl. Phys. **80**, 5955-5962 (1996).

8. Katsumata, H., Makita, Y., Kobayashi, N., Shibata, H., Hasegawa, M., and Uekusa, S., Jpn. J. Appl. Phys. **36**, No.5 (1997) in press.

9. Libezny, M., Poortmans, J., Vermeulen,T., Nijs, J., Amesz, P.H., Herz, K., and Powalla,M., *Proc. of 13th European Photovoltaic Solar Energy Conf.* p. 1326 (1995).

10. Groß, E., Riffel, M., and Stöhrer, U., *J. Mater. Res.* **10**, 34 (1995).

High-Volume Manufacturing Issues: Toxicity, Materials Supply, Yield Management, and Marketing

D. E. Carlson

Solarex, a Business Unit of Amoco/Enron Solar
826 Newtown-Yardley Road
Newtown, PA 18940

Abstract. As new photovoltaic technologies are developed and then scaled up to high volume production, a number of issues such as toxicity, materials supply, yield management, rapid processing and characterization, equipment reliability, product performance and reliability, and marketing should be addressed in the commercialization plan.

INTRODUCTION

While crystalline silicon has been the dominant photovoltaic technology since the development of the first solar cell at Bell Laboratories in 1954 (1), there are currently several new photovoltaic (PV) technologies that are being ramped up for large-scale production. When a new technology is evaluated for commercialization, the company should consider a number of issues such as toxicity, materials supply, yield management, marketing, etc. Some of the issues that should be considered before management commits a company to high-volume manufacturing with a new technology are listed in Table 1.

Since Solarex has just completed the construction of a large (10 MW/yr) thin film PV module manufacturing plant in Virginia, we will review each of these manufacturing isssues in the context of Solarex's commercialization of multijunction amorphous silicon (a-Si) photovoltaics. The device structure is: glass/tin oxide/p-i-n/p-i-n/ZnO/Al/EVA/glass where the first p-i-n junction contains an a-Si i-layer and where the i-layer of the second junction consists of an amorphous silicon-germanium (a-SiGe) alloy (see Fig. 1). The fabrication process was first developed in the laboratory using 1 ft^2 research modules and was then scaled up to 4 ft^2 in a pilot line operation [2]. The manufacturing process

CP404, *Future Generation Photovoltaic Technologies: First NREL Conference*, edited by McConnell
© 1997 The American Institute of Physics 1-56396-704-9/97/$10.00

was qualified on the pilot line by demonstrating a yield greater than 70% for 8% stable modules produced in a run of 40 consecutive modules.

TABLE 1. Commercialization Issues

Issues	Impact
Toxicity of Materials	Environment, health and safety
Materials Supply	Cost
Yield Management	Throughput, yields, cost
Rapid Processing and Characterization	Throughput, yields, cost
Equipment Reliability	Throughput, yields, cost
Product Performance and Reliability	Performance, cost
Markets	Sales, profits

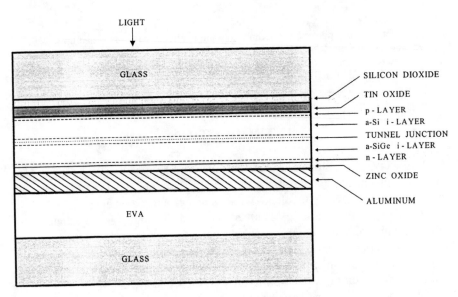

Fig. 1. A schematic of the Solarex tandem device structure.

TOXICITY ISSUES

It is important to understand all toxicity issues related to new technologies even in the early stages of research so that safeguards can be taken to protect all technical personnel. Thus, all feedstock materials and processing materials must be evaluated with regard to toxicity, and appropriate safeguard and training systems must be set up to deal with hazardous materials. All companies that supply materials will provide Material Safety Data Sheets (MSDS) that summarize all known information regarding toxicity issues. However, very little information may be available on recently synthesized materials or those that are rarely used.

One prudent course of action is to avoid highly toxic materials in the development of manufacturing processes whenever possible. Solarex took this course of action several years ago when the decision was made to use tin tetrachloride rather than tetramethyltin in developing the tin oxide front contact.

If the PV module processing requires the use of toxic materials, then systems and procedures must be established to minimize the risk to employees. In the manufacture of amorphous silicon PV devices, toxic doping gases such as diborane and phosphine are used only in a diluted form (~ 1 - 20 vol.% in silane). Trimethylboron (~ 1 - 5 % in silane) is also used as a p-type dopant source and is less toxic than diborane. Silane is pyrophoric so if a leak develops, the dopant gas will be oxidized in the flame, and a silicate glass powder will be formed thus reducing the toxicity hazard.

The silane and germane feedstock gases at the Solarex TF1 plant in Virginia are stored in an outdoor holding area, and are fed into the facility through stainless steel pipes. All exhaust gases are passed through a burn box and the powder is collected in a bag house for disposal. The powder consists mainly of silicon dioxide compounded with small amounts of oxides of germanium, boron and phosphorus.

If toxic materials would be present in the final product, then there must be an evaluation of the hazards associated with fires and long-term disposal in landfills. Amorphous silicon PV modules do not contain any toxic materials that would adversely affect the environment.

MATERIALS SUPPLY

The availability and cost of both feedstock and processing materials are major factors in determining the viability of any new PV technology. Many materials are expensive since they are rare or require costly processing to produce. Other materials are available only in limited quantities since they are produced as by-products of mining or other manufacturing operations. The main source of silicon feedstock for crystalline silicon solar cells is scrap or waste material from the semiconductor industry, and in recent years the cost of this scrap has risen as the semiconductor industry has minimized waste.

The materials that are the major cost drivers in an amorphous silicon multijunction PV module are glass, encapsulation and germane feedstock. (The silane feedstock comprises less than 5% of the total materials cost.) Solarex is currently meeting most of its needs for tin oxide coated glass by purchasing it from major glass companies. This item plus the backsheet glass constitute about 38% of the total materials cost of the module. The cost of the ethylene vinyl acetate (EVA) encapsulation represents about 12% of the total cost while the cost of the germane feedstock gas represents about 25%. Thus, these items together constitute about 75% of the total materials cost of a tandem junction a-Si PV module.

Germane is the only material in the module that is not widely used and that is relatively expensive (~ \$4/g), but since only a very thin layer (~ 180 nm) of a-SiGe is required in the module, the glass sheets constitute an even greater % of the total materials cost. Moreover, the cost of the a-SiGe layer may be reduced significantly in the future with improvements in the utilization of the germane feedstock in the plasma reactors. There are other advantages to increasing the utilization of the feedstock gases; one is that there is a concomitant decrease in the powder produced by burning the exhaust from the reactors so there is less waste removal. Another consequence is that there is a reduced requirement for the handling and storage of hazardous feedstock gases. In addition, the better utilization of feedstock gases will reduce the maintenance associated with cleaning the deposition systems since more of the semiconductor material is deposited onto the substrate and less onto electrodes and chamber walls.

YIELD MANAGEMENT

The overall yield for manufacturing a PV module must be high in order to assure that the total manufacturing cost is low. A process yield can be defined for each major step in the production sequence, and these yields can be monitored using statistical process control methods. It is important to also develop in-line (as well as off-line) diagnostic measurements that can be used to track certain critical parameters or properties during the manufacturing process. These measurements can be used to determine statistically when the process is starting to drift out of control.

In the manufacturing of thin-film PV products, Solarex routinely measures parameters such as sheet resistance and transmission of the tin oxide, resistance of the isolation scribe, illuminated I-V characteristics of the modules after laser scribing and also after encapsulation. Moreover, off-line diagnostics such as characterization of small-area diodes (I-V's and QE's) are used to help in trouble-shooting when yield or performance problems occur. Some representative run data showing variations in the conversion efficiency are shown in Fig. 2 for 4 ft^2 tandem modules that were made in a pilot run at the Solarex facility in Newtown, PA. While most of the data falls within the control limits, some modules exhibit poor performance due to special causes such as defects caused by handling, errors in laser scribe alignment, etc. There were 5 drop-outs (not shown) during this run period.

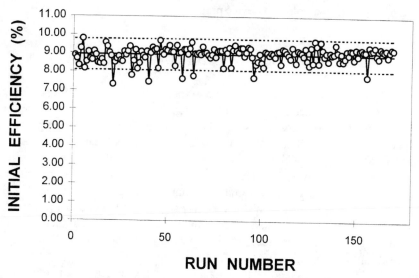

Fig. 2. A run chart showing initial efficiency of 4 ft² tandem modules vs. time.

One factor that plays an important role in assuring the manufacturability of a-Si PV modules is the ability to electrically cure the modules [3]. Since the entire device structure shown in Fig. 1 is about 1.5 microns thick (excluding the glass and the EVA), debris such as dust particles can give rise to pinholes which in turn can lead to either shunts or shorts in the device. Most of these defects can be burned out by subjecting the cell to a reverse bias of several volts. Thus, while it is important to assure clean substrates before each deposition and also to maintain clean conditions inside the deposition systems, the ability to electrically cure defects can assure a high process yield under normal operating conditions. In fact, the Solarex single-junction a-Si PV manufacturing line in Newtown, PA, has been operating for the last few years with electrical yields in excess of 95%.

The yield and performance of a manufacturing process can be continually improved without jeopardizing commitments to customers by using a methodology called evolutionary operations (EVOP). This approach can only be used if the production line is operating within statistical control limits and there is a reasonable gap between the lower control limit and the customer's requirement for performance. Under these conditions, the manufacturing process can be improved by performing small variations in certain process parameters while monitoring the performance.

In the most advanced manufacturing plants, intelligent process control systems couple a detailed mathematical model of the manufacturing process with dynamic feedback controls to keep the entire manufacturing process under control. This approach requires a thorough understanding of how changes in process parameters will influence the material and device properties.

RAPID PROCESSING AND CHARACTERIZATION

There are several reasons why it is important to be able to rapidly process and characterize the product during manufacturing. First, as mentioned above, rapid in-line and off-line diagnostic tools can help to assure high yields. If a processing problem does occur, it is important to identify the root cause and solve the problem as fast as possible since scrapped product can amount to a significant cost penalty. Moreover, if some processing steps are very slow, then a large number of substrates must be handled at that step in order to maintain a high throughput. This means the WIP (work-in-progress) is large, and thus, the total inventory of materials is also large which adds to the manufacturing cost. Another issue is that product quality (performance and reliability) may be adversely affected if partially processed products are stored for long periods of time. Rapid processing and characterization becomes critical if a manufacturing line must be capable of producing a large number of products with changes occurring on short notice (build-to-demand).

The total time for processing a single-junction a-Si PV module (from glass seaming to encapsulation) is about 4 hours in the Solarex Newtown facility. There does not appear to be any adverse effects on module performance or reliability if the partially processed plates are stored at any buffer station for periods up to one week.

EQUIPMENT RELIABLITY

Before committing to large-scale manufacturing of a new technology, a company must develop or have access to reliable, cost-effective processing equipment. If new processing equipment must be developed, then the technical risks associated with the commercialization of a new PV technology are increased significantly. The equipment development costs are one important factor that may strongly influence the overall return on investment. Moreover, the reliability of the equipment will have a signicant effect on the product yields and the plant throughput.

The equipment in the Solarex single-junction a-Si PV line generally operates with at least 90% uptime. While the line is heavily automated, it is functionally separated into five zones so that if equipment in one zone is shut down for unscheduled maintenance, the other zones can continue functioning for some time by using buffers to supply or store partially processed product. A strong preventative maintenance program is essential to assuring high uptimes for processing equipment.

PRODUCT PERFORMANCE AND RELIABILITY

Perhaps the most important factor in determining the viability of a new technology is the long-term performance of the product. The key here is to design reliable performance into the product as part of the product development process.

In the concurrent engineering approach, all key people that are involved in any stage of the product development cycle are actively involved in the defining the product and scoping out the product development plan including the schedule and critical milestones. Concurrent engineering teams will include representatives from research, engineering, finance and marketing and in some cases suppliers and customers will also be involved. The goals are to understand and factor in the customer's requirements for the product at the start of the development process, to build quality and reliability into the product, and to move as rapidly as possible through the development cycle by performing many of the tasks in parallel.

Reliability of a new PV product can be estimated by exposing the product to a battery of accelerated environmental tests [4], but it is also essential to start outdoor field tests in a variety of different climates as soon as possible. One of the critical accelerated tests for thin film PV modules is the wet Hi-Pot test where voltages of about a few thousand volts (actually twice the array voltage + 1000 V) are applied to a module when wet, and the leakage currents must be less than 50 μA. Large leakage currents appear to be correlated with electrocorrosion effects that have been observed in a-Si PV modules [5].

MARKETING

Since a company is taking a technical risk when it launches a new manufacturing technology, most companies will try to minimize the overall business risk by developing a strong marketing plan that can be executed with a high degree of confidence.

Solarex has a significant PV business with most of its products going into various remote applications such as telecommunications, water pumping, cathodic protection, village power, etc. Thus, some of the new multijunction a-Si product will go into markets where Solarex already has a strong worldwide presence (~ 70% of Solarex's sales originate outside the U.S.).

However, Solarex also plans to use this new technology to further develop other potentially large markets such as building-integrated photovoltaics and solar farms. Since the Solarex tandem module consists largely of glass, it can be used as both roofing and cladding for buildings. This type of PV roof is somewhat similar to an atrium or greenhouse roof while glass spandrels (the regions between windows) are widely used in office buildings. It is even possible to pattern thin film PV modules to allow some degree of transparency so that they could be used as windows.

Solarex has recently announced a new utility partnership project, PV-VALUE, where standardized, packaged rooftop systems are being offered to homeowners. These rooftop systems will consist of two and three kilowatt arrays of thin film tandem modules. The project is designed as a market-opening venture that will be self-sustaining within three years and will accelerate the commercialization of grid-tied PV systems.

SUMMARY

Companies planning to commercialize new PV technologies should develop a business plan that addresses a number of issues such as toxicity of materials, materials supply, yield management, rapid processing and characterization, equipment reliability, product performance and reliability, and marketing. These factors can have a large effect on determining the viability of the technology especially when evaluated against alternative technologies.

ACKNOWLEDGEMENT

Some of the work discussed in this paper was partially funded by NREL under Subcontracts ZAN-4-13381-01 and ZM-2-11040-2.

REFERENCES

1. D. M. Chapin, C. S. Fuller and G. L. Pearson, J. Appl. Phys. **25**, 676 (1954).
2. D. E. Carlson, R. R. Arya, M. Bennett, L.-F. Chen, K. Jansen, Y.-M. Li, J. Newton, K. Rajan, R. Romero, D. Talenti, E. Twesme, F. Willling and L. Yang, *Proc. 25th IEEE Photovoltaic Specialists Conf.* (1996) p. 1023.
3. G. E. Nostrand and J. J. Hanak, U. S. Patent No. 4,166,918 (1979).
4. D. E. Carlson, R.R. Arya, L.-F. Chen, R. Oswald, J. Newtown, K. Rajan, R. Romero, F. Willing and L. Yang, *NREL/SNL PV Program Review Meeting*, Lakewood, CO, November 18 - 22, 1996.
5. D. A. Fagnan, R. V. D'Aiello and J. Mongon, *Proc. 19th IEEE Photovoltaic Specialists Conf.* (1987) p. 1508.

The 1997 Calaveras Awards
for "Leapfrog" Technologies

David Carlson

has leapt to a new level with

High Volume Manufacturing Issues: Toxicity,
Materials Supply, Yield Management, and Marketing

the technology that is

The Most Likely to Succeed

The First Conference on Future Generation Photovoltaic Technologies
Sponsored by the National Renewable Energy Laboratory

Feedstock for Crystalline Silicon Solar Cells

Michael G. Mauk, Paul E. Sims, and Robert B. Hall

AstroPower, Inc. Solar Park Newark, DE 19716-2000

Abstract. A mandatory requirement for the development of a large solar photovoltaic power industry is the development a thin silicon layer structure or a cheaper silicon feedstock that does not have the high purity requirements of the semiconductor wafer industry. In the latter case the solar grade silicon feedstock supply *must* be decoupled from the semiconductor silicon feedstock industry. If one of these these events do not occur, then solar energy will remain a small and specialized industry that will start to shrink as the world power grid capacity increases.

BACKGROUND

The poly-Si supply of feedstock for silicon photovoltaics is presently based entirely on the by-product stream from high purity poly-Si generated for the semiconductor device industry. The method for generating poly-Si for the semiconductor industry (SG-Si) is based on the Siemens Process which converts metallurgical-grade silicon (MG-Si) into purified poly-Si. The price of SG-Si is $45/ kg to $60/ kg. The total by-product stream available for PV amounts to about 10% of the SG-Si production rate and is priced at less than $15/ kg. FIGURE 1 graphs data representing 10% of the world-wide annual SG-Si production rate since 1962. The total 1995 production rate was 12,000 MT (metric tons), from which approximately 1200 MT was available for the PV industry.

FIGURE 1 additionally presents the data for the annual production rate of new PV power for the years 1980 to 1994. For the purposes of analyzing these data it is assumed that 1 kg of feedstock material produces 100 watts of PV power. This assumption allows the two sets of data to be compared directly on the same graph. At present a figure of about 50 to 60 watts/ kilogram is applicable for ingot technologies— 15% efficient solar cells, 450-micron thick wafers, and 45% yield of feedstock material to module. Therefore, the difference between supply and demand at present may be significantly less than FIGURE 1 indicates. Figure 1 also forecasts future growth of the production rate of by-product silicon and PV production. For by-product production a 12% growth rate is forecast for the years beyond 1995. A 20% growth rate is shown for PV production. The PV production numbers are similar to those employed by Strategies Unlimited for their Business as Usual case (1), their accelerated forecast (not shown) is for a 25% growth.

CP404, *Future Generation Photovoltaic Technologies: First NREL Conference*, edited by McConnell
© 1997 The American Institute of Physics 1-56396-704-9/97/$10.00

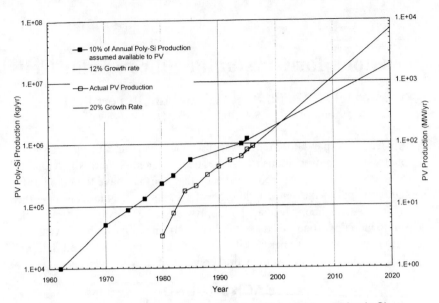

FIGURE 1. Historic growth and projections for PV production and poly-Si available for PV.

Inspection of FIGURE 1 suggests that the PV-Si supply becomes a significant problem by 2002. The problem point may actually move closer in time if poly-Si production increases at a rate less than 12%, or if PV production increases at a rate greater than 20%, or if the watts/kg number does not reach 100. Accordingly, it is prudent for the silicon PV industry to assume control of its feedstock well before the turn of the century.

PRESENT SUPPLY PROCESS

FIGURE 2 illustrates the flow of product (and by-products) starting with MG-Si and ending up with wafers for the semiconductor industry. MG-Si is 98 - 99% silicon and is priced at about $1.80/kg. The product of the Fluidized Reactor Process is a volatile silicon compound (silicon tetrachloride [$SiCl_4$], trichlorosilane [$SiHCl_3$], dichlorosilane [SiH_2Cl_2], silane [SiH_4]). The price for the silicon content in $SiCl_4$ is $15.00 / kg. The product of the CVD Deposition Process (Siemens Process) is semiconductor-grade silicon (SG-Si) that is priced at $45.00/kg. The total 1995 production of polycrystalline silicon for semiconductor industry is about 12,000 MT (equivalent to about 1 GW of PV power).

FIGURE 2 further shows the subsequent process steps employed to yield on-spec silicon boule material and wafers, as well as off-spec by-products. These by-products —tops and tails ($8.00 to $15.00 / kg) and rejected wafers ($45.00 to

$75.00/kg) —are employed by the PV industry in making solar cells. The volume of these by-products as a percent of the material produced for the semiconductor industry is about 6% and 2%, for the tops and tails and rejected wafers, respectively (2). Another 40 to 50% of the volume available as kerf remains.

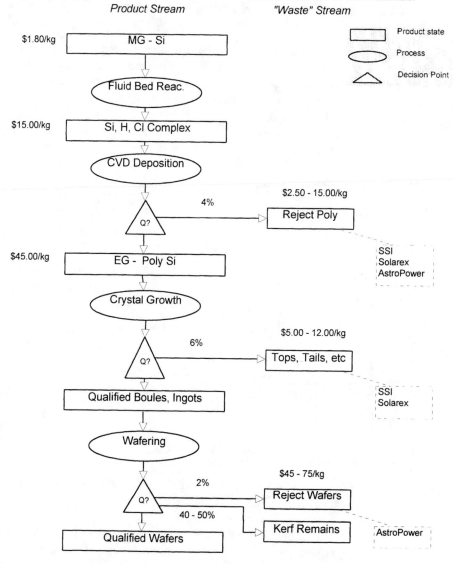

FIGURE 2.   Process for Semiconductor-Grade Silicon

ALTERNATIVE SUPPLY PROCESSES

FIGURE 3 details the standard production of metallurgical-grade silicon by the carbothermic process, and indicates several approaches used to convert metallurgical-grade silicon into solar-grade silicon. A decision point is employed to qualify an approach based on a specification for solar grade silicon.

Paths 1, 2, and 3 all begin with the carbothermic reduction of quartzite (sand). The product of this process step is called metallurgical-grade silicon (MG-Si). Path 1 capitalizes on the idea of starting with pure starting materials for the carbothermic reduction process. Path 2 is presently the one employed by all the manufacturers of Poly-Si for the semiconductor industry (3). This path uses the *chemical approach* to successfully purify MG-Si, and is effective at removing boron. However, the cost of energy to effect these processes is typically high.

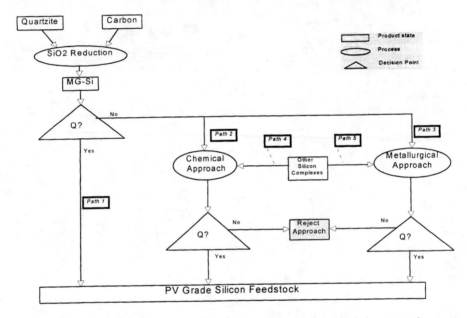

FIGURE 3.   Possible raw materials to solar (PV) grade materials processing methods.

Path 3 is the *metallurgical approach* (4) which has the potential to result in a lower cost product (i.e., lower energy expenditure), but does not deal effectively with boron. Paths 4 and 5 represent alternative non-carbothermic approaches for generating silicon to be subsequently upgraded using either the chemical approach or metallurgical approach.

The available approaches for the manufacture of solar grade silicon are the development of a low boron starting material that is either in the form of synthetic silica or MG-silicon, and the subsequent removal of other impurities with either a chemical approach or a metallurgical approach.

Chemical Approach

The chemical approach involves two separate major process steps. The first step starts with some form of metallurgical silicon generated by the carbothermic (Path 2) or other (Path 4) process and produces a volatile silicon compound—usually trichlorosilane [$SiHCl_3$]. Successive distillations are performed to achieve high purity; this is where the boron is removed. The second step starts with the trichlorosilane (or one of its derivatives, dichlorosilane [SiH_2Cl_2] or silane [SiH_4]) and by thermal decomposition (i.e., a CVD process) deposits semiconductor grade silicon. The distillation and deposition process steps are very energy intensive.

The cost associated with the Chemical Vapor Deposition (CVD) process for silicon deposition is dominated by the growth rate, which is in the range of 5 to 15 microns per minute. This property is particularly limiting for the Siemens Process. The Siemens Process is achieved by decomposing trichlorosilane on a silicon seed rod held at 1200°C. A run begins with a 0.3-cm diameter silicon rod that acquires a diameter of 15 to 20-cm after several days of deposition. However, as practiced by the industry, the temperature of the growing rod is maintained by applying an electric current along the length of the growing cylinder. Since $P=I^2R$, and $R=\rho L/A$, the electrical current requirement for heating the cylinder (which is a square term to start) increases drastically as the diameter increases due to the lowering of the resistance of the silicon rod with the square of the diameter and the linear increase in radiative heat transfer efficiency with the diameter. Since convective heat losses also play a role in this process, the power cost of this type of silicon deposition process is supralinear with diameter.

Given a deposition rate of 10 microns / min (a typical good "reported" rate), a first order analysis predicts that it takes 95 kWhr to generate one kilogram of silicon. This is in agreement with the published results (5), which also report an energy cost of 300 kWhr / kg for the raffination of $SiHCl_3$ step. Thus, a total of 390 kWhr is necessary to generate one kilogram of silicon using the chemical approach; at a rate of $0.03 / kWhr the electricity cost alone is $11.70 / kg. The Siemens process is inherently a batch process.

There is an alternative CVD deposition process, the Fluidized Bed Reactor (FBR), that generates granules of SG-Si. This process has lower energy costs and has demonstrated the capability of running in a continuous mode. However, due to the increase in surface area predicated by this technology, there is a disadvantage to this feedstock in the single crystal semiconductor silicon wafer industry. Energy costs are reduced in the FBR approach even though the silicon growth rate is the same (i.e., 5 to 15 microns / min). This is accomplished by providing very large reaction areas for deposition (i.e., a multitude of sphere-like particles). Simply, the process works by having a stream of small silicon particles fed into a reactor against a counter stream of a halosilane gas. Owing to the large deposition area available, the buildup of mass (i.e., product) is significantly greater than for the same energy expenditure in a Siemens-based process.

The FBR approach was identified as the most promising candidate to produce a low-cost silicon feedstock for the PV industry during the $280,000,000 JPL "Flat Plate Solar Array Project" (1975 to 1985). The bottom line result for the project was that the semiconductor electronics industry achieved higher performance and lower cost silicon wafers using the Siemens-based technology (which was already being employed before the JPL project). Union Carbide was awarded government support to develop the FBR technology, but it was Ethyl (now MEMC) who finally brought it on-line in 1987 and now sells material at $45 / kg.

There are other approaches (Path 4) for providing silicon as feedstock to the chemical branch in FIGURE 3. Primarily they involve replacing chlorine by fluorine, bromine or iodine in the halosilane or halide. There does not appear to be any significant advantage to this replacement. One particular source of interest is the hexafluorosilicic acid (H_2SiF_6) by-product of the phosphate fertilizer industry that is used to generate SiF_4 (6).

Chemical Approach -- Summary

The chemical approach has been well developed for the semiconductor industry but, with the possible exception of FBR, does not meet the requirements for the production of large amounts of cheap, moderate purity, solar grade solid-silicon feedstock. The cost for solar grade silicon and semiconductor grade silicon using chemical processes is essentially the same.

Metallurgical Approach

The metallurgical approach for refining silicon has the potential to result in a lower cost product (i.e., lower energy expenditure). The foremost issue that must be addressed to qualify any metallurgical sequence is dealing with boron.

TABLE 1 lists specific methods to achieve low boron silicon feedstock, and indicates the silicon phase to which the methods are applied. Implicit in this plan

26

is the need to employ the methods in a rational sequence to effect the required purification.

TABLE 1. Purification Methods - Metallurgical Approach

Silicon Phase	Purification Method
Liquid	Starting Materials
	Liquid-Liquid Extraction
	Liquid-Gas Extraction
Liquid/Solid	Segregation
Solid	Solid-Liquid Extraction
	Gettering

There are other approaches for supplying silicon for the metallurgical refinement route. Metallothermic reduction using aluminum has been investigated at Wacker (7), but requires dealing with the Al_2O_3 by-product in order to recycle the aluminum.

Metallothermic reduction of SiF_4 by Na and $SiCl_4$ by Zn have both been employed to produce elemental silicon. The SiF_4 or H_2SiF_6 (a by-product from the phosphate fertilizer industry) reduction by Na has achieved low boron concentration. The Zn reduction of $SiCl_4$ has been extensively investigated. This process was used to generate the first available "electronic" grade silicon in 1952 (8). It has been subsequently replaced by the more economic Siemens Process.

Electrodeposition of MG-Si to produce a photovoltaic grade silicon has been investigated. In general the approach is plagued by the same low linear growth rate (typically 1 to 2 micron / min) that limits the Siemens-based CVD approach, thus requiring very large reaction areas in order to achieve meaningful mass generation rates.

Metallurgical - Summary

The metallurgical approach shows the greatest promise for producing solar-grade silicon at the target price of $8 / kg. A paradigm for achieving this goal is provided by the steel industry that over the last forty years has developed the methods to employ a comparable process to produce steel with impurity control at the ppma level. This approach is presently being pursued by Kawasaki Steel (9). The necessary event is the accomplishment of a metallurgical grade silicon with the proper boron concentration level to enable purification by subsequent metallurgical means.

SUMMARY

The chemical approach for manufacturing silicon feedstock has been well developed for the semiconductor industry, but does not meet the requirements for the production of large amounts of cheap, moderate purity, photovoltaic grade solid-silicon feedstock. The cost for PV grade silicon and semiconductor grade silicon using chemical processes is essentially the same ($45 / kg). Silicon solar feedstock technology has been piggybacking on the semiconductor industry for the past 30 years. A mandatory requirement for the development of a large solar power industry is the development of a cheaper silicon feedstock that does not have the high purity requirements of the semiconductor wafer industry -- the PV grade silicon feedstock supply *must* be decoupled from the semiconductor silicon feedstock industry. For even a moderate growth of the solar industry (10% per year), the future supply of silicon feedstock is in question. As the production of semiconductor grade silicon feedstock begins to level off, a major opportunity-cost driven price restructuring will occur for the whole silicon feedstock industry, and the feedstock cost for silicon solar cells will increase drastically.

REFERENCES

1. STRATEGIES UNLIMITED, Private Communication, (Fall, 1994).

2. STRATEGIES UNLIMITED, *Photovoltaic Industry Information Service*, Report M-34, "Silicon Materials for Photovoltaics," (July, 1991).

3. MCCORMICK, J. R.. "Polycrystalline Silicon Technology Requirements for Photovoltaic Applications," Silicon Processing for Photovoltaics I, Chapter 1, C. P. Khattak and C. P. Ravi, eds., Elsevier Science Publ. B. V., 1985 1.

4. DIETL J., "Metallurgical Route to Solar-Grade Silicon," *8th E.C. PVEC*, Florence, Italy (May, 1988) 599-605.

5. AULICH, H. A., SCHULZE, F. W. AND STRAKE, B., in *Proc. 18th IEEE Photovoltaics Specialists Conference*, Las Vegas, Nevada (October, 1985) 1213-1218.

6. SANJURJO, A., NANIS, L., SAUCIER, K., BARTLETT, R. AND KAPUR, V., *Journal of the Electrochemical Society*, Vol. 128, No. 1, 179-184 (1981).

7. DIETL, J AND HOLM, C., *Technical Digest of the International PVSEC-3*, Tokoyo, Japan (November, 1987) 533-536.

8. LYON, D.W., OLSEN, C. M. AND LEWIS, E. D., *Journal of the Electrochemical Society*, Vol. 96, No. 6, 359-363 (1949).

9. Y. SAKAGUCHI, N. YUGE, H. BABA, K. NISHIKAWA, H. TERASHIMA, and F. ARATANI, "Metallurgical Purification of Metal Grade Silicon Up to Solar Grade" *Proc. 12th Eur. Photovoltaic Solar Energy Conf.*, (1994) 971-974.

Photovoltaics Characterization: Beyond the Horizon

Lawrence L. Kazmerski

National Renewable Energy Laboratory
1617 Cole Boulevard
Golden, Colorado 80401

Abstract. This paper examines current photovoltaic test, measurement, and characterization techniques and makes evaluations and predictions of the next-generation technologies needed to meet the evolving requirements of photovoltaics. The range of support and research areas, from array through atomic-level analysis, are cited. The specific requirements of research and manufacturing sectors are addressed, including the need for more rapid response, new and photovoltaic-specific measurement techniques, manufacturing-environment measurement capabilities, and electronic-based centralized facilities. The integration and cohesion of analytical services with the evolving capabilities of the information highway are discussed and anticipated. To ensure the security of both intellectual and product property, the increased demands of protection of data are emphasized. Trends toward greater accuracy, precision, smaller- and larger-area analysis, and more-versatile measurement technologies are discussed.

INTRODUCTION AND STATUS

"Characterization" has been an essential partner in advancing and realizing photovoltaics technology [1-4]. The boundaries of photovoltaic characterization mirror the technology itself, from *arrays to atoms*, as represented in Fig. 1. Although we sometimes confine ourselves to areas of special technical interest, the technology as a whole owes its scientific development to an ensemble of creative contributions from its parts—covering thought, planning, realization, evaluation, and verification. Theory, materials science, processing, device development, modeling, testing, and measurement are co-dependent. None of these areas can stand alone; together, they provide for the successes and future of the technology. The future of photovoltaics requires nothing less than this synergy. This paper examines test and measurement technologies for photovoltaics—first surveying what is now available, then making some evaluations and predictions of where characterization support and science is heading. As photovoltaics pushes the frontiers of materials science, device engineering, and performance, measurements and characterization

CP404, *Future Generation Photovoltaic Technologies: First NREL Conference*, edited by McConnell
© 1997 The American Institute of Physics 1-56396-704-9/97/$10.00

must match and lead with cutting-edge advancements and technology to support and guide progress.

Table 1 presents a summary of selected, current techniques, covering a wide-range of materials (composition, chemistry, and structure) and device methods (electro-optical, performance, and reliability) [5-9]. An examination of these techniques can lead to some judgments of what "analysis" is required in the near-term and long-term to ensure the progress of photovoltaics. Photovoltaics has benefited greatly from its sister technologies in semiconductor electronics (i.e., microelectronics, displays, and optoelectronics). Most of the techniques cited in Table 1 serve (and perhaps were developed primarily for) the semiconductor industry. Photovoltaics has been able to leverage much of its technical advancement from this related industry. Our laboratories, for example, offer some 57 major and distinct techniques for materials and device evaluation. This leveraging continues, but as photovoltaics matures and evolves into new areas, it begins to stipulate analytical needs that are specific to itself.

Certainly, some degree of this specificity has already been manifested. The technology of terrestrial cell and module performance evaluation (i.e., determining efficiency) has developed directly in response to the needs of photovoltaics. Special electro-optical characterization methods, such as minority-carrier parameter determinations, chemical defect resolution, service-lifetime testing, and certain scanning optical and electron-beam techniques have become routine laboratory and industry-area tools. Research and industry laboratories have established their cadre of analysis methods aimed at their specific needs. Some specifics are indicated in the following section. The approach (centralized or dispersed) to characterization support for the large, directed programs has been guided by service to customer and by economics. Central, multi-technique laboratories have the advantage of being able to select the correct analysis method for the problem from their analysis menu and/or to use a number of complementary techniques to unambiguously solve problems. They also offer the cost advantage over investment in a large number of redundant analytical capabilities. Finally, they can provide the measurement expertise and experience that recognizes problems common to several technologies or clients. The decentralized approach offers immediacy of access (potential advantage in analysis time) and local control over the analysis priorities and data.

The operations for central laboratory facilities include a number of very important components. The first is the most-cited requirement for such a facility: *measurement support and collaborative research*. The support function provides service for the client and should involve extensive interaction between the measurement scientist and the person who knows the most about the sample. This focuses the analysis and distinguishes the established program facility from a commercial analytical laboratory. The analysis is more than just "sample in, data out". A logical extension is to establish a collaborative research activity between the characterization and material/device research entities. This project objective would likely include a range of complementary analytical functions to address the problem. This mode of operation enhances "buy in" from both sides and usually leads to some additional output such as a joint publication in the scientific literature. Not every analysis fits into this category; some materials and device projects only need some chemical or electrical information to establish a process. However, those projects that aim at extended development or problem solution can benefit from the collaborative approach

Table 1. Compilation of selected common methods for analyzing specific structural, chemical, physical, and electro-optical properties of materials and devices.

Property	Technique
Element identification; composition; purity	Auger electron spectroscopy (AES) X-ray photoelectron spectroscopy (XPS) Secondary ion mass spectrometry (SIMS) Time of flight SIMS (TOFSIMS or static SIMS) Ion scattering spectroscopy (ISS) Scanning neutral ion mass spectrometry (SNMS) Surface analysis by laser ionization (SALI) Surface analysis by resonance ionization of sputtered atoms (SARISA) Laser ionization mass spectrometry (LIMS) Laser microprobe mass analysis (LMMA) Rutherford backscattering spectroscopy (RBS) Electron energy loss spectroscopy (EELS) Atom probe field ion microscopy (APFIM) Nuclear reaction analysis (NRA) Particle-induced X-ray emission (PIXE) Glow discharge optical spectrocopy (GDOS) Glow discharge mass spectrometry (GDMS) Surface compositonal analysis by neutral and ion impact radiation (SCANIIR) Energy-dispersive spectroscopy (EDS) Wavelength-dispersive spectroscopy (WDS) Electron probe microanalysis (EPMA)
Chemical state characterization	Infrared and Raman spectroscopy Fourier-transform infrared spectroscopy (FTIR) Electron energy loss spectroscopy (EELS) X-ray photoelectron spectroscopy (XPS) Secondary ion mass spectrometry (SIMS) Electron-stimulated desorption ion angular distribution (ESDIAD) Ellipsometry X-ray fluorescence (XRF) Laser fluorescence (LF) Electro- and cathodoluminescence (EL and CL)
Structural characterization	Low-energy electron diffraction (LEED) High-energy electron diffraction (HEED) Reflection high-energy diffraction (RHEED) Field ion microscopy (FIM) Field emission microscopy (FEM) Transmission electron diffraction (TED) X-ray diffraction (XRD) High-voltage electron microscopy HVEM) Analytical electron microscopy (AEM) Extended X-ray fine structure spectros. (EXAFS) Ion scattering spectroscopy (ISS) Ion and electron channeling Electron-stimulated desorption ion angular distribution (ESDIAD) Ultraviolet photoelectron spectroscopy (UPS) Surface extended X-ray fine structrue (SEXAFS) Vibrational EELS Raman spectroscopy Neutral scattering spectroscopy (NSS)

Topography	Scanning electron microscopy (SEM) Electron microscopy (EM) Optical microscopy X-ray topography Interferometry (including spectroscopic) Profilometry (mechanical stylus) Scanning tunneling microscopy (STM) Atomic force microscopy (AFM) Near-field scanning optical microscopy (NSOM)
Film growth/material absorbed or deposited	Microgravimetric techniques Ellipsometric techniques Radiotracer techniques Laser/optical techniques Quartz crystal (microbalance) Electron stimulated desorption (ESD)
Surface area/roughness	Microgravimetric and volumetric absorption Electron microscopy (EM) Optical microscopy Proximal probe microscopies Optical-beam induced-current spectrocopy (OBIC)
Inclusions/defects	Scanning Auger microscopy (SAM) Secondary ion mass spectroscopy Electron dispersive spectroscopy with TEM Scanning transmission electron microscopy (STEM) Optical-beam induced-current spectrocopy (OBIC) Electron-beam induced-current (EBIC) Electron-beam induced-voltage (EBIV)
Microdefects/features	Electron microscopy (EM) Electron diffraction (ED) Small-angle X-ray spectroscopy (SAXS) Scanning transmission electron microscopy (STEM) Analytical electron microscopy (AEM) Proximal probe microscopies Near-field scanning optical microscopy (NSOM)
Electron density of states/defect chemistry	Ultraviolet photoelectron spectroscopy (UPS) Photoluminescence spectroscopy (PL) Deep-level transient spectroscopy (DLTS) Scanning tunneling spectroscopy (STS) Near-field scanning optical spectroscopy (NSOS) Electro- and cathodoluminescence (EL and CL) Ballistic electron energy microscopy (BEEM)
Atomic-level Imaging	Field ion microscopy (FIM) Scanning tunneling microscopy (STM) Scanning force microscopy (SFM) Transmission electron microscopy (lattice imaging)

Electro-optical determinations	Spectrophotometry Ellipsometry (fixed wavelength, spectral) Photoluminescence (PL), Photoconductivity (PC) Fourier-transform techniques (FTIR, FTPL) Capacitance, current-voltage spectroscopy Minority-carrier lifetime spectroscopy Mobility/resistivity/carrier measurements
Cell and module performance	Simulator efficiency (continuous, pulsed) Quantum efficiency/spectral response Current-voltage spectroscopy (light/dark) Reference cell calibration Spectral measurements Outdoor performance (exposure, power rating) Reliability and durability Solar/optical radiation measurments Rating, certification, acceptance testing Environmental testing (humidity, UV exposure, high-potential, hail impact)

For reference, about two-thirds of NREL interactions with clients fall into the collaborative category. An extremely important program support function is the second category: *standard test and measurements*. The centralized laboratory facilitates an independent source for evaluating components. No evaluation has been more visible or contributory than that of determining the performance (efficiency) of cells and modules. This determination has not only provided the basis for credibility, but has led to a worldwide network of standards laboratories that ensure the independent and fair comparison of devices (within a given technology and between technologies). The centralization of the standard measurement capability has also helped to disseminate standard test methods and reporting conditions within the photovoltaic community [10-12]. It has led to international intercomparisons to make sure that efficiencies measured in India can be expected to track those in Europe and Japan. Almost all these central and standard performance-evaluation laboratories can trace their standards to common sources [13].

A third category is that of *technique development*. Because of expertise in the particular technology (e.g., photovoltaics), these measurement scientists and engineers are able to adapt techniques from other electronics technologies and develop new methods in response to the needs of the technologies.

The evolution of a technology takes many paths, and the measurements and characterization component has to be ready to respond. In fact, this volume provides some interesting insights about this technology development. Many "future" technologies are resurrected technologies: those that may have been introduced or conceptualized before a technical base (e.g., device processing or materials production) existed to effectively realize them. Although electrochemical cells, thermophotovoltaics, thin-film silicon, and polycrystalline III-V semiconductors had origins before 1980, the knowledge and capabilities to fashion these into efficient conversion technologies had not yet completely evolved. They were concepts before their time and could have been lost were it not for persistence and creativity. In a similar manner, many characterization techniques undergo refinement with technology development and will eventually meet specific requirements as characterization science and engineering progresses.

NEAR-TERM REQUIREMENTS

The more immediate or near-term needs for photovoltaics technology can be displayed in the areas of performance, electro-optical, physical, and chemical analysis. The ability to determine and compare the operational characteristics of solar cells and modules has progressed significantly over the past 15 years, with the adoption of international standards for measuring and reporting device efficiency. The accuracy of these measurements is summarized in Table 2, and international intercomparisons and collaborations have provided traceability and confidence in evaluations worldwide. Measurement uncertainty analysis has been applied to every step of these standard evaluation processes, and work continues to increase precision. However, as photovoltaics continues to evolve, so do the requirements in device evaluation. In the standards area, new and novel cell technologies, such as bifacial cells and holographic concentrators, require rating methodologies. Many alternatives to the standard "accepted" reporting techniques must be evaluated for their utility and accuracy. On the measurement side, a growing number of challenges are being presented. Methods for evaluating ultrahigh efficiencies, approaching 50%, are needed to meet the coming needs for multijunction and complex structures currently being developed in the laboratory. Even issues relating to concentrator cells, such as linearity, must be resolved to adequately and fairly evaluate these technologies. New device types, such as thermophotovoltaic and electrochemical cells, need attention for performance determinations. Spectral effects and light response times are critical, and methods for ensuring that these are adequately addressed

TABLE 2. Accuracies in measuring photovoltaic efficiencies for cells and modules.

	Best current methods (%)	Typical methods (%)	International intercomparisons[+] (%)
Single Junction			
Cells (1 sun)	±1.5 - 2	±5	±5 (±3)*
Concentrator cells	±3	±5	
Modules	±3	±5	±7 (±3)*
Multi-Junction			
Cells (1 sun)	±3	±5	
Concentrator cells	±4	±8	
Modules	±5	±8	

* Single- crystal and multicrystal Si only
[+] Based on PEP85 and PEP87

must be considered. Artifacts and other sensitivities of new device types must be probed and identified before standard measurements are adopted. For these and for existing technologies (e.g., a-Si:H and CuInSe$_2$), stabilization criteria must be established. Changes with light exposure are certainly concerns. The accurate measurement of larger-area devices is still evolving. Pulsed-light simulators used for modules (mainly to minimize heating effects) may have problems with uniformity and accurate spectral determinations. Capacitive and other transient effects in the module can hinder measurement accuracy. Large-area continuous simulators are now being developed that can permit the direct application of spectral correction methods used in cell efficiency measurements. Of course, the module efficiency rating (power vs. efficiency, temperature, etc.) must be addressed to better meet the needs of the industry and the client.

Measurement mandates to develop more versatile techniques are becoming increasingly evident. In the case of a central analysis facility, the ability to measure a large number of different materials or cell types is beneficial from the perspective of cost and effectiveness. Even manufacturing facilities now deal with a diversity of products (e.g., thin-film polycrystalline cells and single-crystal devices). An example of one such versatile method, which has leveraged its development with use in other semiconductor electronic applications, is the minority-carrier lifetime spectrometer developed by Ahrenkiel [14,15]. The system, shown schematically in Fig. 2, provides a non-contact and non-destructive means of evaluating the most fundamental measure of the quality of a semiconductor for use in a solar cell—the lifetime of the minority electrons and holes [16]. The technique is extremely versatile, as illustrated by the data shown in Fig. 3. It can accurately determine the lifetimes in direct and indirect bandgap semiconductors, bulk and large-area materials and sub-micron thick films, for single crystals and polycrystalline semiconductors, and for bandgaps covering the range from about 0.5 eV to more than 3.0 eV. In addition to the versatility, the technique can also be adapted directly into the manufacturing line (see following sections) to provide material evaluation and quality control before or following device processing. Such developments provide an important means of technology transfer from the university or research lab to the manufacturing environment.

From purely functional and credibility considerations, other needs are required for photovoltaics as competition grows between international entities. A key requirement is the ability to provide *proprietary measurements* for clients. The protection of the intellectual property and commercial interests of the client is mandatory. This leads to trust between the parties and to efficiency and effectiveness in measurement time because information important to the analysis can be shared. Obviously, advancing any commercial operation is important to the program. And finally, it is important to *minimize the number of interfaces* between the analysis and the client. Certainly, tracking is an important factor for a program, but clients must be able to work directly with that entity concerned with their sample measurements. Because rapid turnaround time is essential to the research group or manufacturer, this tracking decreases the chance of samples being misplaced, left on the desk or mailbox of someone on travel, or that the data will be released incorrectly. Protected electronic transfer and interaction, cited below, are keys to ensuring and enabling rapid response, archiving, and openness of analysis for these operations.

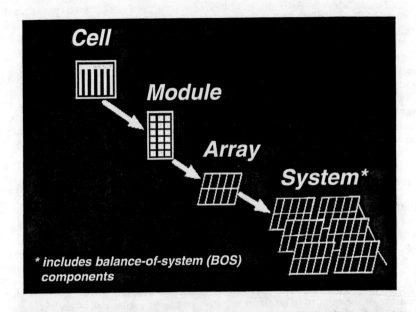

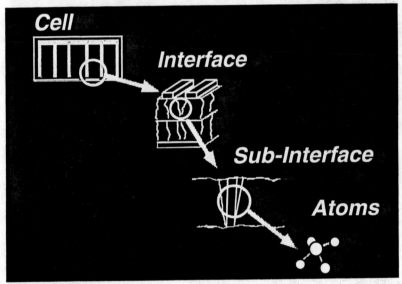

FIGURE 1. Representation of extent of photovoltaic characterization, reflecting the hierarchy of photovoltaic technology from arrays to atoms: (a) arrays through cells; and (b) internal cell considerations.

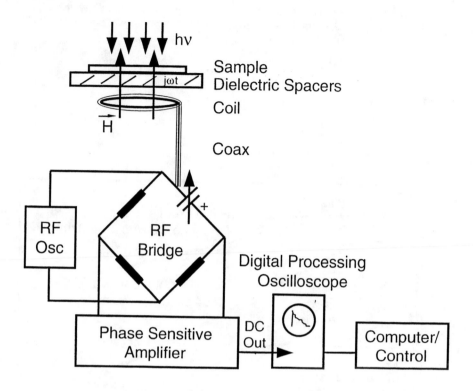

FIGURE 2. Schematic representation of minority-carrier lifetime spectrometer, illustrating technique development of a contactless, versatile, measurement technique for photovoltaic technology.

Reliability is a general area that is closely tied with characterization. It is also an area of growing concern for photovoltaics—a technology that has a 20-30-year operating lifetime tagged to its eventual success. Service-lifetime predictions, lifetime testing, extended exposure to environmental conditions, acceptance testing, certification, and accelerated lifetime testing are all issues that require immediate attention in the near term [17]. It is also an area that has some unique challenges in sensitivity to the industry and real effects on the technology. Reliability, lifetime, and durability programs are important for the wide range of materials and device technologies—both for present commercial products and for those not yet conceived [17].

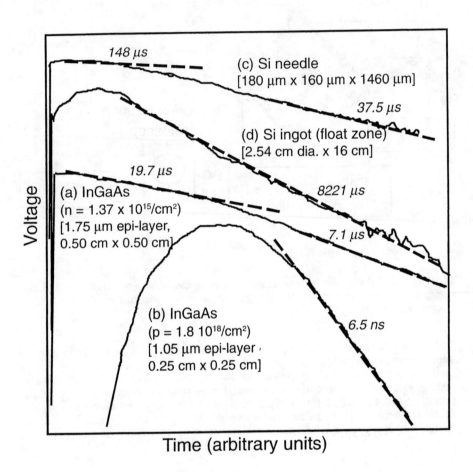

FIGURE 3. Minority-carrier lifetime data from instrument in Fig. 2. These data show the capabilities of the spectrometer to measure from thin films to boules, a wide range of carrier concentrations, and bandgaps from low to high values.

ON AND BEYOND THE HORIZON

Photovoltaics will continue to require measurement technologies that reflect the evolution of the technology. Certainly, this electro-optical technology will continue to benefit from developments for the general semiconductor industry, as well as from those methods contributed specifically from its infrastructure. The trends can be defined in several areas, covering logistical (responsiveness, scheduling, communication) and new areas/novel approaches (larger-area analysis, non-intrusive/contactless/non-destructive measurements, femtosecond-event evaluation, nanoscale area characterization).

The revolution we are now experiencing in communications is already affecting interaction in analytical areas. The ability to transfer results has evolved from mail to fax to electronic means. Many laboratories now have capabilities to transfer data directly from the measurement instrumentation to the client. These electronic schemes have also provided the archiving capabilities, with the clients having access to mailboxes holding their data over a period of time. Mandatory to these electronic interactions is protection of data. The proprietary and intellectual property rights of the client must be protected, and electronic means of encrypting have been made available. These communications steps have significantly improved the capability of the analytical facility to respond to the client's request, and for the client and the analytical scientist/engineer to interact and collaborate. The analytical laboratory is becoming a remote extension of the research laboratory and the manufacturing facility, and this trend will continue to evolve as communications technology improves. A prime example is remote operation of equipment—an operational scheme in its infancy, but one that is becoming more common. For example, the spectroscopist can monitor the data acquisition, change parameters, evaluate data, and download results from locations external to the laboratory. This same next-generation capability should eventually be made available to external clients, providing a new dimension to a "user facility". Specifically for a DOE central facility, this would enhance the outreach of the support portions of the program and benefit the economic and technical objectives by assisting its research and industry partners. There are, of course, some significant cautions. Data interpretation and analysis conditions still need expertise and experience. The remote sample preparation and sample introduction parts of the analysis procedure require coordination with the local facility.

Rapid response, pertinent information, immediacy, and problem identification and solution are paramount to product evolution and market competition. The traditional approach has been to use centralized facilities for sophisticated and more costly measurement requirements, for confirmation of milestones, and to ensure compatibility in the standard performance areas. Currently, there is growing activity in developing manufacturing-environment measurements, primarily non-contact techniques that are incorporated on-site and perhaps directly into the manufacturing line to monitor and ensure product quality. These techniques range from fundamental property determinations to current-voltage spectroscopies, and they represent a growing area of need and investment for the PV industry. Finally, as the industry diversifies (i.e., has multiple PV products under a single ownership), the versatility of the measurement technique becomes important. The availability of a

39

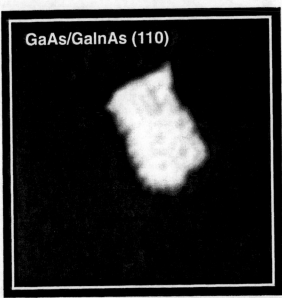

FIGURE 4. GaAs structure grown on InGaAs, using atomic manipulation in scanning tunneling microscope, to provide direct, diagnostic studies of defects and interfaces at this TPV heterointerface.

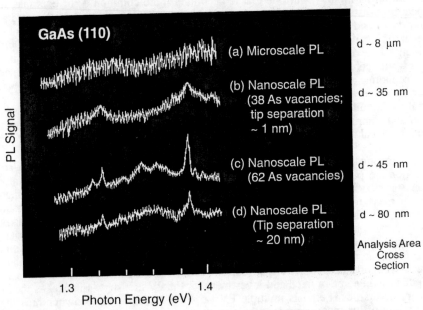

FIGURE 5. Comparison of conventional and nanoscale photoluminescence spectra for atomically engineered GaAs surface, showing evolution of As defect levels.

specific method is desirable for thin films and bulk crystalline materials, for low bandgap absorbers and wide-bandgap windows, for sub-mm^2 as well as cm^2 areas. This is currently an area of growing interest, and one that demands the collaboration with industry of the DOE Program's assets at universities and national laboratories.

Supporting technique development for advancing technology can include contributions on the frontiers of science. An example relating to thermophotovoltaics is in the area of nanoscale science—using the proximal probe methods, i.e., scanning tunneling microscopy (STM), atomic force microscopy (AFM), and near-field scanning tunneling spectroscopy (NSOM), to investigate fundamental issues [18-20]. For example, the STM has been used to manipulate atoms at InGaAs surfaces to create vacancies (point-defect creation) and to directly investigate the electronic levels associated with these defects using NSOM-related methods [21]. Beyond this, the STM has been used to grow atomic dimension interfaces, such as the GaAs-on-InGaAs structure shown in Fig. 4. This technique is being used to characterize and optimize interfaces for these materials, and it provides some direct diagnostic evaluation for potential interface and device optimization. This area represents a growing tool for technology in which events studied on the highest spatial-resolution regimes are used to guide the optimization of macro-area devices.

Certainly, nanoscale characterization complements the evolving capability to image and process on these highest spatial-resolution regimes. Figure 5 shows a series of nanoscale photoluminescence (PL) data for a near-field operation with a specially prepared optical tip. These data show the ability to resolve electronic defect data at higher resolutions than possible using conventional techniques. Figure 5a shows a conventional or macroscale (~8-μm illumination area) PL spectrum, in which no defect levels are discernible. The evolution of the As-defect level is observed in Figs. 5b and c which show the PL spectrum at two As-vacancy levels. Finally, the effect of the near field is represented in Fig. 5d, which shows a decrease in the signal at slightly greater tip separation. These data show the capability of the instrument to image, process, and to directly correlate the results using nanoscale technologies. There is a growing interest in correlating events that occur on the nanoscale with the macroscale operation of devices. This area—integrating processing to address or enhance specific characterization requirements—becoming a vital part of the current semiconductor industry. Just as the introduction of focused ion-beam technologies has redirected failure analysis of microelectronics components [22], the adaption of these submicron ion sources and future use of nanoscale tools will become integral parts of photovoltaics.

SUMMARY

Characterization support and guidance remain integral parts of the photovoltaics program, and the demands on measurement science and technology will grow with the evolution and requirements of photovoltaic materials and devices. These parts will be reflected in demands for more rapid response and needs for manufacturing environment measurements. They will be manifested in developing all-new analytical technology areas, including some that unite the processing with the analysis for technology optimization. Future expectations align with the development of the information highway for improved interactions, collaborations, transfer of data, and even remote facility usage.

ACKNOWLEDGMENTS

The author expresses his gratitude and appreciation to colleagues of the NREL Center for Measurements and Characterization who selflessly helped in providing input to and reviewing this paper. The author would also like to thank the conference Chair for his patience and for the opportunity to present these ideas. This paper was prepared partially through the support of NREL and the U.S. Department of Energy under contract No. DE-AC36-83CH10093, and through the support of the U.S. DOE Applied Energy Projects Program (for TPV and nanoscale portions of this paper).

REFERENCES

1. See, Proc. IEEE Photovoltaic Specialists Conf. (IEEE, New York; 1971-1996); Also, T.J. Coutts and J.D. Meakin, Eds., Current Topic in Photovoltaics (Academic Press, New York; 1986-1989).
2. S.R. Wenham, M.A. Green, and M.E. Watt, Applied Photovoltaics (Centre for Photovoltaic Devices and Systems, University of New South Wales, Australia; 1994) pp. 42-45.
3. L.L. Kazmerski, in K. Boer, Ed., Advances in Solar Energy (Plenum Press, New York; 1986). pp. 1-123.
4. C.R. Brundle, C.A. Evans, Jr., and S. Wilson, Encyclopedia of Materials Characterization (Butterworth-Heinemann, Boston; 1992).
5. F.A. Abulfotuh and L.L. Kazmerski, in Handbook of Thin Films (Plenum Press, New York; 1997) (in-press).
6. L.L. Kazmerski, J. Materials Research (MRS, Pittsburgh, Penn.; 1997), Special issue edited by F. Wald and R. Bell, in-press.
7. See, "From Molecules to Materials", brochure detailing the capabilities of the Center for Measurements and Characterization (NREL, Golden, CO; 1997).
8. Worldwide web site: www.nrel.gov/measurements
9. L.L. Kazmerski, Proc. 23rd IEEE Photovoltaic Specialists Conf. (IEEE, New York; 1993) pp. 1-7.
10. See, for example, K.A. Emery and C.R. Osterwald, in Current Topics in Photovoltaics, Vol. 3, T.J. Coutts and J.D. Meakin, Eds. (Academic Press, New York; 1988).
11. See, for example, M.A. Green, K. Emery, K. Bucher, D.L. King, and S. Igari, Solar Cell Efficiency Tables, published periodically in Progr. in Photovoltaics: Research and Applications (J. Wiley, West Sussex, UK; 1994-present).
12. K.A. Emery and C.R. Osterwald, Solar Cells 17, 253 (1986).
13. H. Ossenbrink, R. Van Steenwinkel, and K. Krebs, Ispra Establishment, PREPRINT EUR 10613 EN, April, 1986; Also, J. Metzdorf, T. Wittchen, K. Heidler, K. Dehne, R. Shimokawa, F. Nagamine, H. Ossenbrink, L. Fornarini, C. Goodbody, M. Davies, K. Emery, and R. DeBlasio, Proc. 21st IEEE Photovoltaic Spec. Conf. (IEEE, New York; 1990) pp. 952-959. Also, Final Report: PTB TPT PTM-Opti31, ISBN 3089429-067-6, Braunschweig, 1990, pp. 174-180.

14. R.K. Ahrenkiel, in <u>Semiconductors and Semimetals</u>, Vol. 39, R.K. Ahrenkiel and M.S. Lundstrom, Eds. (Academic Press, New York; 1993) pp. 39-150.
15. R.K. Ahrenkiel, Patent filed, June 3, 1997.
16. J.W. Orton and P. Blood, <u>The Electrical Characterization of Semiconductors: Measurement of Minority-Carrier Properties</u> (Academic Press, London; 1990).
17. A.W. Czanderna and G.J. Jorgenson, Proc. 14th SNL/NREL Photovoltaics Program Review Meeting, Lakewood, Colorado (AIP, New York; 1997). in press; Also, see A.W. Czanderna, this volume.
18. L.L. Kazmerski, Proc. 23rd IEEE Photovoltaic Specialists Conf. (IEEE, New York; 1993, pp. 1-7. Also, Vacuum 43, 1011 (1992).
19. L.L. Kazmerski, J. Vac. Sci. Technol. B, 9 1549 (1991).
20. L.L. Kazmerski, Proc. European Photovoltaic Solar Energy Conference, Amstermdam (Kluwer Publ., The Netherlands; 1994).
21. L.L. Kazmerski, J. Vac. Sci. Technol. (1997) in press.
22. J.F. Walker, <u>TEM Sample Preparation: Site Specific Search Strategy</u> (FEI Europe, Ltd., Cambridge, UK; 1996).

The 1997 Calaveras Awards
for "Leapfrog" Technologies

Larry Kazmerski

has leapt to a new level by giving

The Most Fun Presentation

on

PV Characterization Beyond the Horizon

The First Conference on Future Generation Photovoltaic Technologies
Sponsored by the National Renewable Energy Laboratory

A Learning Curve Approach to Projecting Cost and Performance for Photovoltaic Technologies

George D. Cody* and Thomas Tiedje**

*Exxon Corporate Research, Clinton Township, Route 22 East, Annandale N.J. 08801,
gdcody@erenj.com
**Physics Department, University of British Columbia, 6224 Agriculture Road, Vancouver,
British Columbia. V6T1Z1, Canada, tiedje@physicsubc.ca

Abstract. The current cost of electricity generated by PV power is still extremely high with respect to power supplied by the utility grid, and there remain questions as to whether PV power can ever be competitive with electricity generated by fossil fuels. An objective approach to this important question was given in a previous paper by the authors which introduced analytical tools to define and project the technical/economic status of PV power from 1988 through the year 2010. In this paper, we apply these same tools to update the conclusions of our earlier study in the context of recent announcements by Amoco/Enron-Solarex of projected sales of PV power at rates significantly less than the US utility average.

EXPERIENCE/LEARNING CURVE PROJECTIONS OF TECHNOLOGICAL PROGRESS

Photovoltaic (PV) power continues to play a significant role in meeting the demands for remote power in undeveloped countries. However the current cost of electricity generated by PV power is still high with respect to utility power, and, despite considerable progress, there remain questions as to when, if ever, PV power can be competitive with electricity generated by fossil fuels and supplied through the utility grid. A previous paper[1] by the authors employed a utility pricing formula and an experience/learning curve, to project cost and performance of PV modules and systems and supplied an objective framework for considering this important question.

Maycock in 1975[2] was the first to use the experience/learning curve to assess progress in PV power and it continues to play a significant role in his surveys of new developments in photovoltaic technology[3]. The concept of the experience/learning curve derives from extensive studies of manufacturing cost reductions in major industries over the last forty years[4], and is a generic description of the observed empirical power law relation between the cost of a product, C(V), as a function of the cumulative quantity, V, of the product that has been produced. In a large number of cases it has been found that the functional relation between the two quantities can be expressed by the power law:

$$\frac{C(V)}{C(V_0)} = \left(\frac{V}{V_0}\right)^b \tag{1}$$

where the exponent b, defined as the learning parameter, is negative and $C(V_0)$ and V_0 correspond to the cost and cumulative volume at an arbitrary initial time. From Eq. (1), an increase in the cumulative production by a factor of 2 leads to a

CP404, *Future Generation Photovoltaic Technologies: First NREL Conference*, edited by McConnell
© 1997 The American Institute of Physics 1-56396-704-9/97/$10.00

reduction in the cost of the product by a progress ratio λ, where $\lambda = 2^b$. The progress ratio, λ, when expressed in percent, is a measure of technological progress that drives the cost reduction, and has been found in a large number of studies to range from 70% to 90%. A progress ratio between 81% and 82%, (reflecting a reduction in cost to 81-82% for every factor of two increase in cumulative production) is the median value resulting from an analysis of many industries, although a 70% learning curve is often found in the mass-production of "high-tech products" such as integrated circuits[4].

As indicated in Figure 1, our data on the average factory price (cost times mark-up) of solar PV modules in ('95$/WattPk) after 1983, agree with recent data of Maycock[3] for all types of PV systems. There is also good agreement between our data for the period 1976-1988, and recent cost and cumulative sales data for PV systems given by Williams and Terzian[5]. Indeed all of the recent data suggest $\lambda \approx 0.8$ for PV systems based on crystalline silicon solar cells.

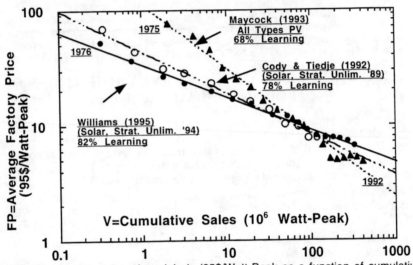

FIGURE 1. Factory price of module in '95$/Watt-Peak as a function of cumulative sales (Megawatt-Peak): All types PV, Maycock[3] 1975-1992; Solar PV from Cody and Tiedje[1]: 1976-1988 and Williams[5]: 1976-1992

A critical condition for the validity of any experience curve projection is that fundamental limits on technical performance or cost do not occur on the projection path. In our earlier paper, we derived absolute limits for the efficiency of amorphous and thin film crystalline solar cells (above 20 and 30% respectively) which clearly do not present fundamental obstacles for system efficiencies of 15% maximum[1]. Are material costs a current fundamental obstacle for progress along the learning curve? To answer this question, we convert a minimum module construction cost, C_m, in units of ($/sqM), to module selling price, S_m in units of ($/Watt-Peak), through Eq. (2) where we have assumed an 80% ratio of system efficiency (SE) to module efficiency,

$$S_m \ (\$/\text{Watt-Peak}) = \left(\frac{C_m(\$/\text{sqM})}{12.5SE(\%)} \right) \qquad (2)$$

The minimum material costs of glass and silicon for a high efficiency thin film cell[6] are about 15('95$/sqM)[1]. An increase of C_m by ≈ 4 to account for other contributions to module cost, leads to $C_m \approx 60$('95$/sqM), and hence, at SE=15%, $S_m \approx 0.30$('95$/Watt-Peak) - a factor of ≈ 10 below the lowest data of Figure 1.

UTILITY PRICING FORMULA FOR PV POWER

As in our previous paper, we use the results of the experience curve analysis to define and project the economic performance of PV power, based on a utility pricing formula originally due to Roger Taylor[7,8]. Our equation for the cost of electrical energy produced by PV systems is given by Eq. (3) where COEPV, is the Cost Of Electricity produced by a PV system expressed *in constant dollars* per kilowatt-hour ($/kWh). In our earlier paper we employed (1986$) as the constant dollar base since the data base was from the period 1976-1988. In the present paper we utilize (1995$) and for reference we note that: 1.00(1986$)=1.34(1995$).

COEPV=

$$\left(\frac{LCC}{8760CF} \right) (1+\delta) \left[1250S_m + \left(\frac{100C_{bos}}{SE} \right) + C_p \right] + \left(\frac{LF}{8760CF} \right) \left(\frac{100C_{om}}{SE} \right) \qquad (3)$$

On the right hand side of Eqs.(3) are the economic and technological factors that determine COEPV: CF = Capacity Factor = Annual Solar Flux Relative to Peak Flux; SE= System Efficiency in percent which is assumed to be 80% of module efficiency; S_m = Photovoltaic Module Cost in ($/Watt-Peak); C_{bos} = Area related Balance of Systems Cost in ($/sqM); C_p = Power Conditioning Cost in ($/Kilowatt-Peak(AC)); C_{om} = Annual Operations and Maintenance Cost in ($/sqM) The economic factors are defined for an "N year" system life, and are derived in our earlier paper[1]: LCC = Levelized Constant Cost of Capital at a given Rate of Return, over N years of System Life; LF= Levelizing Factor for Operations and Maintenance for a given Rate of Return and Inflation Rate over N years; and δ= Indirect Cost Factor.

Reasonable estimates[1] for these quantities in 1995 are: C_{bos} = 70 ('95$/sqM); C_p = 200 ('95$/kW-Peak (AC)); C_{om} = 4('95$/sqM); CF = 0.27 (Southwest Location); LCC=0.0781 (30 year life, 7.5% rate of return); LF=1.48 (30 year life; 3.5% real increase, 7.5% return); δ= 0.5. For independent estimates of the same quantities, the reader is referred to a recent paper of Sabisky[9] that also utilizes Eq. (3) for COEPV. We substitute the above constants, the estimated *current module factory price S_m =5.70('95$/Watt-Peak)* , extrapolated from the Williams data of Figure 1, and System Efficiencies of 7.2% and 15% respectively in Eq. (3), and exhibit the values obtained for COEPV[$\approx$0.40 ('95$/kWh)] in Figure 2. It should not be a surprise, given the dominance of the current module cost, that doubling the system efficiency, while it halves such area related costs as balance of systems (BOS), and operations and maintenance (O&M), has little effect on COEPV.

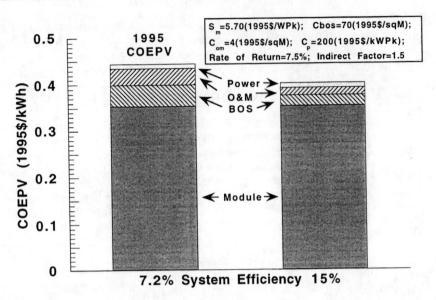

FIGURE 2. Contributions to the cost of electricity produced by PV for a module cost of 5.70('95$/Watt-Peak), System Efficiencies of 7.2 and 15% and estimates of the other cost parameters of Eq. (3) given in the text.

Assuming a constant annual rate of sales, we can project the cost of PV power to any year based on the experience curve and Eq. (3). In Figure 3 we present *two scenarios* with constant *annual growth rates of sales* of *20%* and *40%* respectively, and the 78% learning curve of Figure 1. In 1995 annual sales of PV modules were about 80 Megawatt-Peak. The *present rate of 20% annual growth* projected to 2010, would give a module cost of 1.60('95$/WattPk) in 2010, annual sales of 1700 Megawatt-Peak, 20 times higher than in 1995, and annual sales revenues of 2.7 billion('95$/yr).

From Figure 3, *COEPV in 2010* for a system efficiency of 15%, is projected to be between *0.09 - 0.15('95$/kWh),* above the US utility average of 0.08('95$/kWh), but economically attractive for utilities with day peaking summer power loads.

In Figure 4, we exhibit the components of COEPV for the 40% annual growth projection in the year 2010 [S_m =0.60('95$/Watt-Peak] for System Efficiencies of 7.2 and 15%. We note that under these circumstances, where balance of systems and operations and maintenance costs are comparable to module cost, the magnitude of the efficiency is indeed critical, and that in contrast to Fig. 2, doubling the system efficiency essentially halves the COEPV.

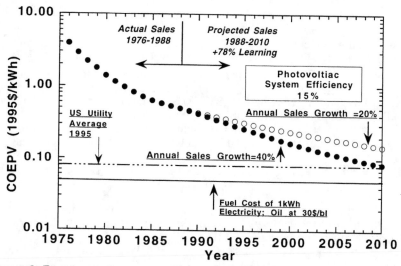

Figure 3. Experience curve projections in time for COEPV in constant dollars for a system efficiency of 15%, for two "annual growth" scenarios of 20 and 40% per year and the technical and economic factors of Eq.(3) given in the text.

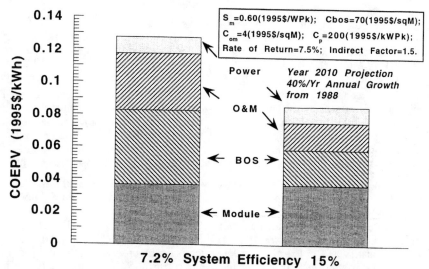

FIGURE 4. Contributions to the cost of electricity produced by PV for a module cost of 0.60('95$/Watt-Peak), system Efficiencies of 7.2 and 15%, and the same estimates of the other cost parameters of Eq. (3) as utilized in Figs 2 and 3.

From Figs. 1-4, we can draw the following conclusions: (1) the Learning/Experience curve can be a useful projection tool - for example from our earlier paper [See Figure 1, *Cody and Tiedje* data, *20% annual growth in sales* and a *78% learning curve*] we projected *in 1988* a *module price of*

4.60('95$/WattPk) *for 1995*, about 20% lower than that used in Figure 2, about 30% higher than the 3.50±40%('95$/WattPk) average module manufacturing cost quoted in a recent review[10] of the PVMaT project of the DOE, and 15% higher than the 4.00('95$/WattPk) quoted in a recent Canadian government report[11]; (2) COEPV is *currently module price limited* and increases in module efficiency without corresponding decreases in module cost in $/WattPk, is not a significant factor in reducing COEPV; (3) Significant reductions in COEPV to 0.0.09('95$/kWhr), about the US average utility rate, might be anticipated to occur by the year 2010 given a 40% annual increase in sales, 78% learning curve, and a system efficiency of 15% - at which point efficiency is critical.

RECENT DEVELOPMENTS IN COST PROJECTIONS FOR PV POWER

In striking contrast to the above "conservative projections", Enron proposed in November of 1994, to sell PV generated electricity *in 1996* to the Department of Energy at *0.055($/kWh)* - 70% below the national average retail price - from a PV power plant starting with 10MW-peak capacity in 1996 and building up to 100MW-peak capacity in 10 years[12]. The PV power plant would be on former nuclear testing grounds in the Nevada desert, powered by thin film a-Si cells produced by a partnership with Amoco-Solarex - Amoco/Enron Solar[13].

In this and subsequent sections we place the Amoco/Enron-Solarex optimistic PV power costs in the context of learning curve projections for the module prices and consider plausible scenarios for the achievement of an *average COEPV of 0.055 ($/kWh)* over the 10 year life of the project. We will assume module efficiencies of the order of 10% for a-Si modules, and system efficiencies of 8%, and a capacity factor, CF=0.27 indicating a Southwest location[1] for the PV system. Our approach has been to start with Eq. (3) and the original technical and economic constants, which we adjust to produce a *net return of zero* between the cost of generating electricity at the COEPV derived from Eq. (3) and the electricity sales of 10 to 100 MWpk over the 10 year manufacturing stage of the project. In what follows we present one self consistent scenario - there are others.

In Figure 5 the scenario starts with an *assumed 1995 module cost of 2.70('95$/Wk)* for thin film amorphous silicon (half that of current Si modules), and shows the dramatic reduction in module cost produced by a *70% learning curve* driven by 10 Megawatt-Peak yearly additions from 1996-2005 starting from an *initial cumulative volume of 2.5MWPk* for thin film a-Si PV solar modules. We may justify a 70% learning curve, which is of the order of that found for "high tech-mass-production", by noting the Amoco/Enron-Solarex project is the first large scale manufacture of a thin film solar modules on low cost, glass substrates

The 70% learning curve scenario projects a module cost of 0.40$/W/Pk in 2005 for cumulative sales of 103MWattPks. It is an interesting coincidence that this module cost is identical to the module cost projected in our earlier paper for the year 2010, which was based on a *40% growth rate for all PV modules*, not just for a-Si modules[1], and cumulative sales by 2010 of 100,000MWPk!

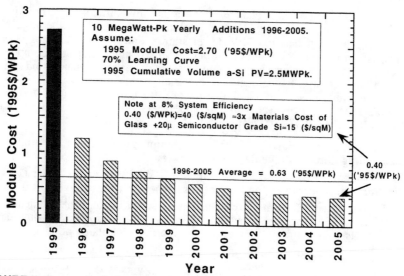

FIGURE 5. One scenario for Enron Amoco a-Si PV modules cost reduction with 70% Learning Curve; 1995 cumulative 2.5MW-P, 1995 Factory Cost S_m=2.70('95$/Wk) and 10 MWPk annual production; 2005 cumulative =103MWPk, 2005 Factory Cost S_m=0.40('95$/Wk)

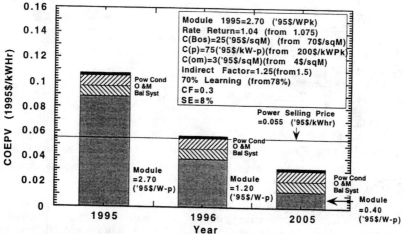

Figure 6. COEPV and its cost contributions for the Enron scenario of Figure 5.

However, additional, but perhaps equally plausible, reductions are still required in the constants of Eq. (3) to fit the ambitious goals of the Amoco/Enron-Solarex proposal to DOE. We have found it necessary to reduce the previous estimates of: "Rate of Return"[4.0% from 7.5%], "C_{bos}" [25 from 70 ('95$/sqM)], "$C_p$"[75 from 200 ('95$/kW-p)], "C_{om}"[3 from 4 ('95$/sqM)] and the "Indirect Factor," "δ" (0.25 from 0.5).

51

The direction of these changes, but not their magnitude, might be expected from the support of the State of Nevada in obtaining financing for the project, the use of DOE land for the site of PV power station, and the vertical integration of PV manufacturing and PV plant operation. Figure 6 summarizes the components of COEPV that meet the cash flow requirements over the 10 year period, for the Amoco/Enron-Solarex PV project with electricity sales at 0.055 '95$/kWHr and average electricity costs that match that figure. *We note from Fig. 6, that by the year 2005 10MW-pk of power is sold at 0.055 ('95$/kWHr), but is generated at a cost of half of that, 0.03 ('95$/kWHr)!*

Figure 7 is derived from an attempt by Al Rose, a pioneer in understanding the fundamental limitations on the cost and performance of photovoltaic energy systems[14,15], to compare "PV Power" to the "Solar Power of Agriculture" that generates food or chemical energy from sunlight. It is an up-date of a similar figure from our earlier paper[1], and illustrates the efficiency/cost tradeoff for PV compared to Photosynthesis, in the context of our earlier projections as well as the COEPV of the Amoco/Enron- Solarex project.

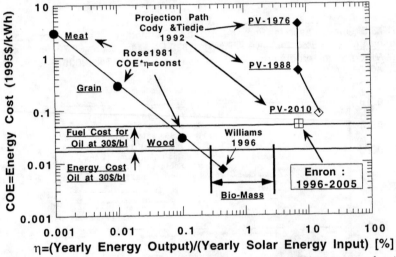

Figure 7. Energy cost of meat, grain, wood and biomass, and PV power as a function of solar efficiency, compared to fuel/energy costs of oil at 30$/bbl.

As illustrated in Figure 7, while PV systems in 1988 had more than 100 times the efficiency in converting sunlight to energy than did trees, the cost of energy produced by PV electricity in 1988 [0.53('95$/kWh)] was 20 times more expensive than the cost of chemical energy obtained from wood [0.027("95$/kWh)] and more than 70 times more expensive than the chemical energy obtained from biomass[16]!

In 1981 Rose predicted that "It would be a gross underestimate of the competence of our scientific community and of the ingenuity of our industry not to believe that they could design a solar cell system significantly cheaper than nature's forests given a hundred-fold advantage in efficiency". Our 1992 paper[1] projected the achievement of that "cheaper system" *eighteen years later* , to the year 2010. The Amoco/Enron-Solarex Nevada project, if successful, will have made a major step in that direction *over the next five years*, by rigorous attention to low cost module manufacturing[17], along the lines driven by NREL's PVMaT program[10], and plausible reductions in balance of systems and capital costs that compensate for the relatively low efficiencies (10%) obtained for current multiple junction amorphous silicon PV modules. Further progress in broadening the utility market for PV in developed countries will require significant increases in system efficiency at the same or lower manufacturing costs levels. Further progress in extending the market for PV in off-grid areas will require equally significant improvements in the cost and performance of electrical energy storage.

REFERENCES

[1]G. D. Cody and T. Tiedje, "The Potential for Utility Scale Photovoltaic Technology in the Developed World: 1900-2010," in *Energy and the Environment*, edited by B. Abeles, A. J. Jacobson, and P. Sheng (World Scientific, Singapore, 1992), pp. 147-217.

[2]P. D. Maycock and G. F. Wakefield, "Business Analysis of Solar Photovoltaic Energy Conversion," Proceedings 11th IEEE PVSC, 1975, pp 252-255.

[3]Paul Maycock, "Photovoltaic Technology, Performance, Cost and Market Forecast to 2010," (1993) and private communication, 2/26/93.

[4]Linda Argote and Dennis Epple, "Learning Curves in Manufacturing," Science **247**, 920 (1990).

[5]R. H. Williams and G. A. Terzian, Center for Energy and Environmental Studies Princeton University, Princeton, New Jersey, Report No. PU/CEES #281, 1995.

[6]T. Tiedje, E. Yablonovitch, G. D. Cody *et al.*, "Limiting Efficiency of Silicon Solar Cells," IEEE Trans. on Electron Devices **ED-31** (5), 711-716 (1984).

[7]Roger Taylor, "Utility Requirements of Photovoltaic Power Systems," Proceedings 161st Meeting of Electrochemical Society, Montreal, Canada, 1982, pp .

[8]Edgar DeMeo and Roger Taylor, "Solar Photovoltaic Power Systems: An Electric Utility R&D Perspective," Science **224**, 245 (1984).

[9]E. S. Sabisky, "A Minimum Achievable PV Electrical Generating Cost," Solar Energy Materials and Solar Cells **40** (1), 55-70 (1996).

[10]R. L. Mitchell, C. E. Witt, H. P. Thomas *et al.*, "Benefits from the US Photovoltaic Manfacturing Technology Project," Proceedings 25th IEEE PVSC, Washington, DC, 1996, pp 1215-1218.

[11]G. Leng, L. Dignard-Bailey, G. Tamizhmani *et al.*, CANMET Energy Diversification Research Laboratory, Natural Resources Canada, Varennes, Quebec, Report No. EDRL 96-41-A1(TR), 1996.

[12]Allen R. Myerson, "Solar Power for Earthly Prices," in *New York Times* 11/15/94, pp. D1-D2.

[13]Press Release, "Amoco Corporation and Enron Corporation Have Formed a General Partnership," 12/19/94

[14]A. Rose, "Solar Energy a Global View, Part 1," CHEMTECH **11**, 566-571 (1981).

[15]A. Rose, "Solar Energy: a Global View, Part 2," CHEMTECH **11**, 694-697 (1981).

[16]Robert H. Williams, Talk at Exxon Corporate Research . April 22, 1996

[17]D. E. Carlson, R.R. Arya, M. Bennett *et al.*, "Commercialization of Multijunction Amorphous Silicon Modules," Proceedings 25th IEEE PVSC, Washington, DC, 1996, pp 1023-1028.

Reliability and Lifetime Issues for New Photovoltaic Technologies

A.W. Czanderna

Center for Performance Engineering and Reliability
National Renewable Energy Laboratory, Golden, CO 80401-3393

Abstract. The purposes of this paper are to elucidate the crucial importance of predicting the service lifetime (SLP) for new photovoltaic technologies (PV) modules and to present an outline for developing a SLP methodology for encapsulated PV cells and minimodules. Specific objectives are (a) to illustrate the essential need and generic nature of SLP for several types of existing solar energy conversion or conservation devices, (b) to elucidate the complexity associated with quantifying the durability of these devices, (c) to define and explain the seven major elements that constitute a generic SLP methodology, (d) to show that implementing the SLP methodology for developing laboratory-scale PV cells and minimodules can reduce the cost of technology development, and (e) to outline an acceptable methodology for relating accelerated life testing to real time testing, using sufficient sample numbers, and applying the methodology in (c) for predicting a service lifetime. The major conclusions are that predicting the service lifetime of PV cells and minimodules should be an essential part of the research and development for developing any future generation PV technology and that using the SLP methodology can be most cost-effectively applied to laboratory scale PV cells and minidevices.

INTRODUCTION

The objectives of this paper, stated in (a) through (e) in the abstract, are all driven by and relate to achieving a goal of more than 30-year service lifetimes (SL) for photovoltaic (PV) systems (1). The reliability and SL goals for developing next generation PV technologies are (i) to identify, understand, and then mitigate the causes of changes in cell and module materials that alter crucial materials properties and reduce the performance or limit the service lifetime, or both, of cells/modules and (ii) to develop new or improved materials that offer greater promise for a module service life expectancy of more than 30 years. These goals are generic for most multilayer, energy efficiency (e.g., conservation) or renewable energy (EERE) conversion devices and can be modified by simply changing "material" in (i) or (ii) to cell, array, or system for other PV specific goals or by changing "module" in (i) or (ii) to some other EERE device such as a solar mirror, electrochromic window, or flat-plate collector. For the service lifetime of other elements, the word "materials" may be changed to be broader, e.g., component or subassembly. In keeping with the generality of the stated goals, we first discuss the general principles behind the requirements for establishing the service lifetime of EERE multilayer devices used for solar energy

CP404, *Future Generation Photovoltaic Technologies: First NREL Conference*, edited by McConnell
© 1997 The American Institute of Physics 1-56396-704-9/97/$10.00

conversion or conservation and then show how these principles are being applied to PV cells and minimodules.

Because emerging advanced multilayer solar energy conversion devices are expected to exhibit service lifetimes of more than 20 to 30 years, technology development cannot be based solely on using real-time weathering for establishing long lifetimes. Typical multilayer SECD include silvered reflectors, PV modules, flat-plate collectors, electrochromic windows, and photoelectrochemical cells. Many U.S. companies are at a critical juncture in marketing improved products, such as protective coatings, interior lighting reflectors, and polymer-based coatings for many uses, whose service lifetime must be predicted. Without confident knowledge from an accurate and reliable service lifetime prediction (SLP) methodology, warranties may be at high risk. Life cycle costs require knowing accurately a service lifetime, as well as the obvious initial cost of the delivered product, cost of initial capital, future costs discounted to present value, and future costs for the operating, maintenance, repair, replacement, demolition and removal activities.

The service lifetime of materials, devices, or systems is the time at which their (time-averaged) performance degrades below a prescribed or required value, i.e., a failure or a failure to perform at the preassigned value. This definition is deduced from the American Society for Testing and Materials (2) definitions for durability, serviceability, and service life. Durability (2) is the capacity of maintaining the serviceability of a product, component, assembly, or construction over a specified period of time. Serviceability (2) is the capability of a product, component, assembly, or construction to perform the function(s) for which it was designed and constructed. For EERE devices, the effective definition of durability is the capability of the device to perform its designed function, i.e., device performance versus time. ("Reliability" is interchangable with this operative definition of "durability".) Service life (2) is the period of time after installation during which all properties exceed the minimum acceptable values when routinely maintained. Thus, service life requires the selection of some minimum performance criteria, e.g., a PV module rated at 50 W at the normal operating temperature condition (NOTC) may be a "failure" when its power output falls below 40 W. *The minimum acceptable performance (i.e., "failure") needs to be defined for a SLP of cells/modules of any PV technology.* SLP is the estimated service life based on criteria and calculated using the protocol outlined later in this paper.

Desired lifetimes of typical EERE devices are as follows: polymeric or glass reflector constructions for mirror applications, > 20 years; PV modules, > 30 years; electrochromic windows, > 20 years; flat-plate collectors, > 10 years; and low-emissivity coated windows, > 20 years. Because the desired lifetimes range from > 10 years to > 30 years, accelerated lifetime testing (ALT) in (simulated) weather environments and a predictive methodology must be used. The lifetimes of EERE devices are not unique in U.S. technology; several first-rate SLP groups have been developed at a few major U.S. corporations. As is the case with EERE

56

devices, U.S. industry (e.g., coatings, lighting, polymeric-based devices) cannot wait for the results from real-time testing (RTT) so ALT and SLP must be used.

We illustrate the vision of being able to predict the service lifetime of an encapsulated PV module in Fig. 1. We have arbitrarily chosen a generic PV module with 100% of its rated output at NOTC. If no loss in performance occurs, the module will produce 100% forever. However, losses in PV *systems* range from a low of 1% per year to 2.5% per year (1), as shown by the solid lines in Fig. 1. The actual losses are shown for the Carissa Plains, CA, 5.2 MW system (3-5), which is the most extreme case of degradation reported. Because some of the modules were removed from the plant after 1991, the projection to seven years is based on the efficiencies of the remaining modules. The losses in real systems are from *all* causes, and not just in the modules. Because it is not known how to project future output from a cell, module, or a system, several possible hypothetical projections are illustrated by the dashed lines for over nine years. These include projections with a simplistic linear extrapolation, with a decreasing rate of loss (perhaps from self-passivating reactions), and with an increasing rate of loss (perhaps from autocatalytic reactions). If the performance could be accurately predicted, the area under the projected curves would permit calculating the predicted output per year until failure is reached, and life cycle costs could then be calculated from the total power that would be produced and from the other life cycle costs, e.g., initial, maintenance, and operating costs. The major issue for any new PV technology is to resolve what (time-averaged) loss in performance, i.e., power output, is permitted until the time of failure (in years) is reached.

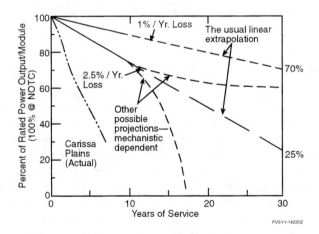

FIGURE 1. Actual and potential percentage losses in efficiency (performance) of current-generation PV *systems.*

Realizing the vision of being able to predict the power output of systems for their operational lifetime is clearly possible, but two significant problems must be resolved. First, the technical reasons for the power losses must be determined. The losses plotted in Fig. 1 are *system* losses; causes of performance losses need to be identified and then mitigated for cells, modules, or any other balance of systems components. No studies have been completed that establishes the relationships between the accelerated degradation of individual modules and RTT. In our prior work, we have clearly demonstrated some of the losses result from EVA browning (1, 6). Secondly, resources need to be increased substantially for a proactive technical approach that will result in improving *current* and *next-generation* PV products, e.g., by (a) monitoring the RTT performance of appropriate statistically-significant sets of individual PV cells, minimodules, and modules, (b) deducing causes of failure in *these* products and (c) by studying new/improved materials and designs at the cell and module level. A SLP then becomes possible by adopting such an approach for identifying and isolating failure modes or degradation mechanisms at the cell/minimodule, module, and other component levels as outlined (7). At present, RTT of *individual* module performance is being monitored at different sites for three cases (8-10), but without complete *initial* characterization before deployment. Eventually, ALT needs to be performed on sets of "identical" modules for accelerating the degradation of design/materials weaknesses and/or for comparing the rates with the RTT results; the RTT data needs to be taken at several environmentally diverse sites. From the work we are now doing, we fully recognize that establishing PV device/module service lifetimes of 30 years or more involves solving many synergistically driven problems. Research and development using ALT at an *early state can reduce the cost of developing advanced PV technologies.*

Collecting Solar Energy

The major problem in solar energy technologies is not discovering how to collect the radiant flux but rather establishing how to collect it at a cost competitive with conventional power generation (11). The latter is one of the reasons ethylene vinyl acetate (EVA) was chosen for use in PV modules rather than other more expensive polymers known to have better properties (1,12). The solar energy reaching the Earth has a typical power density of 500 to 1000 W/m,^{2} which means collection areas of 200 to 400 square miles per quad are required for typical 1-sun solar technologies (12). The cost of the materials used, device production processes, and the operation and maintenance of systems must be held to a minimum. This requires, for example, using multilayered stacks of superstrates, substrates, and the active thin (or thick) films or coatings for various collection schemes, e.g., mirrors, PV systems, electrochromic windows, and flat-plate collectors (as illustrated in Fig. 2). These must be made from inexpensive,

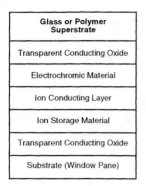

Electrochromic Windows

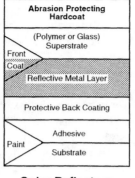

Solar Reflectors

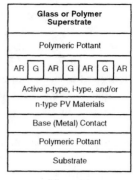

PV Cells

PVSY1-142301

FIGURE 2. Cross sections of typical multilayer stacks used for solar reflectors, electrochromic (EC) windows, and PV cells. The front coat and paint layers are optional additions for solar reflectors. Different detailed mechanisms of failure are expected for the passive reflectors when compared with the active (ion or electron transport) PV or EC devices.

durable, and easily processed materials. For example, if we assume PV devices can be made in 6-foot widths using a continuous process (in vacuum) moving at 1.5 f/s for 24 h/day and 365 days/year, then this manufacturing unit will produce only 10.18 square miles of PV collectors. At 15% efficiency and optimum tracking, the energy output will be 0.025 quads per year and 40 of these manufacturing units will be needed for annually producing about 1% of the current U.S. consumption. A process rate of 1.5 f/s is about 100 times faster than many current production processes as described elsewhere in these proceedings.

The materials chosen not only provide device-specific functions but also environmental protection, which is crucial for the long service lifetimes that will reduce life-cycle costs and increase the market value of the devices. When in use, man-made solar energy conversion systems are subjected to a unique set of in-service degradation factors that may alter their *stability* and, hence, their performance and life cycle costs in addition to the initial cost of the system. Degradation factors include biological (microorganisms, fungi, bacteria), incompatibility (physical or chemical materials interactions), sustained or periodic stresses, use (system design, normal wear, installation and maintenance, abuse) and weathering. For solar energy collecting devices, the weathering factors include irradiance (especially UV radiation), temperature, atmospheric gases (O_2, O_3, CO_2) and pollutants (gases, mists, particulates), diurnal and annual thermal cycles, and, in concentrating systems, a high-intensity solar irradiance. In addition, rain, hail, condensation and evaporation of water, dust, wind, freeze-thaw cycles, and thermal expansion mismatches may impose additional losses in the performance of a solar device. All degradation factors must be considered not only individually, but also collectively for degradative effects that may result from their synergistic action on any part of the system. The first prerequisite is that the bulk properties of the superstrate, substrate, thin film, coating, and other materials be stable. For example, photothermally-induced degradation can be the predominant factor for polymeric or organic-based materials. The activation spectrum depends on the bond strengths and is sensitive to the incident UV wavelengths. Any laboratory-scale, UV testing must correctly simulate the solar irradiance during in-service use. NREL scientists have used xenon arc-lamps with appropriate filters for nearly 20 years for simulating the UV and visible solar irradiance.

After the requisite stability of the "bulk" materials is achieved, interface reactions are known to be thermodynamically driven because of the higher free energy state of atoms at interfaces (13). A further need may then be to choose the different materials carefully to permit achieving a 30-year "stability" (14a) or to modify the interfaces for attaining the same goal (14b). A service lifetime goal of over 20 or 30 years is targeted for all the devices in Fig. 2. For projecting a service lifetime to yield the desired time-dependent level of performance, substantially more SLP-directed work is needed. The detailed application of the SLP methodology will be more challenging for the active (PV and EC) devices than for the passive solar mirror constructions.

The goals cited in (i) and (ii) above are for the type of research needed to develop an understanding of the behavior of low-cost, high-performance, active and encapsulation materials that can be used to extend the service lifetime or to identify materials that offer new options for use in the device. For the conventional triad of requirements that includes low (initial) cost, high performance, and long-term durability (reliability), we substitute service lifetime to replace durability (reliability) as this is what is really desired. A service lifetime *prediction* (SLP) is the ability to project the future time dependence of the performance that defines the durability. Service lifetime must be known to determine the life-cycle cost for using a device of known initial cost and initial performance (i.e., efficiency in PV cells). The cost-effective deployment of any EERE device is partly limited by the durability and life-cycle cost of the materials used. Research on the active and encapsulating materials and studies that address the influence of the materials degradation on device performance are of critical importance, especially to understand soiling of surfaces, degradation of polymeric materials, the effects of oxygen and water vapor permeation, corrosion, the degradation of the active materials, and degradation at interfaces. The ultimate need is to identify materials that will not decrease the performance during exposure to actual use conditions for the desired/required service lifetime of the device. Establishing a service lifetime prediction requires a multidisciplinary team of experts plus supporting diagnostic expertise. These include people knowledgeable in the disciplines of materials science, materials engineering, surface science, corrosion science, polymer science, solid state physics, physics, physical and analytical chemistry, electrochemistry, statistical methods, theorists on lifetime prediction, etc., who have (or can access) sophisticated diagnostic and measuring equipment. Appropriate capabilities for accelerated and real time weathering of devices are also essential. If done properly, predicting a service lifetime of any device requires significant resources but is essential before major investment decisions will be made.

MAJOR ELEMENTS FOR SERVICE LIFETIME PREDICTION

Before implementing a SLP methodology, it is essential to define the problem as exactly as possible. The in-service performance requirements and criteria for the device must be established. The materials in the device must be characterized thoroughly for structure, composition, critical performance characteristics, properties that indicate degradation, range and type of degradation factors, and all possible degradation mechanisms. Postulated degradation mechanisms for a contemporary crystalline silicon PV cell are shown schematically in Fig. 3.

The glass cover plate may or may not contain a UV screen such as cerium dioxide, or a modified polymer may or may not be laminated between it and the pottant. A primer may or may not be used in the EVA formulation or be coated onto the glass substrate. The pottant in nearly all deployed monocrystalline (c-Si), or polycrystalline silicon (pc-Si) systems is EVA, and about 95% of the ca. 500

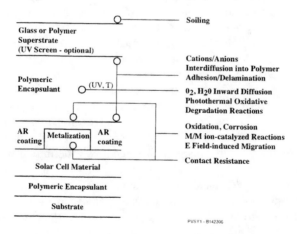

FIGURE 3. Schematic cross-section of an encapsulated PV cell and relevant reactions/processes that may reduce the cell performance and/or service lifetime (adapted from Refs. 1 and 17).

MW of installed PV capacity is pc-Si or c-Si. An antireflection (AR) coating (typically, 50 nm of SiO_2) may or may not be deposited onto the metalization or oxide surface of the Si solar cell(s). The active solar cell material(s) may be several multilayers and have a back or base contact. Another layer of EVA that is shielded from exposure to UV, and the supporting substrate complete the module encapsulation. In a PV module, solar cells (e.g., 36 to 72 or more in a typical module) are joined by interconnects that are also embedded between the two EVA layers. Power output terminals are provided on each module.

As is also illustrated in Fig. 3, degradation can occur by weathering and/or soiling of the cover glass; photothermal, oxidative, or other degradation of the pottant (1, 22, 31); interdiffusion of ions into the pottant; metalization corrosion; electric field-induced ion migration or degradation; and polymer/metal oxide interface reactions or delaminations. Many of these processes may depend on initial impurity concentrations and trapped gases (vapors), and concentration changes during use. We emphasize this complexity of the entire module here because we have to establish which degradative reactions must be mitigated (besides pottant discoloration) and which ones are too slow to impact the performance adversely over 30 years.

Further problem definition involves postulating procedures for accelerating the potential degradation mechanisms (Fig. 3) to be used in the preliminary testing. A number of criteria are necessary for *accelerated testing* to be successful with a *goal* of making service lifetime predictions; these are discussed in some detail by Fischer et al. (15) and outlined in publications from various forums (15, 16), as well as with the PV (17-20) and electrochromic windows (21) communities. These include, for example, that the accelerated test must not alter the degradation mechanism(s); the mechanisms and activation energies of the dominant reaction(s) at normal operating conditions and accelerated test conditions must be the same; both the specimens (including materials and components only) and accelerating parameters (UV, T, RH, product entrapment, etc.) must simulate reality; cells and/or modules that simulate reality must be used in the initial accelerated tests; and the time-dependent performance loss (e.g., power loss for PV modules) must be correlated with the degradative reactions. Ultimately and ideally, the accelerated tests must be made on commercial-scale modules that are the same size as those sold to the consumer, but this ideal *may* not be necessary if predictions from laboratory-scale specimens are reliable predictors of the commercial products. Obviously, a SLP requires a definition of "failure," i.e., what loss in efficiency is acceptable after how many years; failure needs to be defined for a PV module in keeping with the power losses of 1% to 2.5%/yr being observed in systems deployed in the terrestrial environment (1).

We now summarize the seven major elements of a SLP methodology. The first sentence in each of these states the element and subsequent comments clarify the element. The major advantage of the sequence of these elements is that the first four elements can be used to improve multilayer devices until the optimum design and materials are identified. Repeated used of these four elements *is often termed "screening" of design or materials options, or both.* The seven-element, simplified SLP methodology is illustrated in Fig. 4 for c-Si or pc-Si cells or minimodules. For any next generation PV technology, the device construction in element 1 is simply used for the relevant technology. Examples of how some of these elements have been used are available for mirrors (16, 20a, 20b, 23, 24), PV encapsulants (1, 22), and coatings (19).

SLP Element 1. The "final" design and materials selections are needed for the multilayer stack. To improve the durability of the device, each *prototype* design and the materials used can be considered as "final" for elements 1 through 4. When several prototype designs are studied, statistical methods are used to identify a test matrix of the best candidate combinations. Ultimately, a set of materials and a particular design will be identified that permits proceeding to element 5.

SLP Element 2. The degradation factors imposed on the device in real-time use and the same types of factors for ALT need to be identified and quantified. As discussed in the introduction, the factors have been identified for EERE devices used in a solar terrestrial environment. For accelerated environments and for simulating the reality of the solar UV and visible radiation, it is essential that any UV source match the wavelengths reaching the Earth's surface, which means having precise knowledge of the spectral irradiance incident on the EERE device, and that the UV source intensity be a reasonable multiple of the solar intensity. For these reasons, NREL scientists have used filtered Xe-arc lamp sources since 1978 (16, 25) and have rejected other sources, such as fluorescent lamps because they do not simulate reality. Zussman indicates that the solar spectrum cutoff at sea level is 285 nm, and radiation between 290 nm and 300 nm is routinely incident at the Earth's surface (26). UV radiation can severely damage polymers if their activation spectra are at wavelengths from 290 to ca. 380 nm (27). With appropriate filters (28, 29), the Xe-arc light source simulates the solar spectrum very well from 285 to 500 nm. The source intensities usually refer to the number of suns, which are simply multiples of the solar intensity in W/m^2 at the wavelengths of interest. The materials degradation from a Xe-arc light exposure may not match the in-service experience (26). This may result, in part, from the promotion of chemical effects of secondary processes in materials by the synergism of temperature, humidity, O_2, and other weathering factors (27). Similar detailed considerations are required for all imposed stresses unless it is shown that the degradation in performance is not related to a particular stress.

SLP Element 3. The complete devices are subjected to ALT and RTT to determine their durability and *the most sensitive measurement(s) of the performance loss* (or of a parameter that can be correlated to the performance) is measured. Typically, the device performance is evaluated periodically with time from measurements made by moving the samples to the instrument(s). Ideally, the measurement(s) should be made *in situ* either by using probes so the sample is never removed from its test location, (i.e., an outdoor exposure rack or accelerated test chamber), or by using portable measuring equipment at the sample test location. Success in correlating ALT and RTT results depends crucially on

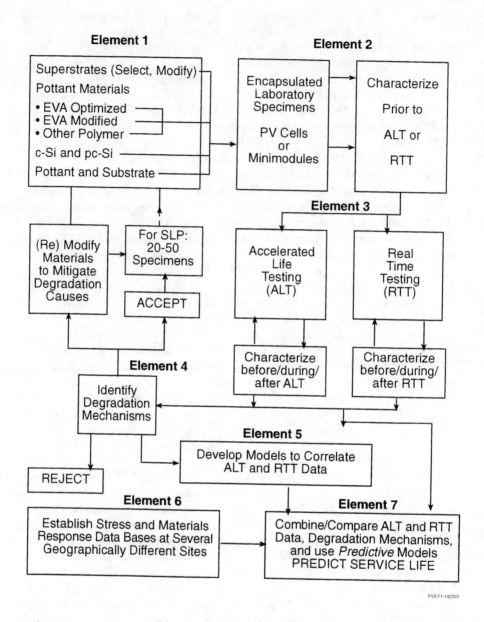

Element 1

Superstrates (Select, Modify)

Pottant Materials
• EVA Optimized
• EVA Modified
• Other Polymer

c-Si and pc-Si

Pottant and Substrate

Element 2

Encapsulated Laboratory Specimens

PV Cells or Minimodules

Characterize Prior to ALT or RTT

(Re) Modify Materials to Mitigate Degradation Causes

For SLP: 20-50 Specimens

ACCEPT

Element 3

Accelerated Life Testing (ALT)

Real Time Testing (RTT)

Characterize before/during/after ALT

Characterize before/during/after RTT

Element 4

Identify Degradation Mechanisms

REJECT

Element 5

Develop Models to Correlate ALT and RTT Data

Element 6

Establish Stress and Materials Response Data Bases at Several Geographically Different Sites

Element 7

Combine/Compare ALT and RTT Data, Degradation Mechanisms, and use *Predictive* Models PREDICT SERVICE LIFE

PVSY1-142303

FIGURE 4. Technical approach for specifically applying SLP Elements 1 through 7 for PV cells and minimodules, but the scheme can be used for other PV components (e.g., modules) and EERE multilayered devices.

64

the sensitivity, accuracy, and reproducibility of the measurement of the performance parameter. For example, if a device performance is degrading at 1% per year, a measurement of the changes in the performance of 0.1% or even less is needed if the ALT data are to be correlated with the RTT data for "reasonable" RTT exposure times. For solar mirrors, specular reflectance is correlated to loss in performance, and changes can be measured accurately and reproducibly (20a). *A measurement of PV performance with a sensitivity comparable to the specular reflectance of solar mirrors needs to be identified for PV cells or modules.*

SLP Element 4. The mechanisms of degradation of bulk materials and reactions at interfaces must be identified and understood. The degradation mechanism must result in a loss in performance of the device, or compromise the materials function, or both, to be of concern. If the rate of performance loss from the degradation is fast relative to the expected service lifetime, the cause of degradation must be mitigated, and the sequence of elements 1 through 4 must be repeated for the new or modified materials or design used initially. If the rate of degradation is slow and the activation energy can be determined for the rate-controlling reaction, it is reasonable to proceed to element 5. At present, the design for silvered polymeric mirrors (Fig. 1) is the only EERE multilayer stack that is ready for us to proceed to element 5. Substantial additional efforts are required with *PV cells* so we will be able to proceed to element 5 in 2002.

SLP Element 5. Models need to be developed for correlating ALT data and RTT data taken at several geographic sites with diverse stresses. The rate of degradation is site dependent, because the stresses that cause degradation vary from site to site. For example, the total UV insolation in the sunny southwesternern deserts in the United States is a more aggressive stress than it is in the cloudy northeastern states. The models for correlating the ALT data and RTT data must be able to accommodate different magnitudes of the stresses including time-dependent variations and any synergism of the stresses that occurs. For a successful SLP, it is critical that correct mathematical interpretations be made of the experimental results that relate or correlate the key environmental stresses (e.g., UV, T, RH).

SLP Element 6. Data bases of stresses and materials response must be established that include data from different outdoor sites. This element follows directly from element 5. While some latitude may result from considering similarities in sites, enough data must be accumulated at sites with the climatic extremes and those in between to permit reasonable interpolation to any site for planned deployment of EERE devices.

SLP Element 7. Predictive service lifetime models are then developed from the data obtained in elements 2 through 6 by using statistical approaches and life distribution models. A sufficient number of replicate samples must be part of the test matrix to deduce the life distribution model from the degradation (7). For example, an initial set of samples, which may range from a minimum of about 12 to 15 up to 50 and that all have "identical" performance, will degrade into a distribution of performances during use or aging. The Gaussian distribution, which is a special case of several types of distributions (15), can be used to illustrate this point. Initially, the Gaussian distribution is characterized by a full width at half maximum (FWHM) that is only limited by the uncertainty in measuring the initial performance parameter. As the sample set ages, the FWHM broadens because the performance of each individual device will degrade differently in comparison to others in the set (30). Thus, the distribution for aged samples will be the superposition of the distribution itself and that imposed by the uncertainty in measurement of the performance parameter(s). With the definition of "failure," the distribution of aged samples yields the time-dependence of failures. Various types of models can be applied to describe the aged distribution (7, 15). Large sample sets and ultrasensitive measurements of the performance parameter are required, if we are to achieve the best prediction results. Both of these requirements increase the cost of making a SLP. Obviously costs increase with increasing sample numbers. The performance parameter may require several measurements or developing a beyond-the-state-of-the-art measurement to achieve the desired result; in either case, the cost for making a SLP is increased. Therefore, it is critical to use efficient, statistical, experimental designs.

Obtaining a SLP for performance may be difficult for several reasons. These include the challenges of dealing with a large variability in failure times, determining the appropriate stresses causing performance degradation, extrapolating the results from ALT at elevated stress levels to the normal stress level, defining what is a "failure" of material(s) or system(s), having to use small lifetime data sets for economic reasons, and demonstrating that the degradation mechanism in ALT is the same as in RTT.

CONCLUSIONS

A methodology for predicting the service lifetime of multilayered EERE devices has been outlined and related specifically to next-generation PV technologies. The SLP methodology is not limited to PV and EERE devices but can also be applied to U.S. industrial needs. Developing the technology base for predicting 30-year PV module lifetimes requires a multiyear research effort. A "failure" in the performance level (efficiency) needs to be defined for PV modules, and is necessary for making a SLP. Furthermore, an extremely sensitive measurement of a PV cell or module performance or one that is directly correlated to the performance also needs to be identified. The multiyear effort must also result in an understanding of degradative reaction mechanisms and their relative importance, establishing the expected levels of degradation, and utilizing the most appropriate experimental methods.

SLP should be an essential element for developing any next-generation PV technology. A durability (reliability) and SLP expert should be a part of any PV technology team to assist in defining a complete programmatic approach. ALT, establishing degradation causes, and improving the new PV technology should be a part of the *initial* research and development program to narrow the materials/design options through preliminary testing. Using the complete SLP methodology that includes ALT, RTT, modeling, and analysis for obtaining a SLP can be done most cost effectively as a research and development effort on *laboratory-scale but complete PV minidevices.* a SLP is an essential element for calculating life cycle costs of any next-generation technology.

ACKNOWLEDGMENTS

The author is pleased to thank G. Jorgensen and F.J. Pern for their technical insight and careful review of the manuscript. The author is grateful to R. Hulstrom for his encouragement, C. Gay and B. Marshall for their support from the Directors Development Fund for developing SLP methodology, and the U.S. Department of Energy for their support of this work under Contract No. DE-AC36-83CH10093.

REFERENCES

1. Czanderna, A.W., and Pern, F.J., "Encapsulation of PV Modules Using Ethylene Vinyl Acetate Copolymer as a Pottant: A Critical Review," *Solar Energy Materials and Solar Cells*, **43**, pp. 101-183 (1996).

2. "ASTM Book of Definitions," American Society for Testing and Materials, W. Conshohocken, PA, 1996.

3. Gay, C.F., and Berman, E., "Performance of Large Photovoltaic Systems," *Chemtech*, pp. 182-186, (March 1990).

4. Rosenthal, A.L., and Lane, C.G., "Field Test Results for the 6 MW Carrizo Solar Photovoltaic Power Plant" *Solar Cells: Their Science, Technology, Applications and Economics*, Elsevier Sequoia, **30**, pp. 563-571, (1991).

5. Wenger, H.J., Schaefer, J., Rosenthal, A., Hammond, R., and Schlueter, L., "Decline of the Carrisa Plains PV Power Plant: The Impact of Concentrating Sunlight on Flat Plates," *Proc. 22nd IEEE Photovoltaic Specialists Conference (PVSC)*, New York, IEEE: 1991, pp. 586-591.

6. Pern, F.J., "A Comparative Study of Solar Cell Performance Under Thermal and Photothermal Tests," *Proc. PV Performance and Reliability Workshop*, L. Mrig, ed., NREL/CP-411-5148, Golden, CO: National Renewable Energy Laboratory, Sept. 1992, pp. 327-344.

7. Jorgensen, G.J., "Accelerated Exposure Testing for Screening and Lifetime Prediction", *Photovoltaic Performance and Reliability Workshop*, B. Kroposki, ed., NREL/TP-411-21760, Golden, CO: National Renewable Energy Laboratory, October 1996, pp. 193-216.

8. Rosenthal, A., and Durand, S. "Long Term Effects on Roof-Mounted Photovoltaic Modules," in S. Smoller, coordinator, *NREL Photovoltaic Program FY 1995 Annual Report*, NREL/TP-410-21101, June 1996, pp. 377-380.

9. Berman, D., Biryakov, S., and Faiman, D., "Efficiency Loss Associated with EVA Laminate Browning Observed in the Negev Desert," *Solar Energy Materials and Solar Cells*, **36**, pp. 421-433 (1995).

10. Mrig, L., Strand, T., Kroposki, B., Hansen, R., and van Dyck, E., "Photovoltaic Module and System Technology Validation," S. Smoller, coordinator, *NREL Photovoltaic Program FY 1995 Annual Report*, NREL/TP-410-21101, June 1996, pp. 350-355.

11. Claasen, R.S., and Butler, B.L., "Introduction to Solar Materials Science," *Solar Materials Science*, L.E. Murr, ed., New York: Academic, 1980, pp. 3-51.

12. Cuddihy, E., Coulbert, C., Gupta, A., and Liang, R., "Electricity from Photovoltaic Solar Cells," *Flat-Plate Solar Array Project, Final Report, Volume VII: Module Encapsulation*, JPL Publication 86-31, Pasadena, CA: Jet Propulsion Laboratory, (October 1986).

13. Czanderna, A.W., "Surface and Interface Studies and the Stability of Solid Energy Materials," Solar Materials Science, L.E. Murr, ed., New York: Academic, 1980, pp. 93-143.

14. a. Czanderna, A.W., and Gottschall, R., eds., "Basic Research Needs and Opportunities for Interfaces in Solar Energy Materials," *Materials Science and Engineering*, **53**, pp. 1-168, (1982); b. Czanderna, A.W., and Landgrebe, A.R., eds., *Current Status, Research Needs and Opportunities in Applications of Surface Processing to Transportation and Utilities Technologies*," NREL/CP-412-5007, Sept. 1992; 444 p., also *Critical Reviews in Surface Chemistry*, guest editors, **2**, Nos. 1-4, **3**, No. 1 (1993).

15. Fisher, R.M., Ketola, W.M., Martin, J., Jorgensen, G., Mertzel, E., Pernisz, U., and Zerlaut, G., "Accelerated Life Testing of Devices with S/S, S/L, and S/G Interfaces," *Critical Reviews Surface Chemistry*, **2**, pp. 317-330 (1993).

16. Masterson, K., Czanderna, A.W., Blea, J., Goggin, R., Guiterrez, M., Jorgensen, G., and McFadden, J.D.O., *A Matrix Approach for Testing Mirrors-Part II*, Golden, CO: Solar Energy Research Institute, SERI/TP-255-1627, July 1983.

17. Czanderna, A.W., "Overview of Possible Causes of EVA Degradation in PV Modules," *PV Module Reliability Workshop*, L. Mrig, ed., Golden, CO: Solar Energy Research Institute, SERI/CP-4079, 26 Oct. 1990, pp. 159-215.

18. Czanderna, A.W., "EVA Degradation Mechanisms: A Review of What is and is not Known," *Proceedings of a PV Module Reliability Workshop*, L. Mrig, ed., Golden, CO: National Renewable Energy Laboratory, NREL/CP-410-6033, 8-10 Sept. 1993, pp. 311-357.

19. Wineburg, J.P., "Can You Believe Lifetime Predictions from Accelerated Lifetime Testing," *Proceedings of the Photovoltaic Performance and Reliability Workshop*, 16-18 Sept. 1992, L. Mrig, ed., Golden, CO: National Renewable Energy Laboratory, NREL/CP-411-5184, September, 1992, pp. 365-375.

20. a. Jorgensen, G.J., "Durability Testing of Silvered Polymer Reflectors for Solar Concentrator Applications," *Photovoltaic Performance and Reliability Workshop*, L. Mrig, ed., Solar Energy Research Institute, SERI/CP-411-5184, Sept. 1992, pp. 345-364; b. Jorgensen, G.J., "An Overview of Service Lifetime Prediction (SLP):, *Photovoltaic Performance and Reliability Workshop*, L. Mrig, ed., NREL/CP-411-20379, Golden, CO: National Renewable Energy Laboratory, November, 1995, pp. 151-171; c. Putman, W., "Outdoor and Indoor UV Exposure Testing," op. cit. ref. 12a., pp. 279-310.

21. Czanderna, A.W., and Lampert, C., "Evaluation Criteria and Test Methods for Electrochromic Windows," SERI/PR-255-3537, July, 1990. Solar Energy Research Institute, Golden, CO.

22. Czanderna, A.W., Pern, F.J., "Estimating Lifetimes of a Polymer Encapsulant for Photovoltaic Modules from Accelerated Testing," *Durability Testing of Nonmetallic Materials*, R.J. Herling, ed., Philadelphia, PA: American Society for Testing and Materials, ASTM STP 1294, pp. 204-225.

23. Kim, H.-M., Jorgensen, G.J., King, D.E., and Czanderna, A.W., "Development of Methodology for Service Lifetime Prediction of Renewable Energy Devices," *Durability Testing of Nonmetallic Metals*, R. J. Herling, ed., Philadelphia, PA: American Society of Testing and Materials, ASTM STP 1294, 1996, pp. 171-189.

24. Jorgensen, G.J., Kim, H.-M., and Wendelin, T.J., "Durability Studies of Solar Reflector Materials Exposed to Environmental Stresses," *Durability Testing of Nonmetallic Materials*, R.J. Herling, ed., Philadelphia, PA: American Society for Testing and Materials, ASTM STP 1294, 1996, pp. 121-135.

25. Webb, J.D., and Czanderna, A.W., "Dependence of Predicted Outdoor Lifetime of Bisphenol-A Polycarbonates on the Terrestrial UV Irradiance Spectrum," *Solar Energy Materials*, **15**, 1-4, (1987).

26. Zussman, H.W., "Ultraviolet Absorbers for Stabilization of Materials and Screening Purposes," *Plastics Encyclopedia*, Sept., 1959, 1A, p. 372.

27. Searle, N.D., "Wavelength Sensitivity of Polymers," *Advances in the Stabilization and Controlled Degradation of Polymers*, Patsis, A.V., ed., Lancaster, PA: Technomic Pub. Co., 1986, **1**, pp. 62-74.

28. Webb, J.D., and Czanderna, A.W., "End Group Effects on the Wavelength Dependence of Laser Induced Photodegradation in Bisphenol-A Polycarbonate," *Macromolecules*, **19**, 2810-2825 (1986).

29. Webb, J.D., Czanderna, A.W., and Schissel, P., "Photodegradation of Polymer Films from Reflecting Substrates," *Polymer Stabilization and Degradation*, H.H.G. Jellinek, ed., Vol. 2, Amesterdam: Elsevier, 1989, pp. 373-431.

30. Kim, H.M., and Jorgensen, G.J. "The Time-to-Failure Distribution of Renewable Energy Devices: Performance Based Approach", submitted to the <u>American Statistical Society</u>, presented at their meeting in Chicago, IL, August 4-8, 1996.

ROLE OF GOVERNMENT, UNIVERSITIES, AND INDUSTRIES IN DEVELOPING AND FUNDING NEW IDEAS

Why Basic Energy Sciences is Funding Innovation

Jerry J. Smith

U. S. Department of Energy
Division of Materials Sciences
Office of Basic Energy Sciences
Germantown, MD 20874

Abstract. The Department of Energy's Office of Basic Energy Sciences has the responsibility   to provide basic research support for the development of energy technologies and to plan, construct and operate special scientific user facilities. The role of the Office of Basic Energy Sciences in the support of innovative research with emphasis on photovoltaic technology-base work is described.

INTRODUCTION

About two years ago, the Division of Materials Sciences (DMS) of the Office of Basic Energy Sciences (BES), Department of Energy (DOE), established a research thrust in "High Efficiency Photovoltaics". This thrust, under the auspices of the DMS Center of Excellence for the Synthesis and Processing of Advanced Materials, was planned, implemented and supported jointly by the DMS and the Office of Photovoltaic and Wind Technologies, Office of Energy Efficiency and Renewable Energy, DOE. The research performed under this thrust and supported by DMS, with a few exceptions, comprises a part of the Solid State Physics Program of the DMS. It was thus logical that I would serve as the DMS point of contact and monitor for the thrust. It is from this involvement that the invitation for me to present the answer to the implied question embodied in the title of this paper arose. This paper summarizes the presentation that I made to the conference. Included in the discussion to follow is an elaboration of the aforementioned program as well as related research supported by BES as a part of its ongoing science program.

The implied question in, "Why Basic Energy Sciences is Funding Innovation", is, of course, much broader than photovoltaics-related research. However, the topic and spirit of this workshop are "Future Generation Photovoltaic Technologies," so I have emphasized

CP404, *Future Generation Photovoltaic Technologies: First NREL Conference*, edited by McConnell
© 1997 The American Institute of Physics 1-56396-704-9/97/$10.00

photovoltaics in the sections dealing with specifics of the kinds of research that BES funds. Many of the comments, along with the rationale given, in response to the question, are equally applicable to the other areas of research supported by BES.

This paper was presented in the session on the "Role of Government, Universities and Industries in Developing and Funding New Ideas." As stated above, I intend to address the topic in the context of photovoltaic technologies. In addition, I intend to discuss the Office of Basic Energy Sciences in relationship to its parent, the Office of Energy Research, the Department of Energy and the research infrastructure of the United States of America.

THE IMPLIED QUESTION

Up front, let me answer the question posed by the title of this paper. Below, I'll elaborate on the answer. The question can be answered in a couple of ways. From a legal standpoint, the answer can be summed up succinctly as, "that is what we in BES are paid to do!" In fact, The Energy Policy Act of 1992, Public Law 102-486, Section 2203, Supporting Research and Technical Analysis, states for Basic Energy Sciences, " (1) Program Direction - The Secretary shall continue to support a vigorous program of basic energy sciences to provide basic research support for the development of energy technologies. Such program shall focus on the efficient production and use of energy, and the expansion of our knowledge of materials, chemistry, geology, and other related areas of advancing technology development." It goes ahead to assign another responsibility under Basic Energy Sciences, namely, "(2) User Facilities - (A) As part of the program referred to in paragraph (1), the Secretary shall carry out planning, construction, and operation of user facilities to provide special scientific and research capabilities, including technical expertise and support as appropriate, to serve the research needs of our laboratories, and others." So, the Office of Basic Energy Sciences funds innovation because it is a part of our mission.

Of course, the mission to support research in energy-related fields is only the starting point in this discussion and alone, doesn't say much about what drives the research nor the priorities operative in the selection of topics for emphasis. Hopefully, a clearer picture of these interpretive aspects of why BES funds innovation will emerge from an examination of DOE, ER and BES. Before proceeding, just what are some of these interpretive factors driving the support of innovation? Three come immediately to mind.

They are need, opportunity, and stewardship. Need plays a role of ever-increasing importance as resources become scarcer. The Department of Energy is, what is often called, a mission-oriented agency. That is, the Department has the mission of addressing the nation's various energy issues. As such, all elements within the Department are charged with supporting this mission and the mission statements of the various elements reflect this charge. For example, the mission statement for the Office of Basic Energy Sciences reads: To support basic research in the natural sciences leading to new and improved energy technologies and to understanding and mitigating the environmental impacts of energy technologies. Each of the technologies represented by the Department's Energy, National Security and Environmental Management Program Offices, has unique scientific and technical needs which change over time as the technologies advance, mature or are replaced. The Office of Basic Energy Sciences plans and implements its research programs, where possible, within the context of these various needs, attempting, at all times, to maintain the necessary base of knowledge upon which new technologies and improvements in existing technologies can be made. BES's research programs must have a significant innovative component to be responsive to these needs.

Opportunity is the second interpretive driver for the BES research program. Opportunity as used here applies to people, ideas and scientific fields. Most technologies go through a repetitive cycle of intense investigation and progress followed by periods of relative inactivity. Reasons for this cyclic nature are varied, however, often the investigations exhaust the accessible information because of limited materials, techniques, instrumentation or ideas. These technologies then remain relatively unexplored until the state-of-the-art in the limiters advances to a point where progress can again be made. We try to identify these areas of opportunity and stimulate research in them in a timely fashion. The goal is to push the technological base before the need arises and to stimulate new technological directions where possible.

The Department of Energy is aware of its national stewardship role in science and technology. We are a full participant in the activities to ensure stable, essential communities and facilities through the support of scientific and technological research and development, education and facilities construction and operation. Included within this stewardship role is the requirement to fund new ideas and approaches to stimulate and promote the most innovative of people and their efforts.

ORGANIZATION

The Department of Energy, a cabinet level organization of the Executive Branch of the federal government, is headed by the Secretary of Energy, Federico Peña at the present time. A simplified organizational chart for the Department is shown in Fig. 1.

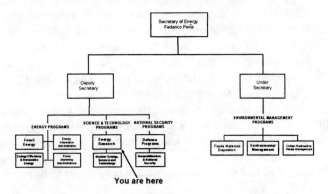

FIGURE 1. Simplified Organizational Chart for the Department of Energy - the Office of Basic Energy Sciences is a part of the Office of Energy Research noted by the arrow.

The Department is organized along program lines into two general categories: technology applications programs consisting of Energy Programs, National Security Programs and Environmental Management Programs and technology-base programs designated Science & Technology Programs. The Office of Basic Energy Sciences is contained within the latter as a part of the Office of Energy Research (ER). Within the context of this conference on photovoltaics, the other major funder of photovoltaics technology, the Office of Photovoltaics and Wind Technology is a component of the Office of Energy Efficiency and Renewable Energy (EE). While components of different programs, both the Director of ER and the Assistant Secretary for EE report to the Secretary through the Deputy Secretary.

A similar organizational chart for the Office of Energy Research is given in Fig. 2.

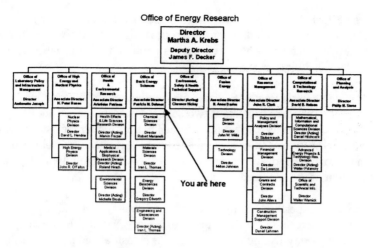

FIGURE 2. Simplified organizational chart for the Office of Energy Research showing the relationship of the Office of Basic Energy Sciences to the other ER components.

The Office of Basic Energy Sciences, along with the other principal offices shown, function as the interface between the Department of Energy and the scientific basic research community. The Office of Basic Energy Sciences is further broken down into four divisions which parallel the areas of emphasis identified in the mission, namely, Materials, Chemical Sciences, Engineering and Geoscience, and Energy Biosciences. Financially, the size of the respective divisions is in this order.

The Department of Energy, primarily through the Office of Energy Research, is one of the major Federal funding agencies for science and technology in the United States. For example, in fiscal year 1996, Energy ranked third in basic research funding with expenditures of $2.1 billion behind only the National Science Foundation (NSF) and the National Institutes of Health (NIH). In the support of research and development facilities, the Department of Energy ranks first, spending, in FY 1996, $1.4 billion. In several disciplines, the Department also ranks high in the level of funding. Energy is first in the funding of the physical sciences, third in environmental sciences, third in mathematics and computing, and third in engineering funding. Summarizing, the Office of Energy Research is the principal U. S. funder of basic research in the physical sciences, with an amount more than three times that of the National Science Foundation, and of the major national scientific user facilities, with funding approximately equal to that of the NSF, NIH and Department of Defense combined. The

Office of Basic Energy Sciences, itself, accounts for a total of seventeen percent of all Federal funding for the physical sciences. It is within this amount that much of the basic research on photovoltaic materials and technology is found.

The Office of Basic Energy Sciences funds approximately two thousand four hundred research projects at two hundred institutions. In addition, BES operates seventeen scientific user facilities, including four synchrotron radiation light sources, four high flux neutron sources and four electron beam microcharacterization centers. Fig. 3 shows the geographical distribution of the facilities and projects.

The research is performed, by the amount of funding, approximately seventy-five percent in the Department of Energy National Laboratory system and twenty-five percent in academic, industrial and not-for-profit laboratories. By any critierion, the Office of Basic Energy Sciences program is successful as well as innovative. The products of the program, which include publications in peer-reviewed journals, patents, and awards and other forms of recognition, attest to the high quality and quantity of scientific research pursued by the principal investigators on BES funded efforts. Publications are routinely featured articles in journals, often cited by selection as the cover illustrations for journals having cover photographs, e.g. Science or Nature. Specific examples of recognition for research performed under the auspices of the Office of Basic Energy Sciences include: five Nobel Prizes since 1985, thirty R&D 100 awards since 1991, all of the first ten American Chemical Society Awards in Organometallic Chemistry and forty-eight major national and international awards, prizes, medals and citations in 1996 alone.

I hope that these figures illustrate that BES succeeds in its fundamental tenets of excellence in basic research, relevance to the Nation's energy future, and stewardship of the Nation's research performers and institutions.

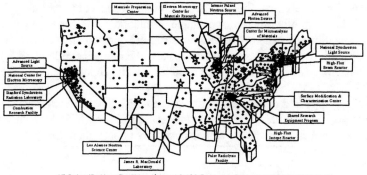

17 Scientific User Facilities (★) and 2,400 Research Projects at 200 Institutions (•)

FIGURE 3. Geographical distribution of BES facilities and research projects

FUNDING

In fiscal year 1997, the Office of Basic Energy Sciences has a budget
appropriation of about $649.7 million. Of this, $232.5 million is for
facilities, $45.7 million for capital equipment and $30.6 million for
construction, and related items. The remainder is the research operating
budget split among the Divisions comprising BES as follows: Materials,
$160.2 million; Chemical Sciences, $111.1 million; Engineering and
Geosciences, $41.2 million; and Energy Biosciences, $28.2 million.

PHOTOVOLTAICS RESEARCH

It is important to remember that photovoltaics is only one of a large
number of technologies that the Office of Basic Energy Sciences supports
scientifically. To the extent that resources permit, needs dictate and
opportunities exist, BES attempts to be responsive to all of the
technologies within the Department's mission. Consequently, the list of
scientific topics within the BES research portfolio is long and extensive.
For a picture of the scope of BES-funded research, refer to one or more of
the references cited below in Bibliography.

The photovoltaics research within BES falls approximately into four broad
categories:
> Generic Core Research
> Technology-Directed Research
> Center of Excellence for the Synthesis and Processing of
> Advanced Materials (CESPRO)
> Small Business Innovation Research.

By generic core research, I refer to research on materials of potential applicability to photovoltaics, for example, semiconducting materials, to characterizations of structure, electronic structure and stability of materials, to investigations of surfaces and interfaces and the influence of light on their behavior and properties, and to facilities which support such research.

Technology-directed research is that research which has a direct relevance to photovoltaic materials, devices and processes and for which the goals of the work are to support the development of new photovoltaic devices or the optimization of existing ones.

The Division of Materials Sciences supports the CESPRO, which has as one of its research thrusts, a project on "High Efficiency Photovoltaics." I will discuss this project in more detail later.

The Department of Energy's Small Business Innovation Research (SBIR) program is administered centrally through the SBIR program office under the formerly Advanced Energy Projects Division. This Division, which is now the Advanced Energy Projects and Technology Research Division, Office of Computational and Technology Research, was until quite recently a part of the Office of Basic Energy Sciences. Consequently, I'll include a brief discussion of the photovoltaic-related work supported out of this Division in this paper. Generally, some photovoltaics research is also found in the SBIR program. The latter has its own annual solicitation and selection process for research. Projects are selected using fundamentally the same criteria used for judging all proposals, namely, scientific or technical quality, energy relevance, and investigator competence. Once selected the projects are normally monitored technically through the appropriate DOE program office.

Now let me speak about some specifics of the photovoltaics research funded by BES. Looking first at the Division of Materials Sciences (DMS). DMS invests over $25 million per year (FY 1996) in generic core research with potential photovoltaic relevance. Specifically, one hundred twelve projects worth approximately $9 million include work on semiconductors; optical-related projects total thirty-five with funding of $2 million; and surfaces and interface research is contained in one hundred forty-two projects at $16 million. The Division has technology-directed research in the form of a research thrust in photovoltaics which supports, primarily, work at the National Renewable Energy Laboratory on materials and structures of direct photovoltaic relevance. The FY 1996 funding for this work is approximately $2 million. One particular effort within this technology-directed category is the research and development of a high efficiency gallium indium phosphide-gallium arsenide tandem photovoltaic cell. This work is being performed at the National Renewable Energy Laboratory. Presently, this cell holds world records for efficiency, having achieved 29.5 percent at one sun in a flat plate cell, 30.2 percent in a 140 - 180 sun concentrator cell, and 25.7 percent in a one sun space cell. This achievement is a good example of integrated research and development as practiced by the Department of Energy and its National Laboratories. The fundamental investigations of the key materials - gallium indium phosphide - properties was supported by the BES Division of Materials Sciences; the cell development work has been funded by the EE Office of Photovoltaic and Wind Technologies. The former represents an investment of about $3 million over the past five years; the latter has invested $6 million over the same period. In addition, I mentioned above the CESPRO whose program includes the "High Efficiency Photovoltaics" project. This project is discussed below. DMS had three relevant SBIR projects funded for a total of $225,000.

The Division of Chemical Sciences' photovoltaics research can be found within the Solar Photochemical program and supported out of the Division's Photochemical and Radiation Sciences Program. The Office of Basic Energy Sciences' photoelectrochemistry research is contained within this program. The photovoltaics-related research is contained within forty-eight university projects funded in fiscal year 1996 at $5.6 million and in four National Laboratories and the Notre Dame Radiation Laboratory at a FY 1996 level of $8.6 million. One specific example of integrated research and development in the DCS is the Photochemical Solar Cell project. This cell, based upon dye-sensitization of nanocrystalline titanium dioxide, is funded in part by the DCS, the EE Office of Photovoltaic and Wind Technologies and the OCTR Advanced Energy Projects and Technology Research Division. Funding levels in FY 1996 were $150,000, $200,000, and $150,000 per year respectively.

The Center of Excellence for the Synthesis and Processing of Advanced Materials is a virtual laboratory concept involving eleven National Laboratories and several universities supported by the Division of Materials Sciences. The objective of the CESPRO is to enhance the science and engineering of materials synthesis and processing to meet the programmatic needs of the Department of Energy and to facilitate the technological exploitation of materials. The Center is coordinated through Sandia National Laboratories, New Mexico. Presently, the Center has seven research projects or thrusts among which is "High Efficiency

Photovoltaics." Projects are selected on the basis of scientific excellence, a clear relationship to energy technologies, multi-laboratory involvement, strong existing or potential partnerships with DOE technologies and strong in-kind industrial partnerships. The Center projects are developed around existing research at the participating institutions and core funding is provided through these existing efforts. Additional funds, often referred to as "glue" funds, are provided to cover expenses associated with the collaborative aspects of the projects, such as travel and workshops. A more detailed description of the Center of Excellence and its projects can be found in the annual brochure, "Research Briefs" published by the center, see Bibliography reference (1).

The CESPRO project on "High Efficiency Photovoltaics" consists of twenty collaborative research projects involving nine National Laboratories and eight universities. Support for the research includes funding from BES, the Office of Energy Efficiency and Renewable Energy and the Electric Power Research Institute. The project has two main themes, silicon-based thin films and next-generation thin film photovoltaics. A more detailed description of this project has already been published(Bibliography reference (2)). Presently, the BES funding level for this project is approximately $2 million.

INFORMATION

The four Divisions within the Office of Basic Energy Sciences each publishs annually, a summary of their respective research programs complete with abstracts and funding levels, Bibliography (3) through (7). The Advanced Energy Projects and Technology Research Division and Small Business Innovation Research Program publish annual summaries as well. In addition, the Energy Materials Coordinating Committee (EMaCC), an internal DOE committee with representatives from the various DOE offices funding materials research, annually publishes a report containing project summaries for the materials projects not given in the Division of Materials Sciences and Division of Chemical Sciences annual summaries, Bibliography (8). Several other publications are available describing the Office of Basic Energy Sciences and the BES scientific user facilities, Bibliography (9) through (11). Copies of some of these reports are available on the World Wide Web through the Office of Energy Research home page at http://www.er.doe.gov. Hard copies can be obtained from the Office of Basic Energy Sciences or the respective Division.

The general mailing address is U. S. Department of Energy, Office of Basic Energy Sciences (or the appropriate Division), 19901 Germantown Road, Germantown MD 20874-1290.

SUMMARY

In summary, the Office of Basic Energy Sciences funds innovation in fulfilling its role in support of the Department of Energy's mission. BES provides the knowledge-base for present and future technologies by anticipating needs and issues and by working cooperatively with the relevant technology offices to identify needs and translating these needs into basic research. BES' programmatic content is defined by these technology needs, scientific opportunities and the requirement to serve as one of the Nation's stewards for scientific expertise and facilities.

BIBLIOGRAPHY

(1) *Research Briefs, The DOE Center of Excellence for the Synthesis and Processing of Advanced Materials*, U.S. Department of Energy, Office of Energy Research, Office of Basic Energy Sciences, Division of Materials Sciences, SAND96-1136, January 1996

(2) S. K. Deb and J. P. Benner, *DOE/OER-Sponsored Basic Research in High-Efficiency Photovoltaics*, Conference Record of the Twenty-Fifth IEEE Photovoltaic Specialists Conference - 1996, May 13-17, 1996, Washington, DC, p. 977

(3) *Materials Sciences Programs Fiscal Year 1995*, U. S. Department of Energy, Office of Energy Research, Office of Basic Energy Sciences, Division of Materials Sciences, DOE/ER-0682, May 1996

(4) *Summaries of FY 1996 Research in the Chemical Sciences*, U. S. Department of Energy, Office of Energy Research, Division of Chemical Sciences, DOE/NBM-1098, September 1996

(5) *Summaries of FY 1996 Geosciences Research*, U. S. Department of Energy, Office of Energy Research, Office of Basic Energy Sciences, Division of Engineering and Geosciences, DOE/MBM-1097, December 1996

(6) *Summaries of FY 1995 Engineering Research*, U. S. Department of Energy, Office of Energy Research, Office of Basic Energy Sciences, Division of Engineering and Geosciences, DOE/ER-0674, March 1996

(7) *Division of Energy Biosciences Annual Report and Summaries of FY 1996 Activities*, U. S. Department of Energy, Office of Energy Research, Office of Basic Energy Sciences, Division of Energy Biosciences, DOE/ER-0698, April 1997

(8) *Energy Materials Coordinating Committee (EMaCC) Fiscal Year 1995 Annual Technical Report*, U. S. Department of Energy, Office of Energy Research, Office of Basic Energy Sciences, Division of Materials Sciences, DOE/EE-0113, December 1996

(9) *Basic Energy Sciences, Research for the Nation's Energy Future*, U. S. Department of Energy, Office of Energy Research, Office of Basic Energy Sciences

(10) *Office of Basic Energy Sciences, Scientific Research Facilities, A National Resource*, Office of Energy Research, U. S. Department of Energy
(11) *Basic Energy Sciences, Serving the Present, Shaping the Future*, Office of Basic Energy Sciences, Office of Energy Research, U. S. Department of Energy

ACKNOWLEDGEMENTS

I wish to thank my colleagues in the Office of Basic Energy Sciences who, in addition to providing data and information to make this presentation as complete as possible, provide an intellectually stimulating and challenging environment in which to work. I particularly thank Cheryl Fee for the figures and Melanie Becker and Christie Ashton for preparing the manuscript in finished form.

Long-Range PV R&D and the Electric Utilities

Terry M. Peterson

Electric Power Research Institute, Palo Alto, California 94303

Abstract. In the short term, photovoltaics will probably continue to enjoy great success in niche markets and non-utility businesses, but see relatively little use within utilities. Deregulation is driving major restructuring of the electric-utility sector, causing great uncertainty among its planners and executives, and leading them to favor cost-cutting over other corporate strategies. However, the competitive motives at the root of that restructuring will ultimately induce resourceful utility executives to seek novel *non-commodity* energy-service businesses to sustain their companies' success in the deregulated industry of the future. In that industry, technology innovation will play a very important role. Specifically, photovoltaics will be highly valued in light of its unsurpassed modularity, extreme siting ease, very low operation and maintenance costs, *and* public popularity. The eventual leaders in wielding that powerful technology likely will be among those who recognize those assets earliest and strive to bring its promises to reality through innovative applications.

INTRODUCTION

Despite the rapid growth of small-scale photovoltaic (PV) applications throughout the world in recent years, energy-significant PV applications remain disappointingly distant from the electric utilities' mainstream businesses, which demand lower cost and higher efficiency than today's PV affords. In the U.S., nationwide electric-utility over-capacity and the present industry turmoil, driven by uncertainties related to deregulation, has dampened utility enthusiasm for new-technology development, particularly higher-cost technologies such as PV. Today, the utility industry's leaders are preoccupied by an urge to cut costs and position their companies for best advantage in the soon-to-be very competitive commodity-electricity business. The most astute of those leaders, however, already recognize that mere cost-cutting—while essential—will not suffice to ensure their future success. They need new ways to offer their customers a wider array of innovative value-added services to stay at the forefront of their industry. From this latter perspective, PV has a number of very attractive characteristics, including great siting ease, low operation and maintenance costs, extreme scalability (important for minimizing capital risk in competitive businesses), lack of fuel-cost escalation risk, and—not least—positive regard in the public's eye.

CP404, *Future Generation Photovoltaic Technologies: First NREL Conference*, edited by McConnell
© 1997 The American Institute of Physics 1-56396-704-9/97/$10.00

TODAY'S MAINSTREAM PV BUSINESSES

Primarily, PV today provides electric service effectively to loads that are not connected directly to the utility grid. These loads fall into the two general application categories, consumer and industrial/commercial, described below.

Remote habitation and other consumer needs. Throughout the United States, dwellings a mile from the grid, or less in some cases, are candidates for PV or hybrid PV-engine generation. Besides powering household appliances, PV frequently runs a domestic water-well pump. Also, PV can be cost-effective in most off-grid battery-charging applications, including electric fences, marine batteries, and radio/TV batteries. Even close to utility power, where PV is typically not the cost-effective choice, it is sometimes the consumer's preferred option because of its proven high reliability or its perceived environmental friendliness (1).

Remote industrial and commercial uses. PV is popular for powering livestock water pumps, providing water more reliably than do windmills. Many megawatts of PV are installed around the world to supply power for telecommunications, telemetry, and radio remote-control systems and their batteries. PV's very high reliability and low maintenance requirements have long made it a favorite for powering remote equipment such as microwave repeaters. In some cases, utility power is actually available nearby, but PV better meets telecommunications' very exacting reliability specifications.

Recent growth in wireless and fiber-optic telecommunications markets has dramatically increased the number of worldwide sites that require small amounts of power. Many of these have proven suitable for PV generation, raising total annual sales in this sector to over 23 MW in 1995. This application group, comprising a vast number of tiny generators, is expected to be in the annual-sales range of 70–100 MW by 2005 (2). Lessons learned in serving the telecommunications premium-power market will also apply in the related, multibillion-dollar power-quality market, as a growing fraction of it opens to ever-lower-priced PV (3).

TODAY'S COST-EFFECTIVE UTILITY PV APPLICATIONS

Within the utility company, EPRI has documented over 80 different cost-effective PV applications, installed at over 50 companies. Typical applications include cathodic protection of underground structures, aircraft obstruction-warning lights, parking-lot and security lighting, and electric distribution-system sectionalizing switches (4).

Looking ahead, the experience that small-scale system users are gaining today with the unique aspects of PV power generation will be key to successfully deploying larger-scale future PV applications. Further, and equally important in the increasingly competitive utility market, early PV applications give companies the opportunity to establish national, or even global, brand-name recognition as purveyors of reliable, environmentally friendly technology. Together, these two factors will be key to success in the international utility business of tomorrow, when PV takes a significant place as one of a small number of clean energy-service technology options capable of bypassing the traditional utility's delivery system.

HISTORICAL PV MODULE PRICE TREND

The downward trend in worldwide average PV-module selling price over the past twenty-some years (2) gives rise to optimism that further substantial reductions are likely. However, a projection of that trend, as in FIGURE 1, shows that a "business as usual" future will only lead to $3/W installed system prices (assuming 50% is module price) when the total PV market has reached about 5 times the 9,000 MW that the Utility PV Group projected for such systems in the U.S. (5) That implies that the domestic market cannot be a strong enough force to "pull us down the learning curve" to cheap PV, and it underscores the importance of jumping to a new curve defined by a fundamentally different manufacturing technology, such as the one hypothesized in FIGURE 1 for amorphous thin-film multijunction modules (6). Other thin-film PV technologies show great promise for similar cost breakthroughs, but they are not as close to commercial reality as amorphous silicon (a-Si), which has been through substantial manufacturing development over the past 10–15 years. Present-day a-Si know-how will take us to $1/W module price, and continued aggressive R&D can yield even lower costs, eventually meeting DOE's and EPRI's cost/performance goals of $50 per square

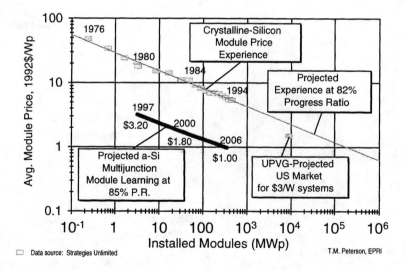

FIGURE 1. Jumping to a new price trend with new technology will be key to accelerating large-scale use of PV.

87

meter and 15% module efficiency (about 30¢/W for the modules) (7, 8).

FLAT-PLATE PV ECONOMICS—THE CASE FOR HIGH-EFFICIENCY

The 50-MW "central station" PV plant referenced in FIGURE 2 is probably not a good model of the near future. Nevertheless, calculation of the levelized cost of electricity (COE) from such a plant (9) serves well to show the extreme importance in overall PV economics of both *module conversion efficiency* and *levelized fixed-charge rate* (which is related, among other things, to the cost of borrowing the capital for plant construction). The first parameter is amenable to technology development, while the second can be influenced by risk sharing; for example, through low-interest loans or other incentives. Smaller-scale plants would have somewhat higher COE. (For example, a 500-kW plant's costs would be roughly twice those in FIGURE 2) But the important message is that successfully pulling on both of these powerful levers will produce very competitive PV power generation—either distributed or centralized—in areas of good sunshine.

Even in less sunny areas, successful science and finance can jointly produce favorable PV electricity prices, as FIGURE 3 illustrates (9). However, it should again be noted that "50-MW Plant" is not a likely deployment mode in the near term and smaller-scale plants will have higher COE, so these figures should be

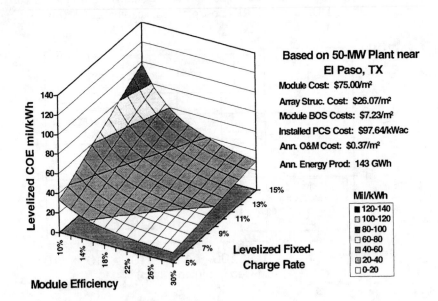

FIGURE 2. Cost of electricity (COE) supplied by a large-scale PV 'farm' as a function of module efficiency and levelized fixed-charge rate.

88

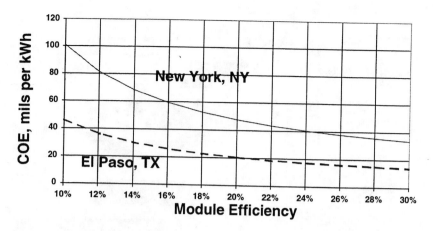

FIGURE 3. Comparison at constant 7% fixed-charge rate of levelized electricity costs (COE) for large-scale (50 MW) plants in Texas and New York.

viewed as merely showing important cost drivers and providing comparative cost trends. Furthermore, in near-term PV applications the system value will likely not simply equate to a busbar COE, because the choice of preferred technology will be many-faceted.

TIMELINE FOR UTILITY PV READINESS

Looking ahead, the waves PV's progress in FIGURE 4 (as in nearly all revolutionary technology changes) are measured in decades. Nevertheless, the seeds of corporate commercial success in these technologies usually must be sown years in advance of full bloom. That is to say, the most likely eventual leaders are not those who choose to sit back and wait for commercially viable technology to appear, but rather those who struggle to make it happen and learn to use it in that process. Following is a brief perspective on each of FIGURE 4's waves in turn.

Flat-Plate Crystalline Silicon

Flat-plate crystalline silicon PV is the workhorse of the fledgling PV industry. It remains the only PV technology available in megawatt commercial quantity, although several thin-film alternatives have threatened to 'go commercial' for the past few years. Despite the enormous development efforts invested, and the continued improvements they have engendered, it is very unlikely that crystalline silicon PV modules will fall below about $2/W.

A variation on the traditional cut-wafer silicon cell is produced by pulling a sheet of silicon from the melt. This approach is sufficiently distinct that it may

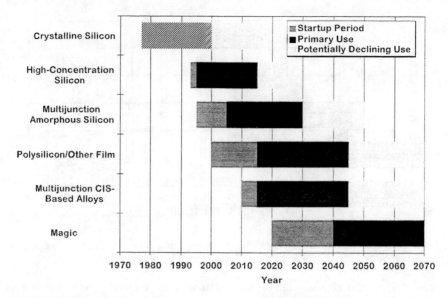

FIGURE 4. Schematic timeline for evolutionary waves of PV technology.

lower the cost floor somewhat below $2/W, although it does not provide the intrinsic scalability of thin-film processing. Nevertheless, it should probably not be regarded as lying on the historic module-price trend line. Moreover, it is now in commercial production—and competitive in today's cut-wafer-dominated market.

High-Concentration Silicon

Thanks to dedicated, creative researchers, support from DOE's Sandia National Laboratory, and sustained EPRI-funded development in the 1970s and '80s, high-concentration PV (HCPV) is now a viable technology option for many good-sunlight locations. Demonstrated array-level solar conversion efficiency is about 20%. However, HCPV presently lacks the commercial infrastructure needed to make it a business success, and already it has missed the leading edge of its opportunity window in FIGURE 4. But given about $15M capitalization and enough people, a two-year HCPV effort on product, manufacturing, and market development could yet provide lower-cost, larger-scale power than can any flat-plate PV system in the near term (10).

Multijunction Amorphous Silicon

The National Renewable Energy Laboratory's track record (11) of best one-of-a-kind single-junction small laboratory PV cells clearly shows that a-Si is not a long-term favorite for top efficiency honors, barring a true materials breakthrough, such as perhaps total elimination of its light-induced degradation. Although research to do that continues, the spotlight in commercial a-Si development has shifted to multijunction devices.

During 1997, three multimegawatt-per-year multijunction amorphous silicon thin-film module manufacturing plants are expected to be producing modules of 7% to 8% stable conversion efficiency. After a few years' production experience, they may be selling products below $2/W_{peak}$, as FIGURE 1 predicts.

Other Thin-Film Materials

Besides a-Si, two other thin-film PV technologies have demonstrated impressive conversion efficiency in the laboratory (11)—copper indium diselenide (CIS, nearly 18%) and cadmium telluride (CdTe, over 15%). Still, scaling up these very promising lab results to large areas and commercial production rates represent significant challenges that will yet require years and dollars to meet.

Another possible contender is "thin silicon film," 10- to 20-micrometer polycrystalline films. High performance has been demonstrated from such films, when supported on a refractory substrate such as crystalline silicon (12). Success in unlocking this candidate's potential most likely lies either in finding a low-cost substrate upon which high-purity films can grow or demonstrating a manufacturable new device design that can deliver high efficiency from lower-purity films.

Multijunction CIS-Based Alloys

Putting together the superior semiconductor properties of thin-film CIS and the advantages of multijunction cell design could produce a breakthrough in thin-film PV module performance. Computer modeling in EPRI's Strategic R&D-funded second-generation thin-film PV module project, called PV-ACIST, indicates that 25% efficiency is well within theoretical reach (13). Such devices, if technological hurdles can be surmounted, should cost little more than single-junction CIS modules on an area basis. Therefore, they would provide a substantial lowering of PV cost per unit power, and offer a new cost-effective option for large-scale power in most of the U.S. and much of the developed world.

'Magic' PV

PV cost and performance are *both* far from fundamental physical limits. An as-yet uninvented device may someday show the way to harvesting well over 50% of sunlight's energy without running afoul of thermodynamic laws. On the cost side, one might imagine PV devices designed to 'grow' like plants, directed by nanomachine 'assemblers' (14), thus reducing manufacturing costs to little more than those of supplying the raw materials and sufficient sunlight. Clearly, such a radical event cannot be planned—but it is much more likely to be found if sought!

CONCLUSIONS

Utility restructuring will open new opportunities for PV applications as the needs of the newly competitive industry are translated into innovative customer-service options. These applications will leverage PV's modularity and reliability, taking full advantage of its ability to build incremental capacity, shave peak loads, and enhance power quality. Of course, utilities will still need low-cost high-efficiency modules for large-scale PV applications. Amorphous thin-film multi-junction modules are likely to be the first to meet utility industry needs for energy-significant applications in the sunniest locations and in specialized applications, such as telecommunications and building-integrated power systems. In the longer term, multijunction modules based on CIS alloys appear a leading candidate for cost-effective large-scale generation in most of the developed world.

REFERENCES

1. EPRI report TR-100711, "Early, Cost-Effective Applications of Photovoltaics in the Electric Utility Industry," 1992.
2. Strategies Unlimited report PM-43, "Five-Year Market Forecast 1995–2000," 1996.
3. EPRI report TR-104372, "Power Quality Market Assessment," 1994.
4. EPRI report TR-102648, "Survey of Cost-Effective Photovoltaic Applications at U.S. Electric Utilities," 1993.
5. Utility PhotoVoltaics Group Summary Report, "Photovoltaics: On the Verge of Commercialization, Report of the UPVG's Phase 1 Efforts," 1994, pp. 35–43.
6. Personal estimate based on available information regarding Solarex and Unisolar large-scale manufacturing plans.
7. DOE report DOE/GO–10096-017, "Photovoltaics: The power of choice. The National Photovoltaics Program Plan for 1996–2000," 1996.
8. EPRI Technical Brief TB-102859, "Amorphous Thin-Film Photovoltaic Module R&D Program," 1993.
9. Calculations based on PVTOOL spreadsheet described in EPRI report TR-101255, "Engineering and Economic Evaluation of Central-Station Photovoltaic Power Plants," 1992.
10. See for example, EPRI reports: TR-106407, "Development of Manufacturing Capability for High-Concentration, High-Efficiency Silicon Solar Cells," 1996; TR-100392, "A Summary of Recent Advances in the EPRI High-Concentration Photovoltaic Program," 1992.
11. Zweibel, K., et al., "Progress and Issues in Polycrystalline Thin-Film PV Technologies," in *Conference Record of the 25th Photovoltaics Specialists' Conference*, 1996, pp. 745–750.
12. Ingram, A.E.., et al., "13% Silicon Film™ Solar Cells on Low-Cost Barrier-Coated Substrates," in *Conference Record of the 25th Photovoltaics Specialists' Conference*, 1996, pp. 477–480.
13. Rockett, A., et al., "Next Generation CIGS for Solar Cells," this conference, 1997.
14. Drexler, K.E., *Engines of Creation*, New York: Anchor Press/Doubleday, 1987.

Technology Development Versus New Ideas Development by Universities

T.W.F. Russell

Institute of Energy Conversion
Department of Chemical Engineering
University of Delaware, Newark, Delaware 19716 USA

Abstract. A logical approach to technology development which stresses the planning and interpretation of experiments useful for the design of commercial scale equipment is presented. The utility of the approach is illustrated by considering the design of a reactor to make copper indium diselenide.

INTRODUCTION

Despite some superb advances in thin film photovoltaic research at the laboratory scale there has not been the same level of achievement in the design, construction and operation of commercial scale facilities to manufacture modules. This is a critical issue and universities have an important role to play in bringing technology from laboratory experiment to profitable production. It is my contention that the new ideas needed in photovoltaic research today should address this critical issue in technology development, particularly in our colleges of engineering.

An engineering presence is required which stresses the planning and interpretation of experiments so that they provide information essential to the design, operation and control of commercial scale equipment. The logic to do so is easy to define and is shown in Figure 1. The effective execution is not so easy to define and requires some very creative engineering with several iterations as critical experiments provide the necessary information.

The iterative process can only be initiated after laboratory scale proof of concept experiments, shown as the block on the top of the diagram, have been completed. The first step in the iterative logic is an engineering analysis which establishes the approximate level of commercial scale production. Since we must deal with very complex problems in making photovoltaic modules, some attempt must be made to quantify the information available at any iteration. A mathematical model is necessary. The model behavior must agree with experiment to be useful but it is essential to have even a crude model if a preliminary commercial design is to be attempted. A preliminary commercial design effectively identifies the critical issues requiring additional experiment. It is a step which is very frequently ignored, but in a resource limited environment we must direct resources to the most important problems.

CP404, *Future Generation Photovoltaic Technologies: First NREL Conference*, edited by McConnell
© 1997 The American Institute of Physics 1-56396-704-9/97/$10.00

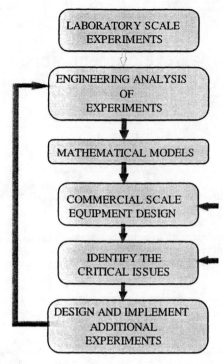

FIGURE 1. Logic diagram.

In this paper we will illustrate the execution of the logic diagram segments by considering the design of a commercial scale reactor to make copper indium diselenide, one of the most promising thin film materials.

LABORATORY SCALE EXPERIMENTS

Copper indium diselenide (CIS) is made in laboratory scale equipment by several different processes. All experiments are operated in a batch or semi-batch mode and make small area CIS or CIGS (CIS with gallium) films.

Physical Vapor Deposition Reactors

These batch reactors deliver the chemical species to make film by evaporating the raw material from some source and depositing by line of sight delivery on the substrate. A typical reactor is shown in Figure 2. An experiment is carried out by loading a substrate into the vacuum chamber, bringing it and the sources up to temperature and then depositing the chemical species on the substrate. It takes about an hour to perform a typical experiment.

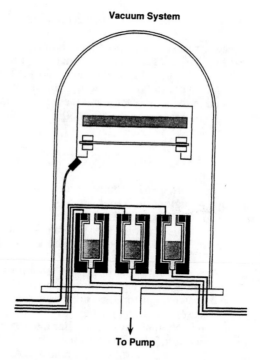

Vacuum System

To Pump

FIGURE 2. PVD batch reactor.

Efficiencies in the 15-18% range on 1 cm^2 cells have been achieved using PVD reactors which have sources for copper, indium, gallium and selenium. The substrate employed is most often molybdenum coated soda lime glass and depositions are carried out at substrate temperatures between 400 and 600°C.

Solar cell efficiencies in the 10-12% range have been reported in PVD reactors which form CIS by delivering selenium to copper indium layers on molybdenum coated soda lime glass. Starting with this substrate avoids the need for high temperature sources of copper and indium.

Chemical Vapor Deposition Reactors

The laboratory scale chemical vapor deposition reactors employed to make CIS cells all employ dilute streams of flowing H$_2$Se to supply selenium species to a substrate with a copper indium layer on molybdenum coated soda lime glass. The film is grown on the substrate in a semi-batch mode, i.e., it is loaded into the reactor, brought up to temperature (400-500°C) and exposed to a flowing H$_2$Se argon mixture for a period of up to about one hour. The H$_2$Se partially decomposes to various selenium species in the gas stream and on the surface of the growing CIS film. Cell efficiencies in the 12-15% range have been obtained in such reactors.

This relatively recent research has produced laboratory scale devices with efficiencies in the 12 to 15% range. The rapid thermal processing is carried out with a stack of copper, indium, gallium and selenium on Mo coated soda lime glass which has special layers to control sodium diffusion into the CIGS film. This cold stack is placed in the RTP reactor and the temperature is rapidly (5 to 10 minutes) brought up to temperatures in the 500-600°C range.

ENGINEERING ANALYSIS OF EXPERIMENTS

Some factor-of-two estimate of the required production is the necessary first step in any preliminary engineering analysis. A commercial scale module manufacturing facility for CIGS on glass should produce a 50 to 100 watt module every one or two minutes to be economically viable. A process to deposit CIGS continuously on a moving web would require a web speed of 1 to 2 meters per minute.

As discussed above, there are a number of proven laboratory scale processes for making CIGS. A complete reaction and reactor analysis would allow equipment design and costing to be carried out for each process and some decision made as to which process or processes should be further developed. In a first iteration all the necessary information is often not available. For a complete reaction analysis species present in the growing film must be identified and their concentration obtained as a function of time and substrate temperature. A reactor analysis must quantify the rate of delivery of raw material to the substrate as a function of reactor geometry, source design and temperature of operation. A complete heat balance on the proposed reactor needs to be carried out. Some thought must be given as to whether the commercial scale reactor is to be run in a batch or continuous mode.

MATHEMATICAL MODELS

Reaction Analysis

Chemical engineering researchers have exploited the capabilities of the CVD reactor to obtain species concentration as a function of time and temperature to develop the chemistry and a reaction model for CIS growth. Some typical experimental data and model predictions are shown in Figure 3 which is a plot of CIS concentration in the growing film as a function of time at reactor substrate temperatures of 400°C (solid circles), 350°C (solid triangles), and 250°C (crosses). The solid and dashed lines are model predictions from the set of first order differential equations describing the reactor behavior. This is the first quantitative reaction analysis that has been done for a thin film photovoltaic material. A complete discussion is available in the literature (1, 2).

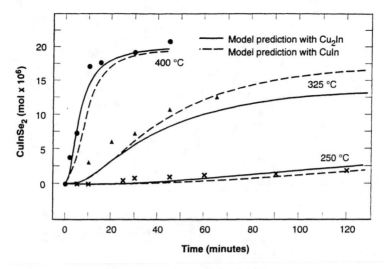

FIGURE 3. Reaction analysis of CIS growth.

Reactor Analysis

Figure 4 presents a simplified summary of the model equations for predicting species delivery rate from a PVD reactor source. The equations allow prediction of rate of effusion from the sources in a PVD reactor. An indication of how it can be used to develop the source design for continuous deposition on a moving web is illustrated on the bottom half of the figure. The model equation derivation and verification of model behavior by experiment in a batch PVD reactor and a pilot scale continuous deposition PVD reactor are discussed in detail elsewhere (3, 4).

COMMERCIAL SCALE EQUIPMENT DESIGN

Figure 5 is a sketch of a preliminary design for a commercial scale four source PVD reactor to make CIGS continuously on a moving plastic web. The model equations shown in Figure 4 were used to size the orifices and to set orifice spacing to achieve uniform CIGS film on the substrate. The model predictions have been verified by experiments on a moving web in a pilot scale reactor. This effective use of the verified mathematical models saved a great deal of expensive trial and error experiment.

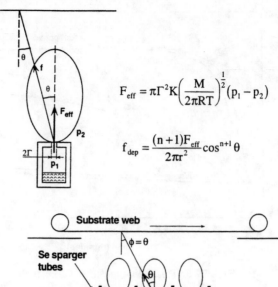

$$F_{eff} = \pi\Gamma^2 K \left(\frac{M}{2\pi RT}\right)^{\frac{1}{2}}(p_1 - p_2)$$

$$f_{dep} = \frac{(n+1)F_{eff}}{2\pi r^2}\cos^{n+1}\theta$$

FIGURE 4. Reactor analysis.

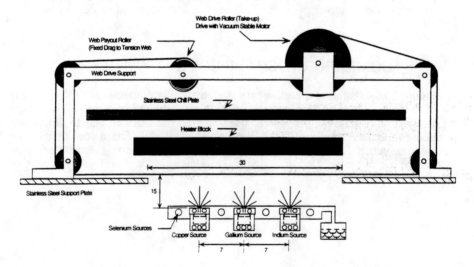

FIGURE 5. Preliminary design of a commercial scale reactor.

IDENTIFY THE CRITICAL ISSUES

To carry out the design of a commercial scale unit for CIGS deposition one must be able to specify reactor holding time for a given temperature. The reaction analysis summarized above supplies this information but the chemical equations, the model equations and the rate constants were based upon experiments in a CVD reactor which employed hydrogen selenide to make CIS films. One would hope that the reaction analysis is not reactor specific, but this needs to be tested if the model equations are to be used to develop a design for each of the proven laboratory scale processes.

DESIGN AND IMPLEMENT ADDITIONAL EXPERIMENTS

The critical experiment in this part of the logic diagram iteration is one in which species concentration as a function of time must be obtained in a PVD reactor. The batch reactor sketched in Figure 2 is presently being modified to contain a cold box into which the substrate can be moved and quickly cooled from 400°C to about 150°C where the reaction is essentially stopped. Experiments to determine concentration time data will then be carried out and compared with model behavior predictions to determine whether or not the reaction analysis is reactor specific.

CONCLUSIONS

A logical approach to the planning and interpretation of experiments which supply useful information for design, operation and control of commercial scale equipment has been proposed. The effectiveness of the logic has been illustrated by considering the design of a physical vapor deposition reactor to make CIGS film.

REFERENCES

1. Verma, S., Orbey, N., Birkmire, R.W., and Russell, T.W.F., *Progress in Photovoltaics* **4**, 341-353 (1996).
2. Orbey, N., Hichri, H., Birkmire, R.W., and Russell, T.W.F., "Reaction Analysis of the Formation of CIS at Temperatures from 250 to 400°C," in *Proceedings of the 25th IEEE Photovoltaic Specialists Conference*, 1996, pp. 981-984.
3. Rocheleau, R.E., Baron, B.N., and Russell, T.W.F., *AIChE Journal* **24**(4), 656 (1982).
4. Jackson, S.C., Baron, B.N., Rocheleau, R.E., and Russell, T.W.F., *Journal of Vacuum Science & Technology* **A3**(5), 1916-1920 (1985).

The Role of Universities in the DOE National Photovoltaics Program

Robert D. McConnell and Jeffrey A. Mazer*

National Renewable Energy Laboratory, Golden, Colorado 80401 and
*U.S. Department of Energy, Washington, D.C. 20585

Abstract: Through subcontracts, universities participate in near-term, mid-term, and long-term technology development in the National Photovoltaics Program of the U.S. Department of Energy. As noted in the DOE Program Plan for 1996-2000, universities' expertise in fundamental science and in materials and device research adds immeasurably to the foundations of science, advancement of technology, and effectiveness of the National Program. However, recent budget cuts have affected university subcontract funding, principally for long-term technology development. Historically, universities funded by DOE concentrated on long-term research, specifically in two subcontract programs entitled "University Participation" and "New Ideas." DOE intends to restart these highly successful programs.

THE DOE NATIONAL PHOTOVOLTAICS PROGRAM

The U.S. Department of Energy's Photovoltaics Program partners with industry, universities, and end-users to make PV a significant part of the domestic economy (1). To place the university portion of the program in perspective, the following summary describes all of the key activities of the National Photovoltaics Program and lists universities under subcontract in each. DOE's program activities are divided into three main categories: Research and Development, Technology Development, and Systems Engineering and Applications (1). The summary presents specific goals and funding that totals $57 million for the U.S. DOE Photovoltaics Program budget for Fiscal Year (FY) 1997. Universities funded within each area (in FY 1996) are listed in square brackets.

Total university subcontract funding is planned at $7.2 million for FY 1997, with another $1.5 million allocated for two special experiment stations managed by universities. This funding is down from an estimated $11.4 million for universities in FY 1995. Another 31 universities received small grants in FY 1997 of about $2000 each for support of highly successful and highly visible university projects for building PV-powered cars for SunRayce. Two universities, Georgia Institute of Technology and the Institute for Energy Conversion (IEC) at the University of Delaware, were designated in 1993 as University Centers of Excellence in Photovoltaics by the Department of Energy.

CP404, *Future Generation Photovoltaic Technologies: First NREL Conference*, edited by McConnell
© 1997 The American Institute of Physics 1-56396-704-9/97/$10.00

Research and Development

Fundamental and Exploratory Research

The goals of this area are to develop solid-state theories for novel photovoltaic materials and devices, perform solid-state spectroscopy studies of thin-film photovoltaic materials, develop and optimize photochemical solar cells, and support universities in their pursuit of long-term, fundamental, PV R&D. ($1.8M) [Johns Hopkins University, North Carolina State University, University of Illinois, University of South Florida, University of Utah, University of Washington, and seven schools under DOE's Historically Black Colleges and Universities Program; Hampton University, Wilberforce University, Clark Atlanta University, Central State University, Southern University, Texas Southern University, and Mississippi Valley State University]

Crystalline Silicon R&D

The goals here are to improve fundamental understanding, develop next-generation technologies, improve present-generation cells and modules, and coordinate national laboratory, university, and industry research. ($3.5M) [University of California, Duke University, North Carolina State University, University of South Florida, Texas Tech University, Massachusetts Institute of Technology, Georgia Institute of Technology's University Center of Excellence in Photovoltaics, University of New Mexico, and Harvard University]

Thin-Film PV Technologies

This is a major mid-term activity to support near-term transition to first-time manufacturing and scale-up of a-Si, CIS, and CdTe thin films, build a technology base upon which these advanced PV technologies can 1) be manufactured successfully and 2) continue to progress in terms of performance, reliability, cost, and manufacturability, and continue to sustain innovation to support progress toward long-term PV cost and performance goals (i.e., 15% modules at under $50/m^2, capable of lasting 30 years). ($13.8M) [Harvard University, Iowa State University, Pennsylvania State University, Syracuse University, University of California at Los Angeles, University of Oregon, University of North Carolina, Colorado School of Mines, Colorado State University, Florida Solar Energy Center (affiliated with the University of Central Florida), University of Delaware's Center of Excellence in Photovoltaics (IEC), University of Florida, University of South Florida, University of Toledo, and Washington State University]

High-Efficiency Concepts and Concentrators

The goals are to push efficiencies beyond 30%, develop concentrator solar cells, and explore high efficiency ((20%) on low-cost substrates such as glass. ($1.3M) [University of California at Los Angeles, and California Institute of Technology]

102

Measurements and Characterization

This is centered at the National Renewable Energy Laboratory to provide technical leadership in standard testing of solar cells and modules, assist R&D and industry efforts through characterizing electro-optical properties, evaluating compositional, structural, and physical properties, conducting surface-analysis measurements to ascertain chemical and compositional information, developing new measurement and characterization techniques, and facilitating collaboration and information exchange through a variety of communication activities. ($3.8M)

Environment, Safety, and Health/Capital Equipment

These are important but small areas centered at Brookhaven National Laboratories to identify critical ES&H issues for PV ($300K) and to support critical capital equipment needs ($400K).

Technology Development

PV Manufacturing Technology

Through government/industry cost-shared R&D, the goals are to improve PV manufacturing processes and products, accelerate PV manufacturing cost reduction, and lay the foundation for significantly increased production capacity, thus assisting the U.S. industry in retaining and enhancing its world leadership role. ($7.9M)

Module Performance and Reliability

The goals in this area are to accelerate the development and improvement of PV module technologies to achieve module service lifetimes of 30 years or more, reduce uncertainty in PV radiometric measurements, and disseminate research results. ($3.0M) [New Mexico State University, Pennsylvania State University, and University of Colorado]

Balance-of-Systems Components Development

The goals are to improve system reliability, reduce life-cycle cost, provide technical support of the PV industry, and support increased standardization of balance-of-systems building blocks. ($1.4M) [Florida Solar Energy Center (affiliated with the University of Central Florida)]

Systems Engineering and Applications

System Performance and Engineering

Goals in this area are to conduct photovoltaic system performance studies through technical assistance, field engineering, outdoor and environmental testing, standards and codes development, performance criteria, test method development, and validation, testing certification, and facility accreditation. ($4.9M) [Southwest Technical Development Institute (affiliated with New Mexico State University), and Texas Southern University]

PV Markets and Applications

The goals are to accelerate acceptance of photovoltaics into existing and new domestic markets through education, training, technical assistance, and regulatory and policy development, and identify non-DOE funding for international projects. ($3.3M) [Clark Atlanta University and University of the Virgin Islands]

Solar Resource Characterization

The goals are to develop accurate solar data bases and disseminate solar resource information. ($500K)

Market Projects

This activity provides support out of DOE Field Offices for markets likely to be cost effective in the near term, such as for buildings (PV:BONUS), cost-effective utility projects (UPVG), PVUSA, PV4U, etc. ($2.8M) [Georgia Institute of Technology, State University of New York at Albany, University of Delaware, North Carolina State University, and Pace University Law School]

Program Management and Communications

Program Management

This task provides program management, analysis, coordination, integration, reporting, and information dissemination necessary to conduct efficient and effective projects at NREL and Sandia. ($2.6M)

DOE Headquarters Special Projects

The purpose is to support Small Business Innovative Research (SBIR) projects, special projects, and DOE-subcontracted studies of PV in the context of other energy technologies. ($5.7M)

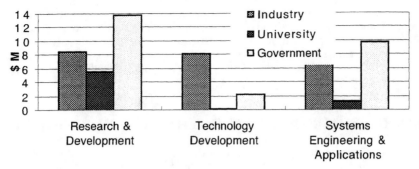

Figure 1. DOE FY 1997 funds for industry, university, and government labs.

THE NEW PARADIGM FOR BASIC RESEARCH

A recent *Basic Research White Paper* prepared by IBM Research, SRI International, Microsoft Engineering, Lucent Technologies, National Science Foundation, and the U.S. Department of Energy presents a new paradigm for basic research, describing it as research projects more than 10 years in duration (2). The old paradigm of basic research was first outlined by Vannevar Bush, President Truman's science adviser, in a 1945 paper entitled *Science--The Endless Frontier* as "research performed without thought of practical ends." That concept may have been appropriate for that period when technologies were believed to progress slowly from basic research to applied research to developmental research. Today, much research is done concurrently, including even production research, to save money and time. The new paradigm places basic research at the end of a spectrum of research ranging from long-term, high-risk research to short-term, low-risk work. Paradigms for applied research are mid-term research (5 years duration) and technology development and applications as near-term research (2 years duration).

There are three observations for this paradigm of basic research within the DOE PV Program. First, some 51 of the 55 universities funded by DOE are now performing mid-term and near-term research projects for the DOE program. Second, this de-emphasis on long-term research is consistent with the findings in the *Basic Research White Paper*, where both U.S. industrial and government basic-research spending have declined throughout the 1990s. Third, as noted in the *White Paper* industry and government labs have been performing less long-term R&D than universities. The danger resulting from a lack of support for basic research performed by universities or others is that the ideas and technologies that industry are going to need in 10 to 20 years from now are likely to be unavailable.

Yet, a balance is needed. In the early 1990s, DOE made the judgment that there was a serious problem in not funding the near-term research needed for PV manufacturing. The outcome was the highly successful PVMaT program (3). Recently a funding gap has been identified between ideas once they have been invented but before they can be tested and demonstrated in production. This mid-term research activity has resulted in a proposal called PV GAP to help bridge the laboratory and early production phases of thin-film photovoltaics (4). Successful technologies require program balance because pumping money into only long-term or mid-term or near-term research isn't a guarantee of success (5). That a balance is needed has also been described in detail in a book entitled *Third Generation R&D* (6).

LONG-TERM RESEARCH IN THE DOE PV PROGRAM

DOE and NREL initiated a Universities Participation Program in the 1980s. The last University Participation solicitation appeared in 1992 with 53 proposals and 5 awards. Five contracts, with the University of Illinois, University of South Florida, Johns Hopkins University, North Carolina State University, and University of Utah are nearing completion in 1997.

A New Ideas Program, first started in 1979, had its last solicitation in 1990, resulting in 104 proposals and only 3 awards. The last of the New Ideas contracts has recently been completed. An assessment of the New Ideas program, conducted in 1986, concluded that more than half of the contracts arising from two separate 1980s procurements were judged to be "successful" (7). About 20% of the New Ideas contracts previous to 1986 went to universities.

Both of these subcontract programs had funded high-risk, long-term research. In the 1990s, the DOE PV program has funded more mid-term and near-term activities. This has been a natural consequence of research progress in the 1980s and the development of a group of stakeholders, such as newly created PV companies and interested electric utilities, for mid-term and near-term research. With decreasing overall program budgets, less funding has been available for the long-term research activities. Nevertheless, as the University Participation and New Ideas contracts are being completed, DOE still sees a need for long-term research in its portfolio, and universities are expected to continue to play an important role. We anticipate that DOE, through NREL, will be soliciting proposals for long-term research in photovoltaics during the next fiscal year.

CONCLUSION

Historically, universities in the DOE PV program have concentrated on long-term, fundamental research. Today much of the university research is centered around two Centers of Excellence and the two experimental stations. Universities also strongly participate in DOE's Thin Film PV Partnership. Their participation has primarily been in mid-term and near-term research projects. This is a consequence of some of the earlier long-term research funded by DOE in the 1980s. The New Ideas and University Participation Programs have been highly successful in the past and DOE looks forward to restarting these efforts that support long-term research leading to future-generation photovoltaic technologies.

ACKNOWLEDGMENTS

We would like to thank John Benner for discussions on *Third Generation R&D* and insights into early procurements for universities and innovative concepts. We also thank Tom Surek for useful conversations about the role of universities in the DOE PV program. This work was completed under Contract No. DE-AC36-83CH10093.

REFERENCES

1) U.S. Department of Energy, *Photovoltaics: The Power of Choice, The National Photovoltaics Program Plan for 1996-2000,* National Technical Information Service, Springfield, VA, DOE/GO-10096-017, 1996, p. 1.

2) IBM Research, SRI International, Microsoft Engineering, Lucent Technologies, U.S. Department of Energy, and National Science Foundation, *Basic Research White Paper, Defining Our Path to the Future,* Des Plaines, Illinois, R&D Magazine, 1997, p. 7.

3) Mitchell, R. L.; Witt, C. E.; Thomas, H. P.; Herwig, L. O.; Ruby. D. S.; Aldrich, C. C., "Benefits From the U.S. Photovoltaic Manufacturing Technology Project," in *Proceedings of the 25th IEEE Photovoltaic Specialists Conference,* 1996, 1215-1218.

4) K. Zweibel, *personal communication*

5) Op. cit., p. 14.

6) Roussel, P. A.; Saad, K. N.; Erickson, T. J., *Third Generation R&D,* Boston, Massachusetts, Harvard Business School Press, 1991, pp. 102-104.

7) W. Luft, *Assessment of the Program for Innovative Concepts and New Ideas for Photovoltaic Conversion,* SERI/MR-211-2793, 1986, p. 21.

Pushing the Frontiers of Silicon PV Technologies: Novel Approaches to High-Efficiency, Manufacturable Silicon Cells

A. Rohatgi, P. Doshi, and T. Krygowski

University Center for Excellence inPhotovoltaic Research and Education
Department of Electrical and Computer Engineering
Georgia Institute of Technology, Atlanta, GA 30332-0250

Abstract. Crystalline silicon is the dominant photovoltaic (PV) material today. Although laboratory cell efficiencies have reached 24%, production cell efficiencies are still in the range of 11-15% at a market price of ˜$4/watt for modules. The real challenge is how to incorporate high efficiency features in the industrial cells in a cost-effective manner. This paper highlights the key high efficiency attributes that can bridge the gap between laboratory and industrial cells. It summarizes some of the recent developments and emerging technologies that have the potential of reaching the cost and efficiency targets simultaneously. Further development of these technologies, along with the push to increase the market size towards 500 MW/yr can lead to next generation silicon cells that can reduce the silicon PV module cost down to $1-2/watt.

INTRODUCTION

Crystalline silicon dominates the world-wide solar cell production, which reached a record level of 90 MW in 1996. Crystalline silicon solar cells accounted for almost 80% of the photovoltaic modules shipped in 1996 at a market price in the range of $3.50 - $4.50/ watt. The cost of PV modules need to be in the range of $1-$2/watt for it to become cost effective for large scale terrestrial and utility scale applications. Cost break down of current Si PV modules reveals that wafer, cell processing, and module assembly account for approximately 45%, 25% and 30% of the module cost, respectively. The cost of silicon wafer can be reduced by low-cost solar grade polysilicon feedstock material, increased substrate size, reduced kerf losses during slicing, and thinner substrates. Cell processing cost can be reduced by integrating multiple high efficiency features in each processing step, reducing the number of photomask and high temperature steps, and increasing the yield and throughput. Higher cell efficiency can reduce the cost of PV by reducing material consumption, module assembly and balance of system cost.

Even though laboratory cell efficiencies have reached 24% (Fig. 1a), the production cell efficiencies are still in the range of 11-15%. This efficiency gap

CP404, *Future Generation Photovoltaic Technologies: First NREL Conference*, edited by McConnell
© 1997 The American Institute of Physics 1-56396-704-9/97/$10.00

can be reduced by cost effective implementation of advanced cell design features such as, (1) surface texturing, (2) reduced contact recombination, heavy doping effects and contact resistance by formation of selective emitter, (3) effective front surface passivation, (4) reduced shadow losses by fine grid line printing, (5) increased diffusion-length/cell-thickness ratio, improved back surface passivation by a back surface field (BSF), point contacts or floating junction, and (6) improved optical design with front surface texturing high back surface reflectance and internal light trapping. Reasonable efficiency targets for industrial cells are 18-20% for monocrystalline Si and 16-18% for multicrystalline Si. This paper highlights some of the promising developments and emerging technologies for the next generation of industrial Si cells which have the potential of reaching the goal of ¯18% efficiency at $1-2 /watt.

2. Next Generation Silicon Solar Cells

2.1   Buried Contact Solar Cells

Buried contact solar cell (BCSC) technology is well documented in the literature [1]. Laboratory scale efficiency as high as 21.3% have been reported on hybrid small area FZ cell with buried contact on front and local back surface field (LBSF) on the rear. Large area (100 cm^2) efficiencies in production have already reached a level of 17-18% using a conventional cell design with full metal back. As shown in (Fig. 1a), this conventional single sided buried contact cell has several high efficiency features including: selective emitter with reduced contact recombination and heavy doping effects, fine-line metallization with high aspect ratio of grid lines, and effective front and back surface passivation. A typical process sequence involves texturing and light n$^+$ diffusion over the whole surface followed by growth of a thick passivating thermal oxide on front and back. A 40-50 μm deep mechanical or laser grooving is performed to define the grid regions. A heavy n^{++} diffusion is performed only in the grooves, since the thick oxide serves as a diffusion mask. An Al back-surface-field is formed, followed by electroless plating of nickel, copper and silver while the thick oxide between the grid lines serves as a plating mask. Even though conventional buried contact technology does not require any photomask steps it involves several lengthy processing steps including three to four high temperature steps, groove formation, etching, cleaning and multilayer plating. Therefore, a simplified buried contact process is being developed and optimized, which involves grooving, texturing, a single P diffusion, rear BSF, AR coating, and plating. This process reduces the number of high temperature steps but sacrifices V_{oc} which decreases to 630 mV.

Another buried contact cell technology has recently been proposed (Fig.

1b), which involves double sided buried contact (DSBC). This technology uses a floating junction for rear passivation along with heavily boron-diffused grooves on the back. The heavily diffused p^+ grooves provide low contact resistance and reduce contact recombination contribution to the base saturation current density (J_{ob}). In addition, a floating junction passivation is superior to Al BSF. These features gave a very impressive V_{oc} of 685 mV, which is ~45 mV greater than the conventional BCSC [2]. However, because of shunting problems with the floating junction, cell efficiencies were ~17%. Even though the floating junction was formed at the same time as the front n^+ diffusion, this DSBC process involves 3 to 4 high temperature steps. More recently a new scheme is proposed to simplify the process sequence of DSBC and reduce the number of high temperature steps down to two [3]. This process scheme involves top phosphorus diffusion followed by simultaneous front and back oxidation. Then front and rear grooves are formed, followed by boron and phosphorus spin-on on the front and rear. Finally a drive-in is performed at ~1000°C followed by Ni/Cu plating. On an untextured surface, this simplified DSBC technology produced 15% efficient CZ cells and is expected to give 17% efficiency for textured cells.

Thus, buried contact technology offers considerable promise for next-generation industrial cells. However, capital investment, process complexity, multiple high temperature steps, and environmental impact because extensive plating chemicals need to be considered and resolved.

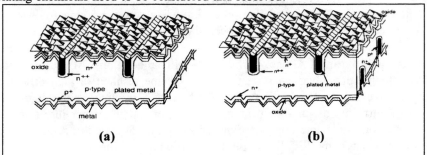

(a)   **(b)**

Fig. 1. Cell structures for (a) conventional buried contact, and (b) double-sided-buried-contact.

2.2   Screen Printed Solar Cells

The advantages of screen printed (SP) cells include: simplicity, high throughput, low-cost, and a proven technology base. In fact, fully automated screen printing production lines are operational in many companies. A typical industrial SP process involves saw damage removal and surface texturing (optional), a 30-50 $\Omega/\square$ phosphorus diffusion for the emitter, followed by TiO_2

or SiN antireflection coating. Al paste is screened and fired to form BSF and back contact on the rear. Grid lines are screen printed using Ag paste and fired through the AR coating. This technology produces CZ Si cell efficiencies in the range of 13-15% [4]. The main draw-backs of the current industrial SP process include performance losses resulting from (1) excessive contact shading (150-200 µm wide grid lines with ˉ10% shading), (2) low fill factors (<0.76) because of high contact resistance and low metal conductivity, and (3) low short-wavelength response because of heavy emitter doping effects and poor surface passivation.

Several groups are now working toward next generation screen printed cells to overcome the above losses, which could amount to about 2% in absolute efficiency. This is being done by using improved pastes, screens, and screen printers that can reduce line width and form selective emitters. Recent development in fine-line printing in the range of 50-80 µm allows the reduction in grid spacing and facilitates the use of more lightly diffused emitter, in addition to reduced shading. This could also improve short wavelength response, provided the emitter surface is well passivated. IMEC has demonstrated 16.6% large area cells using a combination of fine line printing, somewhat lightly doped homogenous emitter, and oxide surface passivation [4]. Fig. 2 shows the schematic of a selective emitter SP cell fabricated by IMEC, which has recently produced 17.3% large area (96cm^2) CZ cells [4]. This process scheme involves (1) front surface texturing & cleaning, (2) selective emitter formation by an grid line protection using screen printing followed by emitter etch back to 100 Ω/□ between the grid, (3) 60µm wide fingers with ˉ6% shadow loss, (4) effective front surface passivation by thermal oxidation, and (5) screen printed Al back surface field with an additional Al gettering step. Even though this advanced screen printing scheme resulted in high efficiency cells, the cost-effectiveness of etch-resist printing/emitter-etch-back, manufacturability of fine-line printing, and use of 3 high temperature steps are of some concern. Some groups are investigating the formation of screen printed selective emitters by PECVD etch-back between the grid lines.

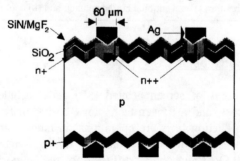

Fig. 2. Structure of IMEC's 17.3%-efficient large area screen-printed Cz silicon cell.

2.3   Rapid Thermal Processing (RTP) of Silicon Cells

Reduction in cell processing time and steps not only enhances throughput and yield but also reduces capital equipment and cell fabrication cost. RTP utilizes banks of tungsten-halogen lamps which radiate from the UV to IR regions. Higher energy photons in RTP enhance the diffusion and oxidation processes to reduce the cell processing time. Because of the reduced exposure to high temperature processing, the cleaning requirements are less stringent and fewer chemicals and gases are consumed compared to conventional furnace processing (CFP). The real challenge in RTP is to prevent quenching-induced lifetime degradation and to obtain surface passivation and appropriate front and back doping profiles compatible with high efficiency devices in a very short time. Until recently, RTP Si cell efficiencies were ˜15% while the CFP cells with comparable cell design are ˜18%. At Georgia Tech, we have been developing RTP technology with the goal of bridging the gap between the RTP and CFP cells through fundamental understanding of RTP-induced effects.

A simple CFP process sequence at Georgia Tech takes about 16 hours due to photolithography, prolonged wafer cleaning, and P and Al diffusions. By virtue of emitter oxide passivation, appropriate doping profiles, lifetime preservation, and photolithography, this CFP process produces ˜18% cells on modest quality untextured Czochralski silicon, ˜19% cells on low-resistivity ($<$ 1 Ω-cm) FZ silicon and $\geq$ 18% cells on HEM mc-Si cells.

An RTP process sequence developed at GT reduces the P and Al diffusion times from 330 min to 3 minutes. A high quality front surface passivation is achieved by a 5 min rapid thermal oxidation (RTO) after phosglass removal. To prevent lifetime degradation and achieve appropriate doping profiles, a short *in-situ* slow cooling is used during the RTD and RTO thermal cycles (Fig. 3). This combination of RTD and RTO resulted in 100 $\Omega/\square$ sheet resistance with a junction depth of 0.25 μm and a reasonably deep Al BSF. The cell data in Fig. 4 shows there was virtually no difference in the CFP and RTP cell performance, in spite of 5.5 hour reduction in high temperature processing time.

The above cells were processed by RTP without any high-temperature furnace steps. However, photolithography was used to define grid lines, which consumed 7 hours (counting the mesa isolation) out of the 8.5 hr process sequence. Therefore, in the next generation RTP cells, photolithography was replaced with screen printing. The contact firing conditions and the RTP diffusion cycle were modified to produce 0.3 μm deep screen-printable junction with a sheet resistance of 40 $\Omega/\square$. This RTP cycle had a peak temperature of 920°C and a short *in-situ* slow cooling, resulting in a 6 min thermal cycle. In an attempt to make these cells more manufacturable, evaporated ZnS/MgF_2

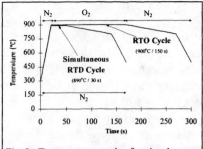

Fig. 3. Temperature cycles for simultaneous emitter/BSF formation and rapid thermal oxidation.

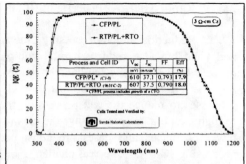

Fig. 4. Comparison of IQE and cell data between RTP and CFP reveals that virtually identical performance results from both processes.

coating was replaced by PECVD SiN AR coating. Finally screen printed contacts were punched through the PECVD AR coating. No RTO was grown in this sequence.

The above 1.5 hour RTP/SP process sequence resulted in ⁻16% efficiency on CZ Si [5]. This was about 2% lower than the RTP/RTO/PL cells. Detailed modeling and analysis of the two cells revealed that the 2% loss in efficiency can be attributed to (1) 0.4% loss in efficiency due to poor surface passivation (2) 0.3% loss due to heavy doping effects in the emitter because of higher N_s (3) 0.2% loss due to higher contact resistance (4) 0.2% loss due to lower conductivity of the screen printed Ag (5) 0.5% loss due to higher shading (6) 0.1% loss in the sheet due to large grid spacing and (7) 0.3% loss due to AR coating properties and absorption in the SiN film.

Based on the above analysis, the next generation RTP/SP cells will include selective emitter by screen printing, surface passivation by RTO, and surface texturing. These modifications should produce ≥ 18% next generation RTP/SP cells without any photomask or lengthy conventional furnace steps. However, before RTP can become commercially viable, a batch or continuous RTP system needs to be built to translate the short RTP cell processing time into high throughout.

2.4   STAR Cells

The STAR process is being developed at Georgia Tech, which incorporates a <u>S</u>imultaneously diffused emitter and back surface field (BSF) on a <u>T</u>extured silicon wafer, with an in-situ thermal oxide for surface passivation and <u>A</u>nti-<u>R</u>eflection coating. In a *single* high temperature step, the STAR process provides five important efficiency enhancement features: (1) front and back doping profiles compatible with high efficiency cell design (2) emitter oxide passivation (3) back surface passivation via a passivated boron back surface field (4) a low reflectance single layer SiO_2 AR coating on the textured surface and (5) Al-SiO_2-Si back surface reflector for light trapping. The star process involves in-house fabrication of doping sources using silicon wafers, coated with P and B compounds. The phosphorus sources are prepared by a 30 min $POCl_3$ diffusion or simply by a spin-on process followed by a short bake. Boron sources are formed by a liquid spin-on process. By controlling the concentration of P_2O_5 and B_2O_3 on the same wafers, a wide range of emitter and BSF profiles can be obtained simultaneously. The resulting P diffusion glass on the solar cell wafer is only 50 A°, which allows in-situ growth of high quality oxide without appreciably increasing the reflectance of the device. It is found that the use of separate source wafers not only gives more flexibility in controlling front and back sheet resistances for the same thermal cycle, but also act as a powerful contamination filter, especially in the case of boron. After each simultaneous diffusion cycle, the source wafers are recycled by a brief dip in 10% HF followed by the short spin-on process.

Fig. 5 shows the source-sample arrangement in the STAR process which is performed in a conventional diffusion furnace [6]. After surface texturing, P and B diffusions are performed simultaneously at 1000 C/60 min followed by a 60 min in-situ oxidation, resulting in 90 $\Omega/\square$ sheet resistance on the front, 30 $\Omega/\square$ on the back, and ˜1100A front and back oxide for surface passivation. Front oxide also serves as a single layer AR coating while back oxide forms a back surface reflector. This is followed by two photomask steps: one for the front grid and the other for point contacts to the p+ region. This process has so far produced ˜20% efficient STAR I cells (Fig. 6). However if the boron is spun-on directly onto the sample wafer, the bulk lifetime drops to 6µs from 1 ms, resulting in a cell efficiency of only 15%. This supports the merit of using separate source wafers to filter contaminants in the spin-on solution. Detailed modeling and analysis of these 20% cells showed that further improvements are possible by incorporating double layer AR coating, a flat back surface, and

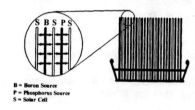

B = Boron Source
P = Phosphorus Source
S = Solar Cell

Fig. 5. Source/sample arrangement for STAR process.

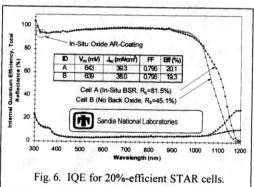

In-Situ Oxide AR-Coating

ID	V_{oc} (mV)	J_{sc} (mA/cm²)	FF	Eff (%)
A	643	39.3	0.796	20.1
B	639	38.0	0.796	19.3

Cell A (In-Situ BSR, R_b=81.5%)
Cell B (No Back Oxide, R_b=45.1%)

Sandia National Laboratories

Fig. 6. IQE for 20%-efficient STAR cells.

making the boron BSF more transparent by increasing its sheet resistance and lowering diffusion temperature (STAR II cell).

In the next generation STAR cells (STAR III), photolithography will be replaced by screen printed contacts. It is also found that the sheet resistance in the textured regions is twice that of the flat regions in the STAR process, which uses limited sources. This will help in the formation of selective emitters without double diffusion, provided a texturing mask can be screen printed in the grid regions. In addition, local BSF will be formed by screen printing Al through the passivating back oxide. Thus, next generation STAR cells will only use one furnace step and three screen printing steps. Preliminary model calculations show that greater than 18% efficient cells are possible on a silicon material with a bulk only lifetime of about 50 μs. STAR II and III cells are currently in the development stage.

3. Conclusions

The current market price of Si PV modules is ~$/4 watt, which is about a factor of 2-4 higher for cost effective utility-scale applications. The cost breakdown of PV module indicates that wafer, cell processing, and module assembly account for 45%, 25% and 35% of the module cost. Since the efficiency influences the production cost at all stages, ranging from Si material to BOS, several promising Si technologies are trying to raise the production cell efficiencies to ~18%. Significant progress has been made in several technologies, including buried contact cells, screen printed cells, RTP cells, and STAR cells, however, there is a strong need for research development to further simplify processing and reduce the number of high temperature steps, while maintaining high efficiency. A recent study in Europe showed that if the market size can grow to 500 MW/yr, then proper combination of promising technologies and Si material can reduce the module cost down to $1-2/watt.

1. Wenham, S.R., et.al, "Silicon Solar Cells," *Progress in Photovoltaics*, vol.4, pp. 3-33, 1996.
2. Hansberg C.B., et.al, "685 mV Open Circuit Voltage Laser Grooved Silicon Solar Cells," *Solar Energy Materials and Solar Cells* vol. 34, pp. 117-123, 1994.
3. Ebong, A.U., et.al, "Fabrication of Double Sided Buried Contact (DSBC) Silicon Solar Cell by Simultaneous Pre-Deposition and Diffusiion of Boron and Phosphorus, *Solar Energy Materials and Solar Cells,* vol. 44, pp. 271-278, 1996.
4. Nijs, J., et.al, "Latest Efficiency Results with the Screen Printing Technology and Comparison with the Buried Contact Structure," *Proceedings of the 1st World Conference on PV EnergyConversion,* pp. 1242-1249, 1994.
5. Doshi, P. et.al, "Modeling and characterization of high-efficiency silicon solar cells fabricated by RTP, screen-printing, and PECVD," accepted for publication, *IEEE Transactions on Electron Devices,* submitted Oct. 1996.
6. Krygowski, T., et.al, "A Novel Technology for the Simultaneous Diffusion of Boron and Phosphorus in Silicon," *Proceedings of the 25th IEEEPhotovoltaic Specialists Conference,*1996.

DYE-SENSITIZED PHOTOCHEMICAL
SOLAR CELLS

Photoelectrochemical Solar Energy Conversion by Dye Sensitization

M. Grätzel

Institut de chimie physique,
Ecole polytechnique fédérale de Lausanne, CH - 1015 Lausanne

Abstract. The lecture desrcibes the salient features of mesoscopic oxide semiconductor film and their electrochemical applications. In particular a molecular photovoltaic device is presented whose overall efficiency for AM 1.5 solar light to electricity at present attains 10-11%. The system is based on the sensitization of nanocrystalline oxide films by molecular charge transfer sensitizers. In analogy to photosynthesis, the new chemical solar cell separates light absorption and charge carrier transport processes. These cells exhibit a remarkable stability making practical applications feasible, the first products being targeted to supply electric power for consumer electronic devices. The mesoscopic oxide semiconductor films offer a number of other attractive research possibilities. Thus, a tandem device based on two superimposed layers with complementary light absorption in the visible range accomplishes the cleavage of water into hydrogen and oxygen with an overall efficiency of 4.5%.

GENERAL FEATURES OF NANOCRYSTALLINE ELECTRONIC JUNCTIONS

Nanocrystalline semiconductor films are constituted by a network of mesoscopic oxide particles, such as TiO_2, ZnO. Nb_2O_5, WO_3 or Ta_2O_5, which are interconnected to allow for electronic conduction to take place.The pores are filled with an electrolyte, or with a solid material such as an amorphous organic hole transmitter or a p-type semiconductor. In this way an electronic junction of extremely large contact area is formed, displaying very interesting and unique properties. Some important features of such mesoporous films are:

i)   an extremely large internal surface area, the roughness factors being in excess of 1000 for a film thickness of 8 microns;

ii)   the ease of charge carrier percolation across the nanoparticle network, making this huge surface electronically addressable;

CP404, *Future Generation Photovoltaic Technologies: First NREL Conference*, edited by McConnell
© 1997 The American Institute of Physics 1-56396-704-9/97/$10.00

iv)   the fast intercalation and release of Li^+ ions into such films.

THE PRINCIPLE OF OPERATION OF THE MOLECULAR PHOTOVOLTAIC DEVICE

The incident photon to current conversion efficiency or "external quantum yield" of such a device is then given by the equation:

$$\eta_i (\lambda) = LHE (\lambda) \times \phi_{inj} \times \eta_e \qquad (1)$$

where $\eta_i (\lambda)$ expresses the ratio of the measured electric current to the incident photon flux for a given wavelength, LHE is the light harvesting efficiency, ϕ_{inj} is the quantum yield for charge injection into the oxide and η_e is the charge collection efficiency. Our solar cell operates on a principle that differentiates the processes of lightabsorption and charge separation, Figure 1. Light is absorbed by a sensitizer (S) attached to the surface of the large band gap oxide. The large internal surface area allows for efficient light harvesting even though dye adsorption is restricted to monolayer coverage. Electron injection from the excited state (S*) into the conduction band of the oxide follows light excitation. The recapture of the electrons by the oxidized dye (S$^+$) is intercepted by transferring the positive charge to a redox mediator R/R^+ present in the electrolyte and hence to the counter electrode. Via this last charge transfer, in which the mediator is returned to its reduced state, the circuit is closed. The system converts light into electricity without permanent chemical transformation. The maximum voltage ΔV that such a device could deliver corresponds to the difference between the redox potential of the mediator and the conduction band position of the semiconductor.

CONVERSION EFFICIENCIES AND STABILITY

The use of mesoporous oxide films to support the sensitizer allows sunlight to be harvested over a broad spectral range in the visible fulfilling the first requirement for efficient photo-conversion. In order for the device to deliver a photocurrent that matches the performance of conventional cells, both the electron injection and charge carrier collection must in addition occur with an efficiency close to unity. The

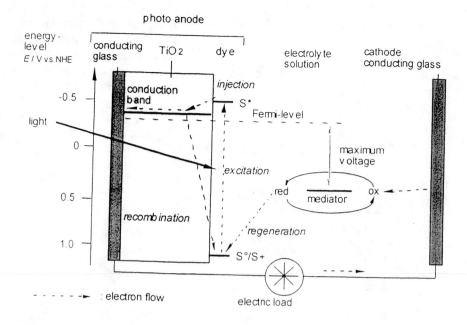

------ : electron flow

Fig. 1 Schematic representation of the principle of the nanocrystalline
injection cell to indicate the energy level in the different phases.

quantum yield of charge injection (ϕ_{inj}) is the fraction of the absorbed
photons which are converted into conduction band electrons. For
sensitizer **1** the injection occurring in the femtosecond domain
dominating largely other competing deactivation reactions of the
excited dye /3/. By contrast, the recapture of the electrons by the
oxidized ruthenium complex is many orders of magnitude slower
favoring light-induced charge separation at the surface of the films.
The rapid regeneration of the sensitizer by the electron donor, i.e.,
iodide:

$$2S^+ \quad + \quad 3\,I^- \quad \longrightarrow \quad 2S \quad + \quad I_3^- \qquad (2)$$

is crucial for obtaining good collection yields and high cycle life of the
sensitizer. In the case of **I** , time resolved laser experiments have
shown the interception to take place with a rate constant of about 10^8
s^{-1} under the conditions applied in the solar cell. This is about one
hundred times faster than the recombination rate and 10^8 times faster

121

than the intrinsic decomposition rate of the oxidized sensitizer in absence of iodide. This, together with the ultrafast nature of the electron injection process from the excited state into the semiconductor, explains the fact that $\underline{I}$ can sustain 100 million turnovers in continuous solar cell operation without loss of performance.

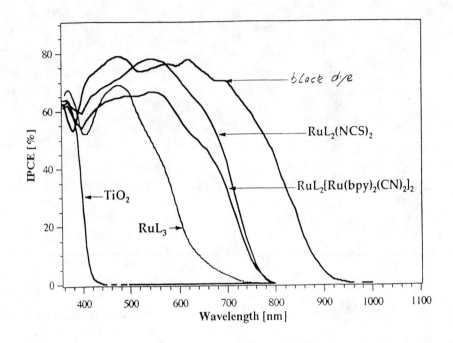

Fig. 2 Photocurrent action spectrum obtained with four different ruthenium complexes attached to the nanocrystalline TiO_2 film. The blank spectrum obtained with the bare TiO_2 surface is shown for comparison.

A graph which presents the conversion efficiency of incident monochromatic light in electric current is shown for several ruthenium complexes in Figure 2 Very high efficiency of current generation, exceeding 75%, were obtained. When corrected for the inevitable reflection and absorption losses in the conducting glass serving to support the nanocrystalline film the yields are practically 100 percent. Historically, RuL_3 (L = 2,2'-bipyridyl-4,4'-dicarboxylate) was the first efficient and stable charge transfer sensitizer to be used in conjunction with high surface area TiO_2 films. A significant enhancement of the light harvesting was achieved in 1990 with the trimeric complex of ruthenium whose two peripheral ruthenium moieties were designed to serve as antennas /4/. The advent of **1** in 1991 marked a further improvement since it extended the light absorption over a broad range in the visible /5/. Its perfomance was only superseeded recently by the discovery of a new black dye having a spectral onset at 900 nm which is optimal for the conversion of AM 1.5 solar radiation to electric power in a single junction photovoltaic cell.

Figure 3 gives an example for the current-voltage characteristics nanocrystalline injection cells based on sensitizer **I**. Overall conversion efficiency for standard AM 1.5 sunlight between 10 and 11% have been confirmed.

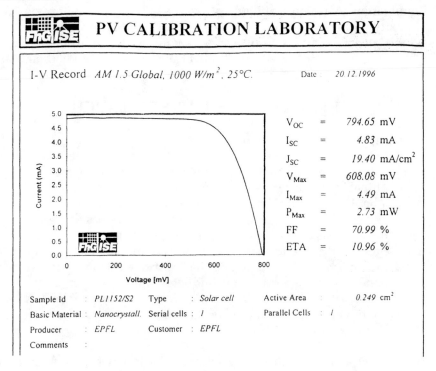

Fig.3   Photocurrent voltage curve for a nanocrystalline photovoltaic cell based on *cis*-Ru(2,2,-bipy-4,4'-dicarboxylate)(SCN)$_2$ as a sensitizer.

An advantage of the nanocrystalline solar cell with respect to solid state devices is that its performance is remarkably insensitive to temperature change. Thus, raising the temperature from 20 to 60 °C has practically no effect on the power conversion efficiency. In contrast, conventional silicon cells exhibit a significant decline over the same temperature range amounting to ca 20 percent. Since the temperature of a solar cell will reach readily 60 °C under full sunlight, this feature of the injection cell is particularly attractive for power generation under natural conditions.

Meanwhile, the stability of all the constituents of the nanocrystalline injection solar cells, that is: the conducting glass, the TiO_2 film, the sensitizer, the electrolyte, the counterelectrode and the sealant have been subjected to close scrutiny. Upon long time illumination, complex **I** sustained 10^8 redox cycles without noticeable loss of performance corresponding to ca 20 years of continuous operation in natural sunlight. Stability tests on sealed cells have progressed significantly over the last few years. These tests are very important, as the redox electrolyte or the sealing, may fail under long term illumination. A recent stability test over 7000 hours of continuous full intensity light exposure has confirmed that this system does not exhibit an inherent instability /6/, in contrast to amorphous silicon which due to the Stabler-Wronski effect undergoes photodegradation.

CURRENT AND FUTURE RESEARCH

Currently research focusses on:
i)   the molecular design and synthesis of new sensitizers having enhanced near infrared light response, examples being phthalocyanines or the black dye discussed above.
ii)   a better understanding of the interface, including experimental and theoretical work on dye adsorption processes.
iii)   the analysis of the dynamics of interfacial electron transfer processes down to the femtosecond time domain.
iv)   the unraveling of the factors that control electron transport in nanocrystalline oxide semiconductor films.
v)   the replacement of the liquid electrolyte by solid materialsvi) the development of tandem cells and their use for the cleavage of water by visible light.

In the following I shall briefly address the last two points. Suitable solid materials to replace the liquid electrolyte are large band- gap p-type semiconductors. First results based on dye sensitized cells using p-type inorganic semiconductors, such as cuprous thiocyanate, CuSCN /7/ or cuprous iodide CuI /8/ have been published. Alternatively, a hole transmitting solid may be employed similar to the amorphous organic compounds used in electroluminescence devices. Light induced electron transfer from the excited state in the conduction band of the oxide semiconductor occurs in the same manner as with liquid electrolytes. However the dye is regenerated by e hole conductor such as the aromatic amine spiro-TAD whose structure is presented in Figure 4. Recent experiments have established that hole injection occurs quantitatively and on a sub-nanosecond time scale /9/. The advantage of this approach is that hole conduction to the counterelectrode is by hopping and does not involve mass transfer. In addition, judicious selection of the organic material allows to match its redox level to the ground state oxidation potential of the sensitizer. This is not the case for the triiodide/ioide redox electrolyte where typically 0.5 eV of driving force are wasted in the regeneration of sensitizers, such as complex **I** . Judicious selection of the type of organic hole transmitting material offers the prospect to avoid this loss resulting in an increased photovoltage that would permit to nearly double the overall efficiency of the device.

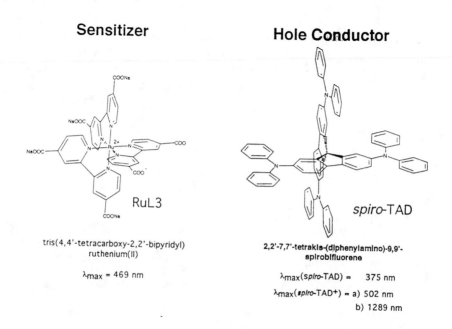

Sensitizer

RuL3

tris(4,4'-tetracarboxy-2,2'-bipyridyl)
ruthenium(II)

λ_{max} = 469 nm

Hole Conductor

spiro-TAD

2,2'-7,7'-tetrakis-(diphenylamino)-9,9'-
spirobifluorene

λ_{max}(*spiro*-TAD) =   375 nm

λ_{max}(*spiro*-TAD⁺) = a) 502 nm

b) 1289 nm

Fig. 4 Molecular structures for a sensitizer and hole conductor
employed in the dye sensitized heterojunction.

A tandem device that achieves the direct cleavage of water into hydrogen and oxygen by visible light was developed in collaboration with two partner groups from the University of Geneva and Bern /10/. This is based on the in series connection of two photosystems. A thin film of nanocrystalline tungsten trioxide absorbs the blue part of the solar spectrum. The valence band holes created by band gap excitation of the WO3 serve to oxidize water to oxygen while the conduction band electrons are fed into the second photosystem It consists of the dye sensitized nanocrystalline TiO2 film. The latter is placed directly behind the WO3 film capturing the green and red part of the solar spectrum that is transmitted through the top electrode. The photovoltage generated by the second photosystem enables the generation of hydrogen by the conduction band electrons. The overall AM 1.5 solar light to chemical conversion efficiency achieved with this device stands at 4.5%.

EMERGING PRODUCTS, SOLAR CELLS AND NANOCRYSTALLINE INTERCALATION BATTERIES

Presently seven industrial organizations are engaged to bring the nanocristalline injection solar cell to market. The near term applications will mainly be in the low power range, the launching of the first product, a solar watch having the photovoltaic cell incorporated in the cover glass being imminent. Considerable progress has also been made in the development of in series interconnected modules and the production of solar tiles will start in Australia at the end of this year.

REFERENCES

1. O'Regan, B., Grätzel, M., Nature (London) 335,1991, pp. 737.
2. Nazeeruddin, M.K., Kay, A., Rodicio, I., Humphry-Baker, R., Müller, E., Liska, P., Vlachopoulos, N., Grätzel, M., J.Am.Chem.Soc., 115, 1993, pp. 6382.
3. Tachibana, Y., Moser, J., Grätzel, M., Klug, D.R., Durrant, J.R., J.Phys.Chem., 100 (1996) 20056.
4. Amadelli, R., Argazzi, R., Bignozzi, C.A., Scandola, F., J.Am.Chem.Soc., 112, 1990, pp. 70293.
5. Nazeeruddin, M.K., Péchy, P., Kohle, O., Zakeeruddin, S.M., Humphry-Baker, R., Grätzel, M., J.Chem.Soc., Chem.Comm., submitted.
6. Kohle, O., Meyer, A., Meyer, T., Grätzel, M., Adv.Mat., submitted.
7. O'Regan, B., Schwarz, D.T., Chem.Mat., 7, 1995, pp. 1349.
8. Tennakone, K., Kumara, G.R.R.A., Kumarasinghe, A.R., Wijayantha, K.G.U., Sirimanne, P.M., Semicon.Sci.Technol., 10, 1995, pp. 1689.
9. Bach, U., Grätzel, M., Lupo, D.W., Salbeck, J., Uebe, J., Weissörtel, F., contributed to the European Conference on Organic Materials, Bayreuth, Germany, April 1997.
10. Augustynski, J., Calzaferri, G., Courvoisier, J.C., Grätzel, M., Ulmann, M., Proc. 10th International Conference Photochem.Stor.Sol. Energy, Interlaken, Switzerland, 1994, pp. 229.
11. Kay, A., Grätzel, M., Sol.Energy Mat.Sol.Cells, 44, 1996, pp. 99.
12. Huang, S.Y., Kavan, L., Exnar, I., Grätzel, M., J.Electrochem.Soc., 142, 1995, L142.

The 1997 Calaveras Awards
for "Leapfrog" Technologies

Michael Grätzel

has leapt to a new level by giving

The Best Oral Presentation

on

Photoelectrochemical Solar Energy Conversion
by Dye Sensitization

The First Conference on Future Generation Photovoltaic Technologies
Sponsored by the National Renewable Energy Laboratory

127

Solid State Photoelectrochemical Cells Based on Dye Sensitization

Brian O'Regan and D. T. Schwartz

University of Washington, Dept. of Chemical Engineering
PO Box 351750 Seattle WA 98195-1750

Abstract. The initial efficiency of dye sensitized heterojunctions (DSHs) made from n-TiO_2, various ruthenium polypyridyl dyes, and p-CuSCN is quite poor. The efficiency increases dramatically when the DSH is exposed to moderate levels of UV illumination. The UV illumination excites electron hole pairs in the TiO_2. The hole moves to the surface and oxidizes some species present at the interface. The rate of the UV enhancement is controlled by the light intensity, the potential, and the TiO_2 thickness. After sufficient UV illumination the efficiency increase is permanent. Possible mechanisms are discussed.

INTRODUCTION

In our previous paper on dye sensitized heterojunctions (DSHs) we demonstrated that the wide-band-gap heterojunction between n-TiO2 (gap 3.2 eV) and p-CuSCN (gap 3.5 eV) could be sensitized to visible light by dyes present at the interface between the two materials.[1] Visible light photons absorbed by the dye cause the injection of an electron into the TiO_2 and a hole into the CuSCN. Both materials are conductive to the majority carrier and the charges can be collected as a photocurrent. We have shown that after certain treatments, the charge injection and collection process can be quite efficient. The necessary treatment was different in the case of two different dyes, rhodamine DSHs responded best to mild heating, whereas ruthenium polypyridyl DSHs responded to best to exposure to UV light. In this paper we will examine in much more detail the characteristics of the enhancement caused by UV light and discuss their implications for possible mechanisms. In this paper we will use the ruthenium dye **Z105** {Ru(II)LL'(NCS) where L is 2,2':6',2"-terpyridine-4'-phosphonic acid and L' is 4,4'-dimethyl-2,2'-bipyridine}.[2] **Z105** has a relatively sharp peak in the visible, which facilitates the measurement of the dye peak above the intrinsic response of the TiO_2 and the measurement of spectral shifts. **Z105** is slightly unstable to UV light while incorporated in a TiO_2/dye/CuSCN DSH. Some other ruthenium dyes are much more stable but their broader peak rendered them less useful for this paper.

CP404, *Future Generation Photovoltaic Technologies: First NREL Conference*, edited by McConnell
© 1997 The American Institute of Physics 1-56396-704-9/97/$10.00

EXPERIMENTAL

Non porous TiO_2 electrodes were fabricated by spray pyrolysis. The thin (50-100 nm) TiO_2 films were deposited on commercial SnO_2 coated glass (Libby Owens Ford) at temperatures of about 500° C. The resultant films had a surface roughness between 1.5 and 2 as determined by dye adsorption. Dyes were adsorbed from ethanol solution by placing the electrode in the solution overnight. CuSCN was usually deposited from ethanol solutions of KSCN and $Cu(BF_4)_2$. x H_2O as previously described. Aqueous solution of KSCN and $CuSO_4$, or ethanol solution of Cu(II)Triflate and KSCN were also employed as deposition solutions where noted.[1]

Photocurrent action spectra were obtained using a 150W xenon lamp (Oriel) focused through glass optics, a 10 cm water filter, and a jobin/yvon 0.20 M monochromator (grating blaze 450 nm). Currents were measured with an potentiostat (Omni 90, Cypress Systems). UV illumination was provided by focused output of the xenon lamp filtered through a UG2 (UV pass) filter. The intensity was approximately 1 mW/cm^2. Photoelectrochemical measurements were carried out using the same lamp and monochromator and various home built cells, using either 2 or 3 electrode system as indicated.

RESULTS

Each of the components of the DSH makes some contribution to the photoresponse of the DSH. Figure 1 shows the photoelectrochemical action spectrum of a spray pyrolysis TiO_2 film and of an electrodeposited CuSCN film. The photocurrent onset for TiO_2 rises steeply in the UV consistent with the anatase band gap of 3.2 eV. The CuSCN photocurrent rises much more slowly, consistent with the reported band gap of 3.5 eV.[3] CuSCN gives a cathodic photocurrent as expected for a p-type semiconductor; the TiO_2 photocurrent is anodic. The magnitude of the photocurrent of CuSCN is small compared to the TiO_2, probably due to the higher doping and thus higher recombination rate in the CuSCN. The as deposited CuSCN has a dark conductivity at least 4 orders of magnitude larger than the TiO_2 films. Figure 1 also shows the photocurrent action spectrum for a TiO_2/CuSCN heterojunction with no dye. The shape of the photocurrent onset for TiO_2/CuSCN is identical with that of the TiO_2 photoelectrochemical response, indicating that most of the UV photocurrent is due to photons absorbed by the TiO_2. In addition, heterojunctions made by depositing CuSCN directly on the conductive SnO_2 layer show ohmic conductivity and no photocurrent or photovoltage.

Figure 2a shows the photocurrent action spectrum of a TiO_2/Z105/CuSCN DSH. The initial action spectrum (solid line, marked "0") shows a large contribution from the TiO_2 and a small peak in the visible that matches the absorption spectrum of the dye. The UV IPCE in Figure 2 is higher than in Figure 1 due to a thicker TiO_2 layer. When the DSH is illuminated with UV light a large change occurs in the action spectrum. Figure 2a shows the evolution of the action spectrum and Figure 2b gives the photocurrent vs time at two different

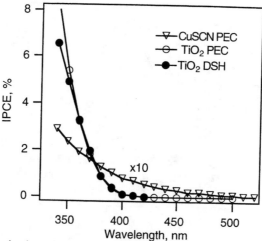

FIGURE 1. Intrinsic photoelectrochemical action spectra for TiO_2 (open circles), CuSCN (triangles) and for a TiO_2/CuSCN heterojunction. The spectra of the TiO_2 and the heterojunction have been normalized to each other at 360 nm to allow comparison of the shape

wavelengths. For the first 30 minutes of UV exposure, the photocurrent peak in the visible increases rapidly. At the same time, the IPCE of photons absorbed by the TiO_2 goes down sharply. During this process the shape of the action spectrum in the visible doesn't change significantly. No changes occur in the visual appearance of the DSH. Beyond 40 minutes, the dye sensitized response plateaus, and then, in the case of Z105, begins to decrease. For some other ruthenium dyes, no decrease is observed.

The rate of increase in the dye sensitized efficiency is dependent on the illumination wavelength. The action spectrum is an onset curve with increasing photon energy which is very similar to the onset of photocurrent in the TiO_2 photoelectrochemical cell. The rate is approximately linear with respect to UV photon flux. Thicker TiO_2 films result in faster increases presumably due to the greater absorption of photons. The rate is strongly affected by the potential at which the DSH is held during exposure. Treating the TiO_2 as ground, positive potentials above 500 mV increase the rate, and negative potentials below -500 mV decrease the rate to near zero. The rate appears to be insensitive to the surrounding atmosphere. Both DSHs sealed with epoxy, and those exposed to UV while in a nitrogen filled chamber, show UV enhancement rates similar to DSHs exposed to the atmosphere. The rate is also not highly sensitive to the deposition conditions. We have deposited the CuSCN from completely aqueous solution and also from ethanol solutions that were as dry as possible. In neither case was the UV enhancement rate strongly affected.

After UV exposure of less than 5 minutes, much of the increase in dye sensitized efficiency is transient. If the DSH is stored in the dark, in air, for 24 hours, the visible photocurrent will decline significantly. A second UV treatment, after dark storage, has the same effect as the first treatment. This "reversion" of the dye sensitized current becomes slower with longer UV

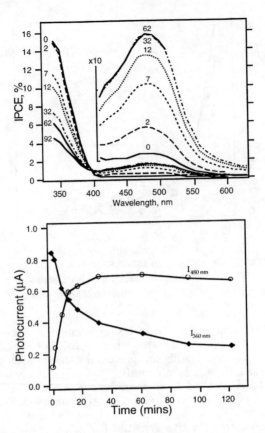

FIGURE 2A: Action spectra under monochromatic illumination, numbers labeling each spectrum are the cumulative UV exposure time. **2B**: Time evolution of the UV and visible photocurrent.

treatments. For UV exposure longer than a few hours, the increase in the dye sensitized efficiency is effectively permanent. The decrease in the TiO_2 intrinsic photocurrent during UV treatment also reverts in the dark. The reversion rate for the TiO_2 photocurrent also decreases with time but it is always faster than the reversion of the dye sensitized photocurrent. . Typically 24 hours in the dark gives an increase in the UV current from about 50% of its original value to about 75%, while at the same time, the dye sensitized photocurrent declines by only about 8%.

The reversion rate of the UV enhancement is not effected by storage in dry nitrogen instead of air. However, dipping the DSH in distilled water for 30 minutes can cause a large increase in the UV current and decrease in the visible. This later may show that the CuSCN layer on the DSH is not completely impermeable to water, or it may be a result of swelling of the CuSCN in water causing delamination of the layers followed by water intrusion. In some cases a few minutes in distilled water has had little effect.

For short UV exposures, if the CuSCN is removed (by soaking the DSH in acetonitrile with LiCl) and new layer of CuSCN is deposited, the effect of the UV exposure disappears. Renewed UV exposure has the same effect as the first treatment. After several hours of UV exposure the same removal and replacement of the CuSCN does not remove all the increase in the dye sensitized efficiency. This fact seems to imply that the longer UV treatments create a layer on the TiO_2 surface which is not removed by the ACN/LICl. Whether this layer causes the increased dye sensitization efficiency, or merely protects another change that has happened at the surface is unclear. In addition, it is possible that the change observed is actually an effect internal to TiO_2.

The UV illumination also causes large changes in the current voltage response of the DSH. Before the UV illumination, the IV response has a definite ohmic component. The initial effect of the UV illumination is to remove this ohmic component causing the IV to have more of a classic diode form. Continued UV illumination causes a steepening of the IV curve, accompanied by a positive shift. The shifts in the voltage for a given current are large. For example, the voltage at 5 mA/cm^2 moves first 300-400 mV negative during the first few minutes of UV exposure, followed by a positive shift of the same magnitude during additional exposure. Beyond 30 minutes the IV response stabilizes. As with the UV component of the current, storage in the dark can cause a reversion of the IV curves, usually to a position similar to that after the first few minutes of UV.

Despite these large changes in the IV response, the photocurrent onset curve does not change dramatically during the UV exposure. The shape of the photocurrent onset steepens as the ohmic component of the IV curve decreases, but the photocurrent onset potential changes only 50-100 mV during both the UV illumination and subsequent reversion in the dark. These results lead us to conclude that the effect of UV exposure on the IV response and on the dye sensitization efficiency are not mediated by the same chemistry. It is possible that the IV response is dominated by the current at a few "short circuits". In other words at a given voltage, not all of the surface may be contributing equally to the dark current. Chemistry that effects the IV response thus only need to occur at the regions of high dark current. The dye however is distributed widely on the surface. Also, at the end point of the increase in dye sensitized efficiency it appears that the APCE (absorbed photon to current efficiency) is near one.[4] This leads to the conclusion that the chemistry increasing the dye sensitization efficiency must occur at all points of the surface.

We have made attempts to "pre-treat" the TiO_2 or the TiO_2/Dye surface to produce the same surface chemistry as the UV exposure, but before CuSCN deposition. Exposure of the TiO_2 to UV illumination before the dye was applied had no effect, nor did exposure of a dyed TiO_2 film in the deposition solution before deposition. Pre-exposure of the TiO_2 to UV light in the presence of KSCN in ethanol did not cause an increase in the initial efficiency of the DSH, however it did result in the appearance of a tail of photocurrent into the visible analogous to the tail that appears during the UV exposure of the TiO_2/CuSCN heterojunction. This onset curve was much larger in magnitude and only somewhat similar in shape to that of the DSH. The dye sensitization efficiency of CuSCN in photoelectrochemical cells was not increased by UV illumination.

133

DISCUSSION

Because of the action spectrum of the UV enhancement rate, we can be reasonably sure that the UV enhancement of the dye sensitized efficiency is mediated by photons absorbed in the TiO_2. Because the photoexcited electron in TiO_2, and the electron injected by the dye, presumably occupy the same states in the conduction band, and dye injection alone does not cause any enhancement in efficiency, it seems that the photoexcited hole is required for the enhancement to occur. The increase in the rate observed under positive bias indicates that the UV enhancement is increased when the ratio of holes vs electrons reaching the TiO_2 surface is increased, or at least, when the products of oxidation are removed from the surface. Taken together, these data limit the mechanism of the UV enhancement to two classes of phenomena. In one, the UV enhancement is caused by photoexcited holes crossing the TiO_2 interface and oxidizing some species present that can not be oxidized by the oxidized dye. In the other, some positive species, trapped in the TiO_2, is made mobile by reacting with the hole, and diffuses out to the surface. We will treat mainly the first hypothesis, but we remark here that Cu(II), present in the electrodeposition solution, is a very small ion and might be able to diffuse into the TiO_2 lattice. During the electrodeposition, the Cu(II) might trap an electron and become immobile until reoxidized by the hole. Why the release of Cu(II) would increase the dye sensitization efficiency is not clear.

For the discussion below, we will assume that the UV enhancement is mediated by the oxidation of some species present at the interface. The main species that are present, and which can be oxidized, are Cu(0), Cu(I), SCN⁻, and water. The presence of Cu(0) is indicated by the formation of copper nodules at pinholes in the TiO_2. Presumably smaller amounts of Cu may be present at the TiO_2 surface. There may also be trapped BF_4^- and ethanol. The participation of ethanol would seem to be eliminated by the fact that the UV enhancement also occurs when the CuSCN is deposited from water. Although the dye is always deposited from ethanol, we presume no ethanol remains at the surface after exposure to a liquid water. We will ignore trapped BF_4^-; making the assumption that it is present very low concentration. To our knowledge BF_4^- does not adsorb to TiO_2 and BF_4^- will be depleted at the interface during electrodeposition due to charge balancing.

Even before the UV illumination, illumination with visible light does create a photocurrent based on dye sensitization. This means that the dye can inject holes into the CuSCN, which is equivalent to the oxidation of surface SCN⁻ or Cu(I). Since dye sensitized photocurrent does not self enhance, whichever species is oxidized by the dye in the process of hole injection is not the same species, which, oxidized by the hole in TiO_2, gives rise to the UV enhancement. Unfortunately, the orbital makeup of the CuSCN valence band is not known, and this information can not be used to eliminate one or both of Cu(I) and SCN⁻. The species $Cu(II)SCN_2$ can be synthesized and is at least metastable. This provides some evidence that the hole in CuSCN exists mainly as Cu(II). However, $Cu(II)SCN_2$ in solution decomposes rapidly to CuSCN and SCN⁺ which may indicate a rough equivalence of the Cu(I) and SCN⁻ "oxidation potentials" in CuSCN and therefore a valence band composed of contributions from both species.

134

If Cu(0) is present near the dye, then presumably it could be oxidized by the oxidized dye after electron injection. Since this process does not cause the enhancement of the dye sensitization efficiency, the oxidation of Cu(O) near the dye is probably not the mechanism of the UV enhancement. Alternatively, Cu(0) far from dye molecules might be oxidized by UV holes but not by the dye. However, it is difficult to see how these copper atoms could lower the quantum efficiency of dye sensitization. We believe that the initial low efficiency of dye sensitization is due to a blockage of charge injection into one phase or both, or an immediate short circuit of the injected charges. If the charges were injected into both phases and recombined at some remote site, then an applied positive bias should increase the collected photocurrent. This latter does not occur. Blocking of injection might occur due to movement of the bands relative to the dye, or by changes in the coupling brought about by shifts in the type of physical contact. An immediate short circuit might be provided by a reductive quenching of the dye excited state by Cu(I) followed by a electron transfer from the reduced dye back to Cu(II). Cu(I) does not result in quenching of the electron transfer to TiO_2 in solution, however Cu(II) does quench the excited state in solution, and in competes with charge injection. Our work with Rhodamine dyes has shown that hole injection may be faster than electron injection for some dyes.

From the above reasoning we are looking for an oxidation reaction that leads to a quasi irreversible change at the interface and which facilitates the charge injection into one of the two semiconductors. Since Cu(II) is a quencher, and we will assume that Cu(I) to Cu(II) oxidation is not the reaction in question. That leaves two other possibilities, the oxidation of SCN⁻ or that of water.

It is known that the photoexcited hole in TiO_2 can oxidize water molecules at the TiO_2 surface. Although our attempts to use dry ethanol and dry salts produced no major changes in the characteristics of the UV enhancement, it is probable that the surface of the TiO_2 was hydrated even in our driest experiments. The ethanol dye solution certainly contains some water, and the electrode was transferred in air between the dye solution and the electrodeposition solution. Even for relatively dry solutions, the amount of water needed to make a few monolayers on 1 cm^2 of TiO_2 is certainly present.

The oxidation of surface water will produce a proton and a hydroxyl radical. The latter is reactive and would presumable attack either a surface thiocyanate or a surface copper ion. The proton could bind to the TiO_2 surface and reduce the band edge energy lowering the barrier to injection increasing the dye sensitization efficiency. Reversion of the dye sensitization efficiency would be due to recombination of the proton and the product of hydroyl attack or perhaps the diffusion of the proton into the TiO_2 lattice. The lack of reversion after longer UV illumination times would be due to a build up of the hydroxylated product such that the number of protonatable sites was smaller than the number of protons created. Since the hole in TiO_2 can also inject a hole directly into the CuSCN (at least energetically), the decrease in the UV current would have to be due to the build up of a blocking layer separating the TiO_2 and the CuSCN, or to an increased recombination in the TiO_2. Reversion of the UV current at a faster rate is harder to explain, but does indicate that the combined reversion can not be a direct recombination of the products back to the starting material. Some possible reactions of the hydroxyl radical are the formation of a Cu(OH) at the surface of the CuSCN, or the formation of a layer of hydroxylated SCN⁻.

Unfortunately, the lack of shift in the photocurrent onset potential argues against this simple mechanism. Given the lack of shift in the onset potential, and the related conclusion that the bands edges did not move, it seems more likely that some form of quenching is removed by the water oxidation. Since water is deleterious to the functioning of photoelectrochemical cells based on dye sensitized TiO_2, it is even possible that it is the removal of water by oxidation, and not the products, that is causing the enhanced dye sensitization efficiency.

A second hypothesis which fits, to some extent, the experimental results is that photoexcited holes in the TiO_2 oxidize SCN^- molecules that are at the surface of the CuSCN. These oxidized thiocyanates combine to form a thiocyanate polymer, a species which has been observed in the oxidation of KSCN melts. This reaction requires, for charge balance, another positive species must leave the interface and reside near, or be reduced by the electron that remains in the TiO_2 after the photoexcited hole crosses the interface. A candidate for the mobile positive species is Cu(I). Another candidate is trapped Cu(II) that is probably present after the electrodeposition. In either case, the Cu species must diffuse to the counter electrode or to some place on the surface where electrons are crossing the TiO_2 interface. Reduction to Cu(0) is then necessary to preserve local charge balance at this location. An arguments against mechanism is that copper mobility in CuSCN at room temperature seems to be quite low. It is also possible that the CuSCN "disproportionates in place" under oxidation by the photohole. In other words, after SCN^- oxidation the an adjacent Cu(I) acts as an electron acceptor and then Cu(0) then coexists with SCN polymer at the interface. Evidence against this latter is that positive bias, which accelerates the UV enhancement, would act against Cu(I) reduction at the TiO_2 interface.

CONCLUSION

We have shown that the UV enhancement is due to photons absorbed by TiO_2. The results of bias strongly point to a mechanism that involves the holes crossing the TiO_2 surface and causing a chemical change at the interface or nearby in the CuSCN. At this time we cannot specify what chemical change is giving rise to the increased efficiency of dye sensitization.

ACKNOWLEDGMENTS

This work was supported by NREL subcontract RAN-5-15234

REFERENCES

1. O'Regan, B., and Schwartz, D. T., *J. Appl. Phys.* **80,** 4749-4754 (1996).
2. Péchy, P., Rotzinger, F. P., Nazeeruddin, M. K., Kohle, O., Zakeeruddin, S. M., Humphry-Baker, R., Grätzel, M., *J. Chem. Soc., Chem. Commun.* **1995,** 65-66 (1995).
3. Tennakone, K., Kumarasinghe, A. R., Sirimanne, P. M., Kumara, G. R. R. A., *Thin Solid Films* **261,** 307-310 (1995).
4. O'Regan, B., and Schwartz, D. T., *manuscript in preparation* .

Electron Transfer in Sensitized TiO$_2$ Photoelectrochemical Cells

Gerald J. Meyer

Department of Chemistry, Johns Hopkins University, Baltimore, MD 21218

Abstract. Light induced electron transfer at Ru(bpy)$_2$(dcb)$^{2+}$ (where bpy is 2,2'-bipyridine and dcb is 4,4'-(COOH)$_2$-bpy) nanocrystalline anatase TiO$_2$ interfaces are presented. Molecular electron transfer processes have been quantified by time resolved spectroscopic methods and correlated with the photoelectrochemical properties of regenerative solar cells based on these same materials. Related sensitizers from the literature are also highlighted and discussed.

Introduction

An order of magnitude increase in solar energy conversion efficiency from regenerative photoelectrochemical cells based on dye sensitized mesoporous nanocrystalline TiO$_2$ materials has recently been achieved.[1] This break through was realized through significant advances in materials chemistry, electrolyte modifications, and synthesis of novel molecular sensitizers. The low production costs, stability, and efficiency of these cells suggest an economically viable alternative to traditional solid state photovoltaics.

In addition to practical applications in regenerative solar cells, sensitized nanocrystalline semiconductor materials allow fundamental interfacial electron transfer processes to be quantified in a manner that was not previously possible. The high surface area, long effective path length, and high optical transparency of the nanocrystalline films allow simultaneous spectroscopic and electrochemical characterization. The dynamics and yield of molecular electron transfer processes that convert (and inhibit) light-to-electrical energy conversion can therefore be quantified spectroscopically and correlated with the photoelectrochemical properties of the device. This is the subject of this report.

Specifically, the sensitizer Ru(bpy)$_2$(dcb)$^{2+}$, where bpy is 2,2'-bipyridine and dcb is 4,4'-(COOH)$_2$-bpy, anchored to nanocrystalline anatase TiO$_2$ films in propylene carbonate electrolyte. This sensitizer-semiconductor combination serve as a representative model wherein all the electron transfer dynamics have been quantified in our labs. Where appropriate, results from other sensitizers and materials will also be discussed. For clarity the electronic

CP404, *Future Generation Photovoltaic Technologies: First NREL Conference*, edited by McConnell
© 1997 The American Institute of Physics 1-56396-704-9/97/$10.00

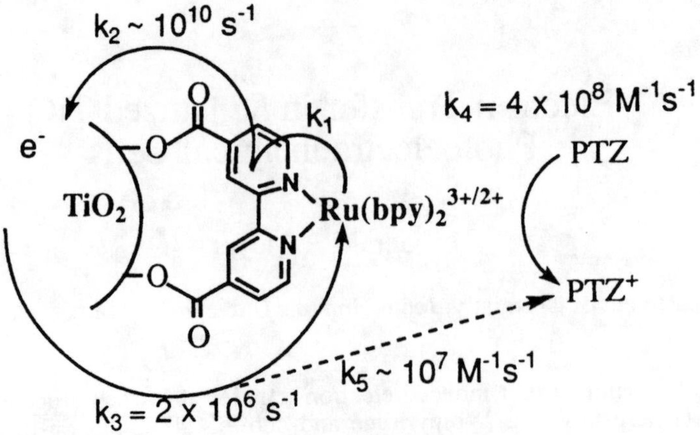

$k_2 \sim 10^{10} \text{ s}^{-1}$

k_1

$k_4 = 4 \times 10^8 \text{ M}^{-1}\text{s}^{-1}$

PTZ

e^-

TiO_2

$Ru(bpy)_2{}^{3+/2+}$

PTZ$^+$

$k_5 \sim 10^7 \text{ M}^{-1}\text{s}^{-1}$

$k_3 = 2 \times 10^6 \text{ s}^{-1}$

Scheme 1

processes will be discussed sequentially as shown in Scheme 1 above: 1. light absorption; 2. charge injection; 3. charge recombination; 4. donor oxidation; and 5. recombination of the electron in the solid with the oxidized donor. The origin of the rates constants given are described below. Important electron transfer processes that will not be presented here include electron transport through the nanocrystalline film.[2]

Light Absorption, k_1

Ruthenium polypyridyl coordination compounds have emerged as the most efficient sensitizers to date. In these sensitizers, a manifold of excited states give rise to broad visible absorption bands which harvest a large fraction of sunlight.[3] Synthetic manipulation of the bipyridine ligands and/or the substitution of non-chromophoric ligands allows the color of these sensitizers to be tuned from yellow to black. When utilized for photovoltaic applications the spectral sensitivity of the solar cell is determined by the absorptance of the sensitizer. Shown in Figure 1 are the absorptance and photocurrent action spectra of $Ru(bpy)_2(dcb)^{2+}/TiO_2$. Negligible photocurrents are observed beyond 700 nm, but the incident photons-to-current (IPCE) efficiency at shorter wavelengths is quite high, IPCE(460 nm) = 0.50 $\pm$ 0.05. The decreased photocurrent at ~ 380 nm observed in the photoaction spectra results from competitive triiodide absorption. The coincidence of these two spectra provides strong evidence that MLCT excitation is followed by electron injection as shown in Scheme 1.

Formally in the MLCT excited state an electron is promoted from the t_{2g} orbitals of the metal to the π^* orbitals of the ligand, Equation I. A large body of

$$Ru^{II}(bpy)_2(dcb)^{2+} \xrightarrow{\ h\nu\ } Ru^{III}(bpy)_2(dcb^-)^{2+}{}^* \qquad\qquad I$$

138

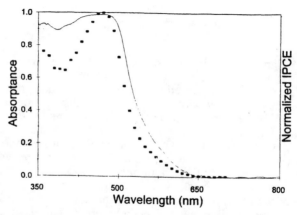

Figure 1. Absorptance and photoaction spectra of $Ru(bpy)_2(dcb)^{2+}$.

data strongly suggests that the electron is localized on one bipyridine ligand, rather than delocalized over all three. For heteroleptic compounds, such as $Ru(bpy)_2(dcb)^{2+}$, the electron is localized on the ligand with the lowest π^* levels.[3] The lifetime of $Ru(bpy)_2(dcb)^{2+*}$ is 300 ns in an argon saturated acetonitrile solution.

Charge Separation, k_2

A key to efficient solar energy conversion is quantitative electron injection from the molecular excited states to the semiconductor, k_2. This process converts the energy stored in the molecular excited state into potential energy in the form of an interfacial charge separated pair. Charge separation is known to occur with a quantum yield near unity under a wide variety of conditions. For Ru(II) polypyridyl sensitizers, electron transfer occurs from the π^* levels of the bipyridine ligand to the semiconductor. We have explored the efficiency and dynamics of this process using photoluminescence (PL) spectroscopy.[4]

Time resolved PL decays from $Ru(bpy)_2(dcb)^{2+*}$ bound to nanocrystalline TiO_2 and insulating ZrO_2 films in operational solar cells were utilized to estimate the rate of electron injection. The PL decay from $Ru(bpy)_2(dcb)^{2+*}$ is dramatically shortened on TiO_2 compared to ZrO_2 consistent with rapid electron injection. The negative conduction band edge of ZrO_2 precludes electron injection. If one assumes that electron transfer is the only quenching mechanism available to the excited sensitizer on TiO_2 not available on ZrO_2 then the difference in lifetimes will yield the electron injection rate, Equation II.

$$k_2 = \frac{1}{\tau_{TiO_2}} - \frac{1}{\tau_{ZrO_2}}$$ II

A significant difficulty in quantifying the observed PL decays is that they are not first order and unique lifetimes therefore, do not exist. Non-exponential kinetics are commonly observed in these materials which presumably stems from interfacial heterogeneity. We have employed a distribution analysis to quantify these complex decays and discovered that symmetric distributions such as Lorentzian, Gaussian, and uniform distributions were unable to model this data. In contrast, Levy or log-normal distributions did accurately model this data. It is the asymmetry or skewness of these distributions that is relevant and it is likely that many other skewed distributions will model this data as well. The difference between these skewed distributions of lifetimes then provides an estimate of the electron injection rates, Equation II. The results from analysis with log-normal distributions is given in Figure 2. This distribution analysis

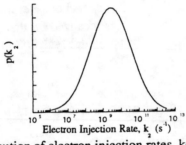

Figure 2. Distribution of electron injection rates, k_2, abstracted from time resolved PL decays observed in operational solar cells.

reveals a log-normal distribution of electron injection rates with peak amplitude $\sim 10^{10}$ s^{-1}. Significant amplitude in the femtosecond and nanosecond time regimes is also revealed. Fast electron injection is consistent with the fact that related sensitizers with short lifetimes efficiently inject electrons into TiO_2.

The rate of electron injection from excited Ru(II) polypyridyl sensitizers to nanocrystalline TiO_2 remains somewhat controversial in the literature. While PL is an indirect method for measuring interfacial electron transfer rates, the distribution shown is consistent with other measurements [5,6] largely because it is so broad. We emphasize that the PL data discussed above were performed in efficient operational solar cells and all the data was included in the analysis.

Charge Recombination, k_3

Recombination of the electron in the solid with the t_{2g} orbitals of the oxidized sensitizer yields ground state products and wastes the energy stored in the interfacial charge separated pairs. Excited state absorption spectroscopy has been used to quantify the kinetics of this important process in the absence of iodide.[7] Shown in Figure 3 are transient absorption signals observed after selective excitation of $Ru(bpy)_2(dcb)^{2+}$ anchored to nanocrystalline TiO_2 in

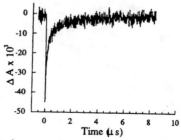

Figure 3. Absorption transient for $Ru(bpy)_2(dcb)^{3+}$ reduction by electrons in TiO_2, k_3 in Scheme 1.

propylene carbonate electrolyte. The kinetics were measured at a wavelength where the formation and loss of the oxidized sensitizers can be cleanly observed. Consistent, with the distribution analysis above, the formation of the oxidized sensitizer cannot be time resolved with our apparatus, $k_2 > 10^7$ s^{-1}. Recombination occurs on a microsecond time scale as is shown. These transients cannot be modeled by first or second order kinetics, but can be fit to distributions of first order rate constants. An average rate constant was calculated for the data shown, $k_3 = 3.2 \times 10^6$ s^{-1}. Therefore, charge separation is approximately 10^3 times faster than charge recombination. The slow recombination rate allows an external donor to intercept the interfacial charge separated pair before recombination.

One strategy for decreasing the charge recombination rate has been to vectorially translate the hole away from the semiconductor interface through intramolecular electron transfer.[8] This was accomplished with the sensitizer shown below in Scheme 2. The strategy in the design of this sensitizer was that after electron injection, k_2, intramolecular electron transfer from the pendant phenothiazine (PTZ) derivative is thermodynamically downhill. Intramolecular electron transfer translates the "hole" from the ruthenium metal center to the pendant phenothiazine group. The rate constant for recombination of the electron in TiO_2 with the oxidized phenothiazine was slowed down by approximately three orders of magnitude, $k_{et} = 3.6 \times 10^3$ s^{-1}. This slow recombination process leads directly to an increased open circuit photovoltage, V_{oc}, in regenerative solar cells based on this sensitizer. Moreover, the measured rate constants applied to the diode equation directly predict the increased V_{oc}. One expects that inhibiting k_3 (and k_5) should be a general strategy for increasing the power output in sensitized regenerative solar cells.

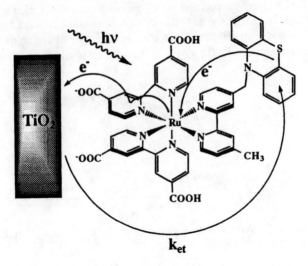

$$k_{et}$$

Scheme 2

Donor Oxidation/Reduction, k_4 and k_5

For efficient solar conversion, electron donors in the external solvent must quantitatively reduce the oxidized sensitizers and transport the charge to the Pt counter electrode. The success of the phenothiazine derivative described above prompted us to explore this chemistry with $Ru(bpy)_2(dcb)^{2+}/TiO_2$ and PTZ as the electron donor in propylene carbonate electrolyte. The formation of the phenothiazine radical cation, PTZ^+, occurs with a rate constant, $k_4 = 4 \times 10^8$ $M^{-1} s^{-1}$. At high PTZ concentrations all the photogenerated oxidized sensitizer, $Ru(bpy)_2(dcb)^{3+}/TiO_2$, are reduced by PTZ. This indicates that PTZ has access to all the surface bound sensitizers.

Recombination of the electron in the solid with PTZ^+, k_5 in Scheme 1, follows a second-order equal-concentration kinetic model. Representative data is shown in Figure 4. Because the appropriate path length is not known, it is not straight forward to convert the change in absorbance observed to a change in concentration. However, for this sensitizer bound to TiO_2 colloids recombination to related phenothiazine acceptors has also been measured, $k_5 =$. An important consequence of second-order kinetics is that the observed rate is fast in nanocrystalline films as a result of the high local concentrations of PTZ^+ and electrons trapped in TiO_2.

When PTZ is employed as the donor in regenerative solar cells negligible photocurrents are observed. If iodide/iodine is substituted for PTZ, milliamp

142

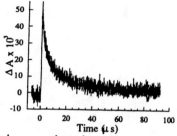

Figure 4. Absorption transient that corresponds to electron transfer from TiO_2 to PTZ^+, k_5 in Scheme 1.

recombination occurs. Therefore no photocurrent is observed. In many respects this is not too surprising. What is surprising is that triiodide, I_3^-, does escape the TiO_2 network when it too has the possibility for recombination with trapped electrons. Clearly recombination of the electrons in TiO_2 with I_3^- is inhibited. However, the reason(s) for this remain unknown. Studies are currently underway to understand this fascinating interfacial electron transfer process.

ACKNOWLEDGEMENTS

The National Renewable Energy Laboratory (NREL XAD-3-12113-04), the National Science Foundation (CHE-9322559, CHE-9402935), and the Division of Chemical Sciences, Office of Basic Energy Sciences, Office of Energy Research, U.S. Department of Energy are gratefully acknowledged for research support.

REFERENCES

1. O'Regan, B., and Grätzel, M. *Nature* **353**, 737, (1991).

2. Cao, F., Oskam, G., Searson, P.C., and Meyer, G.J., *J. Phys. Chem.* **100**, 17021 (1996).

3. Meyer, T.J., *Acc. Chem. Res.* **22**, 364 (1989).

4. Heimer, T.A., and Meyer, G.J., *J. Lumin.* **70**, 468 (1996).

5. Fessenden, R., and Kamat, P.V., *J. Phys. Chem.* **99**, 12902 (1995).

6. Tachinaba, Y., Moser, J.E., Gratzel, M., Klug, D.R., and Durrant, J.R., *J. Phys. Chem.* **100**, 20056 (1996).

7. Argazzi, R., Bignozzi, C.A., Heimer, T.A., Castellano, F.N., and Meyer, G.J., *H. Gerischer volume of J. Phys. Chem. B, in press.*

8. Argazzi, R., Bignozzi, C.A., Heimer, T.A., Castellano, F.N., and Meyer, G.J., *J. Am. Chem. Soc.* **117**, 11815 (1995).

Photochemical Solar Cells Based on Dye-Sensitization of Nanocrystalline TiO$_2$

A.J. Frank, B.A. Gregg, M. Grätzel*, A.J. Nozik, A. Zaban,
S. Ferrere, G. Schlichthörl, and S.Y. Huang

*National Renewable Energy Laboratory, Golden, Colorado 80401, and *The Institut de Chimie Physique, Ecole Polytechnique Fédérale, CH-1015 Lausanne, Switzerland*

Abstract: A new type of photovoltaic cell is described. It is a photoelectrochemical device that is based on the dye sensitization of thin (10-30 μm) films of TiO$_2$ nanoparticles in contact with a non-aqueous liquid electrolyte. The cell is very simple to fabricate and, in principle, its color can be tuned through the visible spectrum, ranging from being completely transparent to black by changing the absorption characteristics of the dye; the photovoltage of the cell is not determined by the threshold energy for light absorption (bandgap) as in conventional photovoltaic cells based on solid-state semiconductors. The highest present efficiency of the dye-sensitized photochemical solar cell is about 11%. The cell has the potential to be a low-cost photovoltaic option. Unique applications include photovoltaic power windows and photoelectrochromic windows.

INTRODUCTION

The application of dye-sensitized semiconductor electrodes to a new type of photovoltaic (PV) cell has generated much interest recently. The photoelectrode of these PV cells consists of a 10-20 μm film of nanocrystalline TiO$_2$ particles (10-30 nm in diameter) which contain a monolayer of adsorbed dye molecules, and which are deposited on a transparent conducting glass substrate (e.g., tin oxide). The pores of the nanocrystalline TiO$_2$ film are filled with a liquid electrolyte containing the iodide/iodine redox couple in acetonitrile. A transparent counter electrode is placed over the nanocrystalline TiO$_2$, and the edges of the cell are sealed. Upon photoexcitation of the cell, the excited dye molecules inject electrons efficiently into the TiO$_2$ conduction band, affecting charge separation. The injected electrons traverse the nanocrystalline film with little loss and are collected at the conducting glass substrate. After passing through the external circuit and delivering power to a load the electrons re-enter the cell at the counter electrode, reduce iodine to iodide, which then diffuses into the pores of the TiO$_2$ film to reduce the photo-oxidized dye back to its original state. These cells are termed either photochemical solar cells, dye-sensitized

CP404, *Future Generation Photovoltaic Technologies: First NREL Conference*, edited by McConnell
© 1997 The American Institute of Physics 1-56396-704-9/97/$10.00

solar cells, or Grätzel cells, the latter because of the breakthrough report (1) by M. Grätzel and B. O'Regan on dye-sensitization of TiO_2. The standard dye used in present cells (2) is $Ru(II)(4,4'\text{-dicarboxy-}2,2'\text{-bipyridine})_2(NCS)_2$ (absorption peak at 550 nm); this system shows conversion efficiencies of 7-9% under standard solar conditions. A recent new "black" dye (4,9,14-tricarboxy 2,2'-6,6'terpyridyl Ruthenium(II) trithiocyanate) has recently been discovered which produces an efficiency of nearly 11%.

The NREL Photochemical Solar Cell Project

The Photochemical Solar Cell project at NREL is an integrated program of basic and applied research that is jointly planned and funded by the U.S. Department of Energy's Office of Energy Research and Office of Energy Efficiency and Renewable Energy. It has three components: (1) applied research and development of dye-sensitized PV cells; (2) basic studies of the dynamics of electron injection from dye molecules into the conduction band of TiO_2; and (3) research and development of dye-sensitized solar cells that split water into hydrogen and oxygen rather than to produce electricity. In this report we only cover work on the applied research and development of dye-sensitized photovoltaic cells; the report is presented in three parts.

IMPROVING THE PHOTOVOLTAGE THROUGH SURFACE MODIFICATION

(S. Y. Huang, G. Schlichthörl, A. J. Nozik, M. Grätzel, and A. J. Frank)

Dye-sensitized nanocrystalline titanium dioxide photochemical solar cell systems are of considerable technological interest because of their demonstrated high power conversion efficiency (7-10% at AM 1.5) (1-6), potential low cost (7) and high semiconductor stability. The most extensively studied cell consists of a monolayer of a Ru-bipyridyl-based charge-transfer dye adsorbed onto the surface of a thin nanocrystalline TiO_2 film (ca. 10 µm) supported on transparent SnO_2 conducting glass. The particles of the film are in contact with an electrolyte solution containing iodide and triiodide ions as a redox relay and are sandwiched by a second plate of glass covered with platinum. Photocurrent is generated when visible light absorption by the dye leads to electron injection into the conduction band of TiO_2. The injected electrons diffuse through the interconnecting network of TiO_2 particles and are collected at the transparent conducting electrode. The resulting oxidized dye molecules are reduced by I^- ions, regenerating the original dye molecule. The oxidized iodide ions diffuse back to the Pt counter electrode as I^- ions, where reduction occurs to complete the cyclic process.

146

A major factor limiting the conversion efficiency of present dye-sensitized TiO$_2$ solar cells is the low photovoltage (8), which is substantially below the theoretical maximum (9-11). Charge recombination at the nanocrystallite/redox electrolyte interface is expected to play a significant role in limiting the photovoltage. There are two likely recombination pathways occurring at the interface. The injected conduction-band electrons may recombine with oxidized dye molecules or react with redox species in the electrolyte. Owing to the rapid rate of reduction of the ionized dye molecules by I$^-$ ions, which are present at high concentration, the contribution of this latter energy-loss channel to the recombination current can usually be ignored (12). The net recombination process, controlling the photovoltage, is represented by the reaction (3):

$$2e^- + I_3^- \rightarrow 3I^-$$

Some suppression of back electron transfer in TiO$_2$, as manifested by a higher open-circuit photovoltage V_{oc}, has been reported (3,4), as a result of chemically treating the surface with 4-tert-butylpyridine.

In this paper, we report on the effect of various surface modifying reagents on V_{oc} and the underlying mechanism (13,14) for their action. An unexpected result is the discovery that the reaction rate for recombination is second order in triiodide ion concentration. The mass-transport theory is also applied to determine whether the nanoporous TiO$_2$ film impedes the diffusion of triiodide ions in the solution phase.

The methodology for preparing the dye-sensitized nanocrystalline solar cells is detailed elsewhere (13). Table 1 summarizes the PV characteristics of cis-di(thiocyanato)-N,N-bis(2,2'-bipyridyl-4,4'-dicarboxylic acid)-ruthenium(II) (RuL$_2$ (NCS)$_2$)-coated nanocrystalline TiO$_2$ electrodes exposed to 4-tert-butylpyridine (TBP), 2-vinylpyridine (VP), poly(2-vinylpyridine) (PVP), or ammonia. The electrodes were immersed in CH$_3$CN/3-methyl-2-oxazolidinone (NMO) (50:50 wt %) containing 0.3 M LiI and 30 mM I$_2$. A comparison of the untreated and treated electrodes shows that the surface treatment does not significantly alter the short-circuit photocurrent J$_{sc}$ and does not therefore affect electron injection into the conduction band by excited dye molecules. The fill factor is also not significantly changed by surface treatment. The major effect of surface treatment is to increase V_{oc} and consequently the cell efficiency. The improved V_{oc} with respect to the untreated surface (V_{oc} = 0.57 V) ranges from 0.64 V, for the VP-treated sample, to 0.73 V, for the PVP-treated sample, corresponding to respective increases of 70 and 160 mV. The largest improvement was for an NH$_3$-treated electrode, which yielded a V_{oc} of 0.81 V, corresponding to an increase of 240 mV, and a conversion efficiency of 7.8 %. The latter efficiency represents a 34% improvement over that of the untreated electrode.

TABLE 1. Photocurrent-Voltage Characteristics of Untreated and Pyridine Derivative-Treated [RuL$_2$ (NCS)$_2$]-Coated Nanocrystalline TiO$_2$ Solar Cells[a,b]

Electrode Treatment	J$_{sc}$ (mA/cm^2)	V$_{oc}$ (mV)	FF	η (%)
Untreated	14.9	570	0.68	5.8
VP[c]	14.8	640	0.70	6.6
TBP[c]	14.7	710	0.72	7.5
PVP[c]	14.5	730	0.71	7.5
NH$_3$	15	810	0.64	7.8

[a]Radiant power: 100 mW/cm^2 (AM 1.5). [b]Redox electrolyte: CH CN/NMO (50:50 wt %), LiI (0.3 M) and I (30 mM). [c]VP (2-vinylpyridine); TBP (4-tert-butylpyridine); PVP (poly (2-vinylpyridine)).

To understand the underlying cause for the improved cell characteristics, resulting from surface treatment, we have investigated the phenomenon using intensity modulated photovoltage spectroscopy (IMVS) (14). IMVS, which is closely related to intensity modulated photocurrent spectroscopy, measures the modulation of the open-circuit photovoltage in response to the modulation of the incident light intensity as function of the modulation frequency. We have developed a theoretical model of IMVS and have derived analytical expressions for determining the kinetic parameters for dye-sensitized nanocrystalline solar cells. The model considers charge trapping/detrapping and electron transfer from the conduction band and surface states of the semiconductor to redox species at the solid/solution interface. Analysis of the IMVS response yielded one time constant from which the accumulated charge (Q) in the TiO$_2$ film was determined. From studies of the V$_{oc}$ dependence on Q, it was shown that the principal mechanism by which surface modification of dye-sensitized nanocrystalline TiO$_2$ electrodes with NH$_3$ and TBP improves the photovoltage is to shift the conduction-band edge to negative potentials and that suppression of recombination by shielding surface sites is relatively not important.

It is generally assumed that the recombination reaction is first order in I$_3^-$ concentration. To investigate the reasonableness of this assumption, we studied the dependence of V$_{oc}$ on the I$_3^-$ concentration and the radiant power. Contrary to conventional wisdom, we find that the recombination reaction in dye-sensitized nanocrystalline TiO$_2$ solar cells is second order in I$_3^-$ concentration. This discovery is consistent with transient absorption studies of I$^-$ ions in colloidal TiO$_2$ solution (15). The second-order process is attributed (15) to dismutation of I$_2^-$ to I$_3^-$ and I$^-$. The presence of dismutation suggests that back electron transfer from TiO$_2$ involves I$_2$ as an electron acceptor.

To determine whether the nanoporous TiO_2 film impedes the diffusion of I_3^- ions in the liquid phase via adsorption processes, the dependence of V_{oc} on the radiant power at low I_3^- concentration was studied and mass transport theory was applied to the experimental data to obtain the diffusion coefficient of I_3^-. The calculated curve coincides closely with the experimental data for an optimized diffusion coefficient of 7.55×10^{-6} cm^2/s for I_3^- ions in CH$_3$ CN/NMO (50:50 wt%)/TiO_2. After correcting for the TiO_2 porosity (0.3) (16), the diffusion coefficient of I_3^- ions in the solution phase was determined to be 2.5×10^{-5} cm^2/s, which is in good agreement with values obtained for I_3^- ions in CH$_3$ CN [(8.5-30) $\times 10^{-6}$ cm^2/s] and NMO (2.8×10^{-6} cm^2/s) (17-19). The similarity of our measured value of the diffusion coefficient with those reported in the literature implies that in the I_3^- concentration range investigated, most of the I_3^- remains in solution and is not adsorbed to the TiO_2 surface. In other words, the porous structure of the TiO_2 films does not significantly retard the diffusion of I_3^- ions in the solution phase.

MECHANISMS
(B. A. Gregg, A. Zaban, S. Ferrere)

Induced pH-Sensitivity in Sensitizing Dyes

The photosensitization of high bandgap semiconductor electrodes by adsorbed sensitizing dyes has been studied for a number of years. Despite the high level of activity, a comprehensive fundamental understanding is lacking and unexpected results are still being encountered. One such result was recently reported (20) where the rate of recombination between the electron in the TiO_2 and the oxidized dye was measured as a function of pH. Since the conduction band potential of the TiO_2 is known to have a Nernstian dependence on pH, while the potentials of the usual family of ruthenium-based sensitizing dyes show little or no pH dependence in solution, it was assumed that varying the pH would vary the driving force for the recombination reaction. Surprisingly, the rate of the electron transfer reaction was independent of pH over a range where it was expected to change by many orders of magnitude.

We report here measurements on a similar dye and show that, although its oxidation potential is independent of pH when the dye is dissolved in solution, its potential becomes pH-dependent when it is adsorbed on the TiO_2. The pH-dependence is close to the 59 mV/unit pH expected theoretically for the flatband potential of an oxide semiconductor. Therefore, in this system there is little or no change in the difference between the TiO_2 flatband potential and the Ru(II)/Ru(III) potential of the adsorbed dye over the range from pH 2.5 to pH 8. This may explain the lack of pH-dependence of the recombination rate observed in reference 20. Therefore, it is not necessary to invoke unusual models of electron transfer to explain the behavior of the dye-sensitized cells. Such a change in pH-dependent behavior of a sensitizing dye upon adsorption has not been previously reported.

We have extended our experiments to include several types of dyes on both semiconducting and insulating surfaces and see similar behavior in all cases. We believe the effect is caused by the dye being inside the Helmholtz plane of the semiconductor, thus it experiences a substantial fraction of the potential experienced by the semiconductor upon adsorption or desorption of ions such as H^+ and OH^-. The induced pH-dependence of the oxidation potential has implications for the design and optimization of dye-sensitized solar cells. Specifically, it is not possible to independently adjust the potentials of the semiconductor and the dye by the use of potential-determining ions. Although our experiments were in aqueous systems, the same effect should occur in the non-aqueous solvents used in the standard PV cell configuration that contains the potential-determining Li^+ ions.

Potential Distribution in Dye-Sensitized Cells

One important factor that has never been clearly understood about the dye-sensitized cells is the distribution of electrical potential through the cell under working conditions. The individual TiO_2 particles are too small to support a space charge layer, but they are sintered together to form an electrically conducting, porous film. Is there then a space charge layer across the film at short circuit and/or at open circuit? We have investigated this problem by impedance spectroscopy measurements and electrochemical dye desorption measurements. The results are unambiguous: because of the porous nature of the TiO_2 film, ions can migrate through the film to neutralize any applied fields over very short distances. Therefore, under normal operating conditions, there are essentially no electric fields of range longer than about 10 nm in the cell. It is clear from these considerations that charge motion through the TiO_2 films occurs entirely by diffusion rather than by drift.

This understanding has important consequences for the design of solid state analogues of the dye-sensitized cells. Since there are essentially no electric fields present in the dark, an electric field is created by the photoinjection process upon illumination and this electric field must oppose charge separation. In the standard, solvent-containing configuration this induced electric field is quickly neutralized by the motion of electrolyte ions, and thus the electrons can be separated from the holes (oxidized ions). However, in the solid state analogues proposed and studied so far, there have been no mobile electrolyte ions. The conversion efficiency in such systems has been uniformly low. We now believe this was caused by the induced, uncompensated electric field in such cells that opposed charge separation. It should be possible, however, to design a

150

solid state version of these cells that contains mobile electrolyte ions that will eliminate this problem. Such experiments are in progress.

PHOTOELECTROCHROMIC WINDOWS
(B. A. Gregg, A. Zaban, S. Ferrere)

We recently invented a new type of self-powered smart window that we called the photoelectrochromic window. Photochromic materials undergo a change in color upon absorption of light, while electrochromic materials undergo a change in color upon a change in oxidation state. Both types of materials are being investigated for potential applications that include displays, imaging devices, and adjustable-transmission ("smart") windows. The new photoelectrochromic (PEC) cells are based on integrating electrochromic films with dye-sensitized photoelectrochemical cells. The observable process occurring in PEC cells is photochromism, but the mechanism is unique and has several potential advantages over conventional photochromic films. The light absorbing process is physically separate from the coloration process, allowing each to be individually optimized. The materials constraints are greatly relaxed compared to single-component photochromic films in which one material must meet all criteria (color change, switching speed, photostability, etc.). Furthermore, since the coloration process in a PEC cell requires an external electron current between the two electrodes, a particular state (transparent, absorbing, or imaged) can either be stored when the electrodes are at open circuit, or can be changed when the electrodes are connected.

The PEC devices appear to have advantages over earlier designs of smart windows. Use of smart windows can significantly decrease the air-conditioning costs in commercial buildings. However, existing smart windows require an external power source to change the transmittance, making the retrofitting of existing buildings difficult and expensive. A self-powered smart window, that uses the energy of the incident sunlight to modulate its own transmissivity, could be easily installed in existing buildings without requiring additional electrical connections. Dye-sensitized electrodes have several distinct advantages over conventional photovoltaic cells, or other semiconductor electrodes, for large-area window or display applications. The PEC cells described here are remarkably simple devices that are easily assembled without electrical shorts. The light-absorbing dye layer can be made optically thin by either reducing the thickness of the TiO_2 layer or by decreasing the concentration of the adsorbed dye. On the other hand, decreasing the optical thickness of inorganic semiconductor films often leads to electrical shorts. Inorganic semiconductors absorb all wavelengths shorter than their bandgap, and thus can be completely transparent only if the bandgap lies in the ultraviolet range. In contrast, sensitizing dyes of any color, even infrared-absorbing dyes, can be employed for dye-sensitized electrodes. In the latter case, the device would be completely transparent to visible light in the "off" state but could be darkened by sunlight or addressed with an infrared diode laser.

FUTURE WORK

We will continue our studies of dye-sensitized cells, concentrating especially on understanding the factors that limit the efficiency of present cells and on the realization of a solid state analogue of the cell. We are also working on a two photon device, or tandem cell, in an effort to increase the photovoltage and use a greater fraction of the incident solar light. Current work on the photoelectrochromic cells is directed toward making an all solid state device and on finding an infrared sensitizing dye to make a completely transparent window.

ACKNOWLEDGMENTS

This work was supported by the Office of Basic Energy Sciences, Division of Chemical Sciences (G.S., B.G.L., B.A.G., A.J.N., and A.J.F.) and the Office of Utility Technologies, Division of Photovoltaics (S.Y.H., A.Z., S.F.), U.S. Department of Energy.

REFERENCES

1. O'Regan, B.; and Grätzel, M., *Nature* **353**, 737 (1991).
2. Nazeeruddin, M. K., Kay, A., Rodicio, I., Humphry Backer, R., Mueller, E., Liska, P., Vlachopoulos, N., Grätzel, M., *J. Am. Chem. Soc.* **115**, 6382 (1993).
3. Grätzel, M., and Kalyanasundaram, K., *Current Science* **66**, 706 (1994).
4. Grätzel, M., *Platinum Metals Rev.* **38**, 151 (1994).
5. Grätzel, M., in *Research Opportunities in Photochemical Sciences*, DOE/BES Workshop Proc., Division of Chemical Sciences, Estes Park, CO, Feb. 5-8, 1996.
6. Hagfeldt, A., Didriksson, B., Palmqvist, T., Lindström, H., Södergren, S., Rensmo, H., and Lindquist, S.-E., *Sol. Energy Mater. Sol. Cells* **31**, 481 (1994).
7. Smestad, G., Bignozzi, C., and Argazzi, R., *Sol. Energy Mater. Sol. Cells* **32**, 259 (1994).
8. Stanley, A., and Matthews, D. *Aust. J. Chem.* **48**, 1294 (1995).
9. Matthews, A., Infelta, P., and Grätzel, M., *Aust. J. Chem.*, in press.
10. Smestad, G., *Sol. Energy Mater. Sol. Cells* **32**, 273 (1994).
11. Liska, P., Ph.D. thesis of Swiss Federal Institute of Technology, No. 1264, 1994.
12. Hagfeldt, A., Lindquist, S. E., Grätzel, M., *Sol. Energy Mater. Sol. Cells* **32**, 245 (1994).
13. Huang, S.Y., Schlichthörl, G., Nozik, A. J., Grätzel, M., and Frank, A.J., *J. Phys. Chem.* **101**, 2576 (1997).

14. Schlichthörl, G., Huang, S.Y., Sprague, J., and Frank, A.J., submitted.
15. Fitzmaurice, D. J., Eschle, M., and Frei, H., *J. Phys. Chem.* **97**, 3806 (1993).
16. Papageorgiou, N., Grätzel, M., and Infelta, P. P. *Sol. Energy Mater. Sol. Cells,* in press.
17. Macagno, V. A., Giordano, M. C., and Arvia, A. J. *Electrochimica Acta* **14**, 335 (1969).
18. Desideri, P. G., Lepri, L., and Heimler, D., in *Encyclopedia of Electrochemistry of the Elements*; Bard, A. J., Ed., Marcel Dekker, Inc., New York, 1973, Vol. 1, p. 91.
19. Papageorgiou, N., Athanassov, Y., Bonhôte, P., Pettersson, H., and Grätzel, M., *J. Electrochem. Soc.,* in press.
20. Yan, S. G., and Hupp, J., *J. Phys. Chem.* **100**, 6867 (1996).

SINGLE-CRYSTAL-LIKE FILMS ON LOW-COST SUBSTRATES

Ion-Assisted Deposition of Textured Thin Films
on Low-Cost Substrates

R. H. Hammond, C. P. Wang, K. B. Do,
M. R. Beasley, T. H. Geballe

Department of Applied Physics
Stanford University, Stanford, California 94305-4085

Abstract. Ion Beam Assisted Deposition (IBAD) can result in biaxial texturing of certain materials on non-textured substrates permitting the subsequent growth of films that require an epitaxial template. Using *in situ* structure monitoring (RHEED) we have shown MgO can be made more than 20 times thinner than YSZ, thereby lowering the cost considerably, and the mosaic spread is 7°, the lowest yet seen.   The results of a crude cost study are given.

INTRODUCTION

There is an increasing number of thin film materials which, in their application to devices of all types, depend on properties dependent on their anisotropic physical properties, or which require low-angle grain boundaries between grains. Thus, these materials must be deposited in special orientations in order to utilize their properties. The present method of achieving this is to epitaxially deposit onto carefully selected single crystal substrates with matching lattice constants and chemical compatibility. These single crystal substrates are expensive, especially in large sizes. The work briefly described here is a method that can start with an amorphous or polycrystalline substrate, and, by the use of a directed ion beam, deposit a biaxial buffer layer upon which the final thin film product can be deposited.

The areas of application that will benefit from this ability include:

- High temperature superconductors
- Semiconductors
- Magnetic thin films
- Photovoltaics

CP404, *Future Generation Photovoltaic Technologies: First NREL Conference*, edited by McConnell
© 1997 The American Institute of Physics 1-56396-704-9/97/$10.00

With respect to the application to GaAs photovoltaic solar cells, a recent study by Kurtz and McConnell[1] of NREL discusses the effect of grain size and crystal orientation in order to achieve an efficiency of 20%. They estimate that a grain size of 20–50 μm may well be needed, and that the need for biaxial GaAs may not be required. However, that is not certain, and may help in the growth of large grains, as well as decrease the inter-grain minority-carrier recombination. Also, inplane low-angle grain boundary would decrease the emitter sheet resistance, thus simplifying the grid structure needed to collect the current.

IBAD DESCRIPTION

Ion beam assisted deposition (IBAD)[2] is a general term. In this particular application the ion beam is used to produce or result in a biaxial texture of the deposited film material.

There is a large body of literature concerning the effect of ion beams on a thin film deposit, both after the deposit and during the deposition. However, in the author's experience, the first consideration of this effect in the present context was by James Harper of IBM-Yorktown, in the summer of 1979. The first publication concerning Nb thin films was in 1985.[3] A theoretical description of the process was published in 1985 by Bradley, Harper, et al.[4] This description proposed that the different rates of sputtering of different orientations of the film deposit results in a gradual evolution and over-growth of that orientation that has the lowest rate of sputtering. It was also suggested that an orientation with the greatest transparency to the ions would have the lowest sputter rate. This is usually called the channeling direction. It was also suggested that a big advantage would result if the preferred texture of the film (here meaning having one orientation perpendicular to the surface—so-called "fiber texture") without the influence of the ion beam was the desired texture—thus requiring that the ion beam only result in the in-plane orientation.

Other mechanisms have been suggested in the meantime. It is thus not clear if the Bradley process is the best or only one. In particular, channeling should not be as effective at the low energies used—300-800 volts—since at these energies the classic channeling concepts are not operative.

The present interest in IBAD for HTSC was initiated by Iijima[5] who independently discovered the process in 1991. It has now expanded to include workers in the U.S.,[6,7,8] Japan,[9] and Germany.[10]

EXPERIMENTAL METHODS

We first describe the methods used in the Stanford research.[11] It is somewhat unique at the present time among the rest of the world wide efforts. It is firstly characterized by using *in situ* real time monitor of the structure during growth:

1. *In situ* RHEED to monitor the surface structure in real time.

2. Adjust rates to be in the "critical nucleation regime," i.e. slight net growth.

3. Amorphous Si_3N_4/Si as the substrate to grow on.

4. TEM, both planar and cross-section (X-TEM) views, to determine the microstructure of the growth.

In common with other efforts, *ex situ* x-ray is used to determine the exact amount of inplane alignment, as well as the overall alignment. The criterion of quality is the FWHM of the ϕ scan (the width of the pole figure), i.e., the mosaic spread. The final test in the case of the HTSC is the critical current, which is a strong function of the grain-to-grain angle.

RESULTS

The material of choice of most of the efforts in the field has been yttrium stabilized zirconia (YSZ). Stanford also chose that in the initial attempts, but changed to MgO, as will be discussed below. YSZ was chosen because it was compatible chemically and lattice-wise with YBCO. CeO_2 was also briefly used for the same reason. The methods used for the depositions have been ion beam sputtering, and PLD of the YSZ. A few groups , including Stanford, have used electron beam evaporation.

Stanford Result on YSZ

The desired texture for obtaining the needed 4-fold symmetry for YBCO is the (001) orientation. The equilibrium orientation is (111) for YSZ, which has 3-fold in plane symmetry. Indeed when deposited with the ion beam along the most open channel direction, good 3-fold orientation is obtained, particularly at an elevated substrate temperature. We have obtained 80% (100) by working at ambient temperature and with an added flux of oxygen directed at the substrate. Note that this possibly explains the success of the methods using sputtering or PLD, as these methods operate with a high flux of particles directed at the growing film. In fact, the Japanese researchers at Sumitomo[9] have used the plume of ablated particles and energetic species in place of the ion beam to create the (100) biaxial orientation.

Our X-TEM reveal a sharp interface between the α–SiN or Ni alloy substrates and the IBAD-ed YSZ. In the case of the (111) growth, it is that structure from the interface upward. In the case of the (100) it is also sharp, but is initially a mixture of the three orientations, (100), (110), and (111) in a columnar morphology. As the thickness increases it gradually becomes (100). This is consistent with the findings of the rest of the groups working with YSZ. The best results are obtained only after about 2000–5000 Angstroms. We believe that this is consistent with Bradley's description.

Stanford Results for MgO

MgO was chosen because it is chemically stable with YBCO, has a fair lattice match, and has the rock salt structure. This structure already deposits in the (001) texture, and thus only needs to be aligned inplane. Using the RHEED it is observed to basically nucleate that way. By the time the deposit is thick enough to diffract (~20 Angstroms) a pattern indication of biaxial 3D growth appears. After about 100 Å the pattern starts to show mixtures of other orientations. Stopping the ion beam, heating to about 600° C, and continuing the MgO deposition—-i.e., homoepitaxy—continues the good pattern. The deposition can be continued until a thickness sufficient for XRD is reached. *Ex situ* pole figure or ϕ scan shows the degree of inplane orientation. The best found is about 7° at FWHM.

These results are new and research is continuing. The aims are to understand the mechanism, to improve the alignment, and to determine the factors that determine it. The key to rapid progress has been the use of in situ RHEED, and ex situ TEM, particularly cross-sectional TEM

FUTURE APPLICATION POTENTIAL

The factors that need consideration in apprising the future success of IBAD biaxial texturing include:

- Mosaic spread—the low angle grain boundary spread.

- Grain size.

- Growth of a "healing" buffer.

- Cost

The grain size of our IBAD MgO is very small, ~50 Å at about 100 Å thickness. When a homoepitaxial MgO is grown without the ion beam at 600°C, there is substantial grain growth to a few hundred Å. To increase the grain size of the buffer to provide 20-50 μm GaAs gain will require a "healing" buffer, referred to above. This term comes from the observation we and others have made that, in the case of overgrowths of YBCO on these MgO films, the X-TEM shows that the unit cell step-edge growth extends over the grain boundary and defects in the MgO, resulting in larger grains of the order of 2 μm. A "healing" buffer material would have this property. CeO_2 seems to have this property, for example. In addition, selected area nucleation to enforce edge growth may provide a means to large grain size.

COSTS

It is too early to make good cost estimates of this process for the photovoltaic application. However, a rough guide can be found from a cost estimate made for the HTSC-YBCO on nickel alloy tape application.[12] The application for this is expected to be for the power industry. A scale-up study and cost estimate has been done for both the IBAD and YBCO deposition aspects. A plant designed for 10 cm wide metal substrate tape producing 6×10^4 sq. meters per year had a capital cost of $6.3M, which, if depreciated at 20% per year for 5 years, yields a cost of $21 per square meter. The cost of the substrate is not included. The cost of materials, labor, etc., are small compared to the capital outlay. The cost of money is also not included in this estimate. (For a yield of 5×10^5 sq. meter per year, the cost is $15 per square meter.)

CONCLUSION

IBAD-induced biaxial buffer on low-cost substrates for photovoltaic cells is a realistic possibility. Basic thin film materials research is just getting started. At the present time MgO seems to be the material of choice, based on the low mosaic spread found, and on the orders of magnitude lower cost due to the much lower thickness required to achieve the best results. An important goal is to find materials and methods for increasing the grain size to values consistent with the needs of the photovoltaic material.

ACKNOWLEDGMENTS

We wish to thank the Electric Power Research Institute for supporting this work, and especially Dr. Paul Grant for his interest and guidance.

REFERENCES

1. Kurtz, S. R., and McConnell, R., "Requirement for a 20% Efficient Polycrystalline GaAs Solar Cell," National Renewable Energy Laboratory (1997).

2. Harper, J. M. E., Cuomo, J.J., and Kaufman, H. R., *J. Vac. Sci. Technol.*, **21**, 737 (1982).

3. Yu, L. S., Harper, J. M. E., Cuomo, J. J., and Smith, D. A., *Appl. Phys. Lett.* **47**, 932 (1985).

4. Bradley, R. M., Harper, J. M. E., and Smith, D. A., *J. Appl. Phys.* **60**, 4160 (1985).

5. Iijima, Y., Tanabe, N., Ikeno, Y., Kohno, O., *Physica C* **185**, 1959 (1991).

6. Reade, R. P., Mao, X. L., and Russo, R. E., *Appl. Phys. Lett.* **59** , 739 (1991).

7. Sonnenberg, N., Longo, A. S., Cima, M. J., Chang, B. P., Ressler, K. G., McIntyre, P. C., and Liu, Y. P., *J. Appl. Phys.* **74**, 1027 (1993).

8. Wu, X. D., Foltyn, S. R., Arendt, P., Townsend, J., Adams, C., Campbell, I. H., Tiwari, P., Coulter, Y., and Peterson, D. E., *Appl. Phys. Lett.*, **65**, 1961 (1994).

9. Fujino, K., Yoshida, N., Okuda, S., Hara, T., Ohkuma, T., and Ishii, H., *Advances in Superconductivity VII. Proc. of 7th International Symposium on Superconductivity*, **2**, 629 (1995).

10. Wiesmann, J., Heinemann, K., and Freyhardt, H.C., *Nuclear Instrument and Methods in Physics Research, Sec. B* **120**, 290 (1996).

11. Do, K. B., Wang, C. P., Marshall, A. F., Geballe, T. H., Beasley, M. R., and Hammond, R. H., "Control of B-Axial Texture of MgO Thin Films by Ion Beam Assisted Deposition (IBAD)," (invited talk), Materials Research Society Fall Meeting, December 1995, Boston, MA; Wang, C. P., Do, K. B., Hammond, R. H., Geballe, T. H., and Beasley, M.R., "Growth Study of Ion-Beam-Assisted Deposition of Bi-axial Textured MgO Buffer Layer of Non-epitaxial Substrates for Large Area YBCO Deposition," to be submitted to *J. Appl. Phys.*

12. Hammond, R.H., *Advances in Superconductivity VIII, Proceedings of the 8th International Symposium on Superconductivity* (1995), Eds. H. Hayakawa, Y. Enomoto (Springer-Verlag Tokyo, 1996), pp.1029-1033. Invited talks: Hammond, R. H. "YBCO Thick Film Manufacturing Issues: Beam Evaporation Controlled by *In situ* Sensors", Materials Research Society, San Fransico (1996); Hammond, R. H. "Optical and Electron Beam Probes for Control of Vapor Phase Manufacturing: YBCO Thick Film Manufacturing Issues," International Conference on Metallurgical Coatings and Thin Films (1996).

Selective Nucleation-based Epitaxy (SENTAXY)

Takao YONEHARA

Canon Inc,. Device Development Center, 6770 Tamura, Hiratsuka, Kanagawa 254, Japan

Abstract. We have proposed a novel approach to form polycrystalline thin films called SENTAXY (Selective Nucleation-based Epitaxy). The location of the crystallites and boundaries with neighbors are generally random in poly-crystalline films over amorphous substrates, because the formation of the films is initiated by the spontaneous nucleation of crystallites. The proposed method introduces artificial nucleation sites at which the crystallites nucleate selectively and grow epitaxially. As a result, it becomes possible to predetermine the location of crystallites and their boundaries. The principle of the method has been demonstrated in the chemical-vapor deposition of Si and the solid-state crystallization of amorphous Si films and has also been applied to other Si.

INTRODUCTION

Thin films formed over amorphous substrates are amorphous or polycrystalline at best due to the lack of long-range order in the substrates surface. Polycrystalline films consist of crystallites that have various orientations and sizes. The location of the crystallites is random, so that the location of the boundaries formed between the adjacent crystallites is also random. The randomness of the structures could limit the performance of devices such as thin film transistors fabricated using these films.

Some attempts have been made to obtain thin films with more controlled structures by graphoepitaxy (1) or diataxy.(2) These methods use patterned substrate surfaces that align the deposited crystalline overlayers. However, they do not control the location of crystallite nucleation nor of the boundaries.

As a novel approach for highly-tailored thin films, we propose a method called SENTAXY (Selective Nucleation-based Epitaxy). The method relies on selective nucleation and growth of a single crystallite at an artificial site. By manipulating such nucleation sites of the uniform crystallites in polycrystalline thin films with predetermined locations for crystallites and their boundaries. In this paper, we present the principle of SENTAXY and its application in the chemical-vapor deposition (CVD) of Si and the solid-state crystallization (SSC) of amorphous Si (a-Si) thin films. Related issues and other recent advances are also described.

PRINCIPLE

The formation of polycrystalline thin films is initiated by nucleation of crystallites. Spontaneous nucleation occurs at random positions and at random times (Fig. 1(a)) because nucleation is essentially a probabilistic kinetic process. The nuclei grow to form crystallites and finally impinge upon their neighbors. Because of this, the

CP404, *Future Generation Photovoltaic Technologies: First NREL Conference*, edited by McConnell
© 1997 The American Institute of Physics 1-56396-704-9/97/$10.00

locations of crystallites and their boundaries are random, and the crystallite size varies widely.

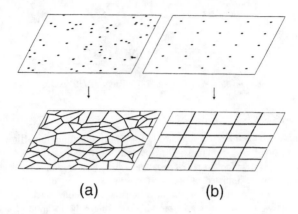

Fig.1. Schematic representation of the relation between positions of nucleation and the resultant polycrystalline film. (a) Random nucleation leads to random location of boundaries and broad size lattice points results in grid-pattern boundaries and uniform size distribution of crystallites.

The basic idea of SENTAXY is to place artificial nucleation sites at predetermined positions. The artificial site is a finite area in which a single crystallite selectively nucleates and grows, and more than one real (atomic) nucleation site can be contained. The probability of nucleation must be higher in the artificial site than outside it. The site must be small enough to select only one crystallite for nucleation. After nucleation, the crystallites must grow beyond the area of the sites. It is also necessary to suppress nucleation outside the sites. If nucleation occurs within a period of time sufficiently shorter than that required for the growth, and if every crystallite grows at nearly the same rate, the crystallites meet near the mid-point between their nucleation sites when they impinge upon each other. For example, if the sites are placed at the points of a square lattice, the boundaries make a grid-like pattern (Fig. 1(b)). Thus one can control the size distribution of the crystallites and the location of both the crystallites and their boundaries.

To produce the artificial nucleation sites, it is essential to control nucleation and /or growth in terms of both space and time. Practical ways of doing this vary for each system.

CVD-Si SENTAXY

CVD-Si SENTAXY is done using a conventional reactor in which single crystalline Si grows epitaxially. Spatial control of the nucleation is achieved using multiple materials simultaneously for the substrate surface. The density of the crystallites nucleating over Si_3N_4 is much higher than that over SiO_2 (3-5) when the ambient gas contains chlorides. SiOx (x<2) and SiNx (x< 4/3), (6) which can be formed by CVD and/or Si-ion implantation, yield crystallite densities higher than that of SiO_2. Figure 2 shows that the saturated density of the nucleating crystallites increases with the surface density of excess Si atoms in SiOx. Si s deposited at 1223 K under a pressure of 2 x 10^5 Pa, using gases SiH_2Cl_2, HCl, and H_2 at flow rates of 0.53, 1.8, and 100 l/min, respectively. If the small portions made of these materials are surrounded by SiO_2, they behave as artificial nucleation sites.

The following is one procedure for preparing the artificial nucleation sites in CVD-Si SENTXY. The substrates is a single-crystalline Si wafer coated with a 0.2-um-thick SiO_2 film formed by thermal oxidation. The artificial nucleation sites are 1.2 x 1.2 um^2 areas of SiOx (x<2) which are placed at the square lattice points of the SiO_2 surface (50 um in period). The Si-rich SiOx areas are formed by irradiating the focused ion beam of Si^{2+} locally onto the SiO_2 at an accelerating energy of 40 KeV and at a dose of 4 x 10^{16} cm^{-2}. To promote recovery from the damage caused by ion irradiation, the substrates is annealed at 1173 K for 10 min in H_2 atmosphere. To expose the most Si-rich layer of SiOx areas, about 0.1 um of the top layer is removed by dipping the substrate in HF solution. Finally, Si is deposited for 30 min under the same CVD conditions as described in Fig. 2 legend. Figure 3 shows the plan-view SEM (scanning electron microscope) micrograph of the deposited substrate. The crystallites are growing in the matrix at the predetermined positions.

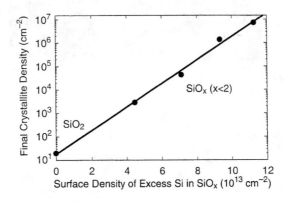

Fig. 2. Dependence of the saturated crystallite density on the surface density of the excess Si atoms in SiOx (x<2).

To use SiNx for artificial nucleation sites, photolithographic techniques are useful to make the thin SiNx films formed over the SiO₂ surface into small islands.

The opposite composition is also possible, i.e., the underlaid SiNx can be exposed by making small openings in the SiO₂ films formed over the SiNx.

As shown in Fig.3, the faceted crystallites grow three-dimensionally over the substrate. They are classified into three types based on their crystal habits. The first type is a part of the single Crystal bounded by 8 {111} facets and 24 {311} facets. The second is the simple mirror-twin crystal of the first type. The third is a part of the multiply twinned icosahedron which is bounded by equivalent {111} facets. Because crystallites of any type grow like hemispheres, the surface of the film is not always flat even after they impinge upon their neighbors. It is possible, however to level the surface by conventional polishing techniques. Moreover, if the artificial nucleation sites are placed in the patterned hollows in the SiO₂ surface and the crystallites overgrowing the hollows are selectively polished, we can obtain discrete islands of Si crystal. Metal-oxide-semiconductor field-effect-transistors fabricated within such polished crystallites demonstrate performance as good as those fabricated on bulk single-crystalline Si wafers. (5)

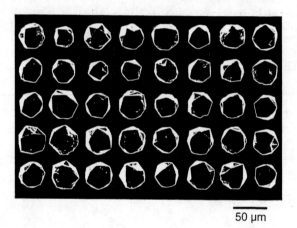

50 μm

Fig. 3. A plan-view SEM micrograph of the typical results of CVD-Si SENTAXY using SiOx (x<2) formed by focused ion beam of Si²⁺.

To understand the control of time for nucleation, it is necessary to know how a single nucleus is selected to grow at an artificial nucleation site. We have previously reported observing a coarsening phenomenon of Si crystallites and importance of this process in the selective nucleation. (7-9) Figure 4 shows the dynamic evolution of crystallites nucleating on an island of SiNx (x=0.56) serving as the artificial nucleation site. After some induction period, many submicron-sized crystallites suddenly nucleate on the SiNx island (Fig. 4(a)). Initially they seem to increase in number but do not grow in size. After a period of time, only one micron-sized crystallite emerges among the small crystallite (Fig. 4(b)). This large crystallite grows rapidly, while the others seem to vanish (Fig. 4(c)). Finally, the large growing crystallite occupies the whole area of the island and selective nucleation is completed.

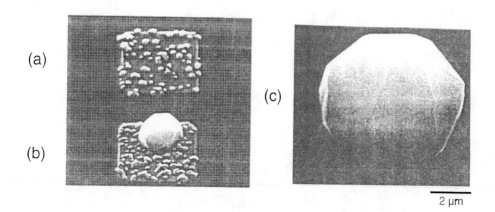

Fig. 4. SEM micrographs showing the coarsening phenomenon of the Si crystallites nucleating on an island of SiNx (x=0.56). The CVD conditions are identical with those described in Fig. 2 legend. (a) At 480s, many small crystallites nucleate exclusively on the SiNx island. (b) At 720s, a large crystallites emerges among the small ones. (c) At 960s, the large crystallite rapidly grows, while the small ones seem to vanish.

The coarsening of Si crystallites is observed not only in the small sites but also over a broad area of the same material. Figure 5 shows the dynamic evolution of the CSD (crystallite size distribution) over a SiOx (x<2) surface. The SiOx is formed by the uniform implantation of Si^+ ions into SiO_2 at an accelerating energy of 20 KeV and at a dose of 4 x 10^{16} cm^{-2}. It is clear that the CSDs of the small crystallites are restricted in the submicron range, while the micron-sized crystallites emerging at 360s grow rapidly. The concentration of the small crystallites decreases by nearly three order of magnitude within the short period from 420s to 480s. Therefore, selective nucleation is attributed to the emergence and rapid growth of the large crystallites, and to the rapid disappearance of the small crystallites.

The mechanisms of the coarsening phenomenon have not been sufficiently clarified. However, under deposition conditions when the nucleus density is relatively low, the small crystallites on some artificial sites disappear entirely although no large grain appears there. (8) This observation suggest that the small crystallites reevaporate into the deposition atmosphere. It is suggested that HCl gas, which etches Si, play a crucial role in detail, we found that the phenomenon could not be explained simply by the previously reported mechanisms such as Ostwald ripening. (7-8)

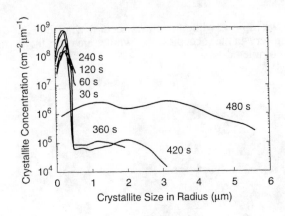

Fig. 5. Dynamic evolution of the CSD observed over a broad are of SiOx (x<2).

SSC-Si SENTAXY

The SSC-Si SENTAXY is demonstrated in an experimental system that the a-Si thin films deposited by LPCVD (low pressure CVD) over SiO_2 are annealed and crystallized in solid state. The crystallites nucleate and grow in the films by rearranging the disordered atomic bonds of the amorphous. They initially grow three dimensionally, but the growth is restricted in the film plane after their size reaches the film thickness. Figure 6(a) shows a plan-view TEM (transmission electron microscope) micrograph of the partially crystallized film without manipulating the nucleation sites. The dendritic crystallites are seen at random locations in the amorphous background. The CSDs are widely distributed with the monotonous decrease at any time (Fig. 6(b)).

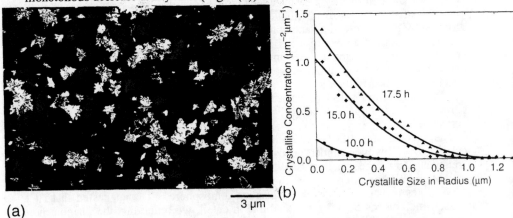

(a)

(b)

Fig.6. Random nucleation in the SSC of 100 nm-thick a-Si films. The a-Si films is deposited over SiO_2 by LPCVD, using SiH_4 gas under 40 Pa at 823K. After Si^+ ions are

168

implanted uniformly onto the as-deposited film at 70 KeV and at 1×10^{15} cm^{-2}, the film is annealed at 873K in nitrogen ambient. (a) A plan-view TEM micrograph of the film partially crystallized for 17.5 hours. Dendritic crystallites are seen at random locations in the amorphous background. (b) Time evolution of the CSD. The CSDs are widely distributed with the monotonous decrease at any time. The solid lines are the theoretical CSDs by eq. (31) of re. 14.

The spatial control of the nucleation is realized by implanting Si ions locally into the amorphous films prior to the annealing (10-12). It is found that the Si-ion implantation suppresses the nuclation of the crystallites in the subsequent crystallization(13,14). figure 7 shows the dependence of the nucleation parameters on the dose of the implangted Si$^+$ ions. The acceleration energy is fixed at 70 KeV. The steady-state nucleation rate drastically decreases and the time lag for nucleation increases as the dose increases. On the other hand, the growth rate hardly changes with the ion implantation. Thus, the artificial nucleation sites can be formed by giving the different conditions of the ion implantation inside and outside the sites' area.

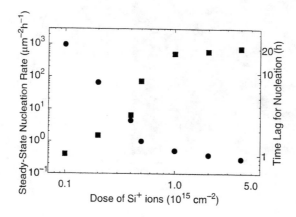

Fig. 7. Dependence of the nucleation paramaters on the dose of the Si$^+$ ions that are implanted into a-Si films prior to the crystallization. The as-deposited films are the annealing temperature are the same of those of Fig. 6. The accelerating energy is fixed at 70 Kev.
The steady-state nucleation rate (marked by the circles) drastically decreases and the time lag for nucleation (marked by the squares) increases as the dose increases. The Si-ion implantation suppresses the solid-state nucleation, while the growth rate hardly changes.

The followings are one of the typical procedures to prepare the artificial nucleation sites in SSC-Si SENTAXY and the result: first, 100 nm-thick a-Si films are deposited over thermally oxidized Si wafers, using SiH4 gas under 40 Pa at 823 K. Si⁺ ions are implanted uniformly into the as-deposited film at the dose of 4 x 10¹⁴ cm⁻² and at the accelerating energy of 70 KeV. Photoresisit masks (0.66 um in diameter) are formed over the a-Si film at the square lattice points (3 um in period). Then the Si⁺ ions are implanted again with the masks at the dose of 2 x 10¹⁵ cm⁻² and at the same accelerating energy. Thus the second (local) implantation affects only the unmasked area. The masked areas become the artificial nucleation sites later. After removing the masks, the film was annealed at 873 K in nitrogen ambient. Figure 8(a) show a plan-view TEM micrograph of the film that has been annealed for 10 hours. The matrix of the crystallites is seen at the dark marks of the artificial nucleation sites. Most of the crystallites have already grown beyond the sites' areas. The selected-area electron diffraction reveals that they are multiply twinned but have the continuous crystalline structures(15). The CSDs show a peak like normal distribution and simply shift with time (Figure 8 (b)), in contrast to those of random nucleation (Figure 6(b)).

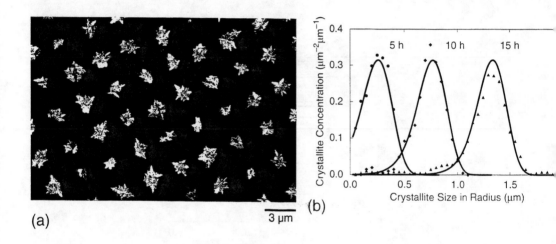

(a)   (b)

Fig. 8. Results of SSC-Si SENTAXY. The as-deposited films are the same of those of Fig.6. Si⁺ ions are implanted uniformly into the as-deposited film at 4 x 10¹⁴ cm⁻² and at 70 Kev. Then the Si⁺ ions are locally implanted again with photoresist masks at 2 x 10¹⁵ cm⁻². The masks (0,66 um in diameter) are placed at the square lattice points (3 um in period). After removing the masks, the film was annealed at 873 K.   (a) A plan-view TEM micrograph of the film partially crystallized for 10 hours. The crystallites are seen at the dark marks of the artificial nucleation sites. (b) Time evolution of the CSD.   The CSDs show a peak and simply shift with time.   The solid lines are the theoretical CSDs by eqs.(31) and (41) of re. 14.

The process of the selective nucleation in SSC-Si SENTAXY is found to be quite simple(12), compared to that in CVD-Si SENTAXY. This is completed if the first crystallite nucleating in the site grows to occupy the whole area of the site before the excess nucleating in the site grows to occupy the whole area of the site before the excess nucleation occurs. In the above procedural, the first (uniform) implantation adjusts the nucleation rate in the sites' area. This controls not only the number of the crystallites nucleating in a site, but also the range of the time that the first crystallites nucleate in the multiple sites. The shape of the CSDs (Fig. 8(b)) directly reflects the range of the time for nucleation(14). Therefore, it is concluded that both the spatial and the time controls of nucleation depend on the precise adjustment of the nucleation rate.

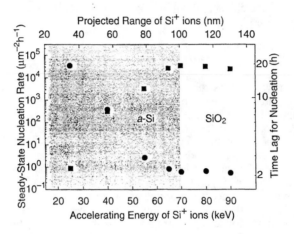

Fig. 9. Dependence of the nucleation parameters on the accelerating energy of the Si+ ions that are implanted into a-Si films prior to the crystallization. The top horizontal axis represents the projected range of Si+ ions that corresponds to the accelerating energy. The as-deposited films and the annealing temperature are the same of those of Fig. 6. The dose is fixed at 1×10^{15} cm^{-2}. The steady-state nucleation rate (marked by the circles) decreases and the time lag for nucleation (marked by the squares) increases with the accelerating energy, but they both saturate beyond 70 keV. At this energy, the projected range exceeds about 100 nm, which is just the thickness of the a-Si film.

It is also interesting why the ion implantation suppresses the solid-state nucleation. Figure 9 show the dependence of the nucleation parameters on the accelerating energy of the ions. The dose is fixed at 1×10^{15} cm^{-2}. the steady-state nucleation rate decreases and the time lag increases as the accelerating energy increases, but they both saturate beyond 70 KeV. At this energy, the projected range of Si+ ions exceeds about 100 nm that is just the thickness of the a-Si films. Comparing these observations with the results of the simulation, we have suggested that the suppression of the nucleation originates in the modification of the interface to the SiO$_2$ underlayer, such as the change of the self-diffusion or of the interfacial energies(14).

Lately, we developed the two methods for the more detailed understanding of the solid-state nucleation and growth(16). One is the method to measure the effective dimension of the fractal crystallites. By applying it to the present system, it is clarified that the dendritic crystallites indicated the irrational effective dimension. The other is the method to measure the free-energy barrier to nucleation directly from the CSDs, taking the fractal nature of the crystallites into account, without invoking any model for nucleation, and independent of the energy barrier to growth. As the initial results, it is found that the magnitude of the free-energy barrier to the solid-state nucleation suppressed by the ion implantation could not be accounted for by the classical theories. We expect that these approaches are useful to make the suppression mechanism of nucleation clear.

SIMILAR APPROACHES

In this section, we discuss the applications of the principle of SENTAXY to the other systems and the similar approaches.

Noguchi and Ikeda(17,18) use the local heating of a-Si thin films to form the artificial nucleation sites in the SSC of Si. The excimer laser is irradiated to the a-Si film on which the reflecting overlayers are formed with the small openings. The patterned overlayer makes the distribution of temperature inhomogeneous. By the laser irradiation, the crystallites grow selectively at the locally heated portions of the film. They are grown beyond the sites by the subsequent isothermal annealing. Yang and Atwarter(19) recently reported an attempt to manipulate nucleation sites in the SSC of Ge thin films. They produce the arrays of metal dots on top of the amorphous Ge films doped with boron or phosphorous. Though the metal-induce (10,21) selective-area(22) nucleation, the metal dots could provide the artificial nucleation sites. Ma *et al.* (23,24) deposited diamond films by CVD over the Si wafers on which the dots of SiO$_2$ films are formed. The diamond crystallites tend to nucleate over the rough surface, particularly at the edge of the steps. The small SiO$_2$ dots work as the artificial nucleation sites. On the other hand, Hirabayashi et al. (25) used the Si wafers on which the smooth surface and the rough surface coexist. The small portions with the rough surface become the artificial sites. Tokunaga et al. (26) reported the selective growth of polycrystalline GaAs films deposited by metalorganic CVD. They choose small islands of the polycrystalline Si films, for the artificial nucleation sites over SiO$_2$ surface. For the sufficient selectivity, it is necessary to add HCl to the ordinary deposition gases composed of AsH$_3$, trimethylgalllium, and H$_2$. As a related approach, we have also demonstrated a method similar to the CVD-Si SENTAXY. This method replaces the artificial nucleation sites with the single seed crystallites formed by agglomerating dots of polycrystalline Si films. (27) The seed epitaxially grow resulting in the similar films.

SUMMARY

We originated a novel approach to the highly-tailored crystalline thin films, which is the method named SENTAXY. The location of both the crystallites and their boundaries is controlled by arranging the artificial nucleation sites. The principle of the method has been demonstrated in chemical-vapor deposition of Si, the solid-state crystallization of a-Si thin films, and applied to the various systems and materials. It is expected that the method spreads more widely and the resultant films provide the material for the higher-performance devices.

ACKNOWLEDGMENTS

The author is grateful to Professors H.A. Atwater, A. Hiraki, H. Ishiwara, T. Itoh, H. Kawarada, I. Ohdomari, R. Reif, F.G. Shi, H. Smith, C. V. Thompson, and Drs. A. Chiang, T. Noguchi, N. Yamauchi, for their valuable suggestons and comments. He also thanks to his colleagues, Y. Nishigaki, T. Noma, M. Ohtsuka, N. Sato, H. Tokunaga, H. Kumomi, K. Sakguchi and K. Yamagata for their enthusiastic contributions.

REFERENCES

1) H. I. Smith, M. W. Geis, C.V.Thompson, and H.A.Atwater: J. Cryst. Growth **63** (1983) 527.

2) N.N.Shefta: *Growth of crystal* **10** (Consultants Bureau, New York, 1976)195.

3) T.Yonehara, Y. Nishigaki, H.Mizutani, S.Kondoh, K. Yamagata, and T. Ichikawa: *Ext. Abst. 19th   Int. Conf. Solid State Device and Materials* (Business Center for Academic Societies Japan,   Tokyo, 1987) 191.

4) T.Yonehara, Y. Nishigaki, H.Mizutani, S.Kondoh, K. Yamagata, T. Noma, and T. Ichikawa: *Mat. Res. Soc. Symp. Proc.* **106,** ed C.Y.Wong, C.V.Thompson, K-N. Tu (Materials Research Society, Pittsburgh, 1988)21.

5) T.Yonehara, Y. Nishigaki, H.Mizutani, S.Kondoh, K. Yamagata, T. Noma and T. Ichikawa: Appl. Phys. Lett. 52 (1988) 1231.

6) N.Sato and T. Yonehara: Appl. Phys. Lett. 55 (1989) 636.

7) H. Kumomi, T. Yonehara, Y. Nishigaki and N. Sato: Appl. Surf. Sci. **41/42** (1989) 638.

8) H. Kumomi, and T. Yonehara: *Mat. Res. Soc. Symp. Proc.* **202**, ed. C.V.Thompson, J.Y.Tsao,   D.J. Srolovotz (Materials Research Society, Pittsburgh, 1991) 83.

9) H. Kumomi, and T. Yonehara: Appl. Phys. Lett. **54** (1989) 2648.

10) H. Kumomi, and T. Yonehara: *Ext. Abst. 22nd Int. conf. Solid State Devices and Materials*(Business Center for Academic Societies Japan, Tokyo, 1990)1159.

11) H. Kumomi, and T. Yonehara: Appl. Phys. Lett. **59**(1991)3565.

12) H. Kumomi, and T. Yonehara: *Mat. Res. Soc. Symp. Proc.* **202**, ed. C.V.Thompson, J.Y.Tsao,   D.J. Srolovotz (Materials Research Society, Pittsburgh, 1991) 645.

13) I.-W.Wu, A. Chiang, M.Fuse, L.Ovecuoglu, and T.Y.Huang: J. Appl. Phys. 65 (1989) 4036.

14) H. Kumomi, and T. Yonehara: J. Appl. Phys. 75(1994)2884.

15) T. Noma, T. Yonehara and H. Kumomi: Appl. Phys. Lett. 59(1991)653.

16) H. Kumomi and F.G.Shi: Phys. Rev. B**52**(1995)16753.

17) T.Noguchi and Y. Ikeda: *Proc. Sony Research Forum* (Sony Corp., Tokyo, 1993) 200.

18) T. Noguchi: *Abst. Mat. Res. Soc. Symp. Fall Meeting* (Materials Research Society, Pittsburgh,   1996) **G3.7**, 124.

19) C.M.Yang and H.A.Atwater: *Abst. Mat. Res. Soc. Symp. Fall Meeting* (Materials Research Society, Pittsburgh, 1996) **G3.6**, 124.

20) S.R.Herd, P. Chaudhari, and M.H.BRODSKY: J. Non-Cryst. Solids 7(1972)309.

21) J.E.Greene and L. Mei: Thin Solid Films **37** (1976)429.

22) G.Liu and S.J.Fonash: Appl. Phys. Lett. 55 (1989(660.

23) J.S.Ma, H.Kawarada, T. Yonehara, J. Suzuki, J. Wei, Y. Yokota, and A. Hiraki: Appl. Phys. Lett.   **55**(1989)1070.

24) J.S.Ma, H.Kawarada, T. Yonehara, J. Suzuki, J. Wei, Y. Yokota, and A. Hiraki: J. Cryst. Growth **99**(1990)1206.

25) K. Hirabayashi, Y. Taniguchi, O.Takamatsu, T. Ikeda. K. Ikoma, and N.Iwasaki-Kurihama: Appl.Phys. Lett. **53**(1988)1815.

26) H.Tokunaga, H.Kawasaki, and Y. Yamazaki: Jpn. J. Appl. Phys. **31**(1992)L1710.

27) K. Yamagata and T. Yonehara: Appl.Phys.Lett. **61**(1992)2557.

Low-Cost Metal Substrates for Films with Aligned Grain Structures

D. P. Norton, J. D. Budai, A. Goyal, D. H. Lowndes, D. M. Kroeger,
D. K. Christen, M. Paranthaman, and E. D. Specht

Oak Ridge National Laboratory, Oak Ridge, TN 37831-6056

Abstract: Polycrystalline metal substrates that possess a significant amount of in-plane and out-of-plane crystallographic texture have recently been developed for high-temperature superconducting film applications. These substrates enable the virtual elimination of large angle grain boundaries in subsequent epitaxial films, having been successfully utilized in various oxide thin film architectures. This paper describes the characteristics of these substrates, and briefly discusses their potential applicability in polycrystalline thin-film photovoltaic applications.

INTRODUCTION

For many thin-film device applications, substrate selection is a primary determinant of film characteristics and device performance, as well as a significant cost factor. As such, significant attention has been given to device structures fabricated on a variety of substrate materials, including inexpensive glass, polycrystalline ceramics and metals, and single crystals. Of course, films deposited on amorphous or polycrystalline substrates are, at best, polycrystalline with a random distribution of grain boundaries that can impose serious limitations on the performance of electronic devices (1). Extended defects at the boundary result in dangling bonds that introduce electronic states in the band gap of semiconductors and insulators. This typically leads to a reduction in both the carrier lifetime and mobility. Grain boundaries also introduce an electronic interface between grains, with grain boundary charging due to permanently trapped charges at the interface. Additional difficulties associated with grain boundaries include impurity segregation and secondary phase formation at the boundary. In many instances, elimination of grain boundaries with the use of

CP404, *Future Generation Photovoltaic Technologies: First NREL Conference*, edited by McConnell
© 1997 The American Institute of Physics 1-56396-704-9/97/$10.00

epitaxial films grown on single crystal substrates results in superior device performance. Epitaxial films reproduce the crystallinity from the substrate, resulting in a single crystal-like film of the desired material. Unfortunately, the cost of single crystal substrates, as well as limitations on the available size, significantly limits the applicability of single crystals in the elimination of grain boundaries in electronic materials for large-area applications.

GRAIN ALIGNMENT WITH BIAXIALLY-TEXTURED SUBSTRATES

Until recently, the manipulation of grain boundaries through a significant reduction of the misorientation angle between adjacent grains has been limited to cases where single crystal substrates could used. However, recent developments in the area of superconductivity have made possible the formation of large-area substrates in which the crystallographic orientation at the substrate surface is highly textured, with only low angle grain boundaries present in the material. When considering the properties of polycrystalline high temperature superconducting oxides, large-angle grain boundaries have a profound and detrimental effect on the ability of these materials to carry significant electrical currents, as indicated by the superconducting critical current density, J_c. Key experiments in which the critical current density was measured across single grain boundaries in $YBa_2Cu_3O_7$ (YBCO) films deposited on $SrTiO_3$ bicrystals clearly indicate that grain boundaries greater than $10°$ dramatically reduce J_c. This observation motivated an effort to develop a means for achieving crystallographic alignment with the virtual elimination of large angle grain boundaries in long-length substrates.

Two different approaches have emerged that are effective in producing a significant degree of in-plane and out-of-plane crystallographic texture in large-area, long-length substrates. One technique involves the use of energetic ions directed at the surface of a growing film to induce crystallographic orientation of the depositing film (2). This technique, known as ion beam-assisted deposition (IBAD), results in highly textured films, with particular success achieved with yttria-stabilized zirconia (YSZ) buffer layers for superconducting oxide film growth on nickel-based alloys. Similar approaches using energetic ions in magnetron sputtering and pulsed-laser deposition have also been developed (3,4). In each case, the impinging energetic ions create crystallographic texture, either through an ion channeling or preferential sputtering mechanism. A competing approach, designated as rolling assisted biaxially-textured substrates (RABiTS), results in a crystallographically textured substrate by thermomechanical deformation of a metal (2). Both the IBAD and RABiTS approach have led to significant enhancements in the critical current density for deposited YBCO films when compared to randomly oriented HTS materials.

The RABiTS Concept

The basic concept of RABiTS, schematically illustrated in Fig. 1, begins with the formation of crystallographic texture in a metal foil by thermomechanical deformation. It is well known that significant in-plane and out-of-plane texture can be induced in most metals simply by rolling and annealing (6). In some cases, a single component of texture can be achieved, with the foil emulating a poor single crystal with a broad mosaic spread in the texture. A good example of this is cold-rolled and annealed nickel. Figure 2 shows the out-of-plane and in-plane texture as determined by four circle x-ray diffraction, in a pure nickel foil that is formed by rolling at room temperature and annealing in a reducing atmosphere. The diffraction pattern shows significant in-plane and out-of-plane texture, with the full width at half maximum (FWHM) of the in-plane and out-of-plane diffraction peaks approximately 7-10°. The grain size of the nickel foil is on the order of 50-100 μm.

RABiTS Process

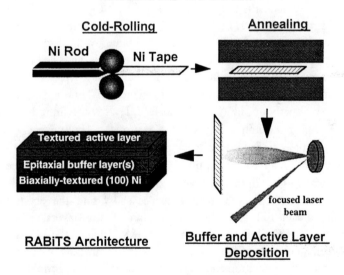

Cold-Rolling

Ni Rod   Ni Tape

Annealing

Textured active layer
Epitaxial buffer layer(s)
Biaxially-textured (100) Ni

focused laser beam

RABiTS Architecture

Buffer and Active Layer Deposition

FIGURE 1. The RABiTS process for fabricating biaxially-textured substrates for polycrystalline film growth. Crystallographic alignment in the substrate can virtually eliminate large-angle grain boundaries in subsequent epitaxial films.

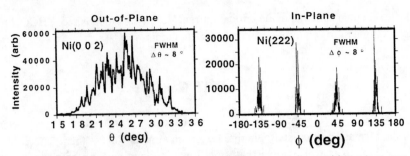

FIGURE 2. Significant in-plane and out-of-plane crystallographic texture can be achieved in pure nickel by simply cold-rolling and annealing. This figure shows the x-ray diffraction θ-scan through the Ni (200), as well as the φ-scan through the Ni (222).

In most cases, the epitaxial growth of electronic materials of interest directly on a metal is not feasible due to chemical interaction of the film with the substrate. An epitaxial buffer layer that chemically isolates the active layer from the metal substrate while transferring the texture from the substrate to the film will be needed. Previous work on HTS conductors based on the RABiTS approach has implemented various oxide buffer layers to separate the superconducting film from a nickel foil. Figure 3 shows a schematic illustration, along with the x-ray diffraction scans, for a YBCO/YSZ/CeO$_2$/Ni structure grown by pulsed-laser deposition (PLD), in which the YSZ/CeO$_2$ buffer layers isolate the nickel substrate from the HTS film. Typically, the epitaxial growth of a (001)-oriented cubic oxide on a (001) Ni surface is inhibited by the formation of (111) NiO at the oxide/metal interface. The Ni substrates were annealed at 900°C in a 4% H$_2$/Ar gas mixture prior to film growth to reduce any NiO on the substrate surface. In order to further suppress the formation of NiO and achieve (001)-oriented epitaxy directly on the (001) Ni surface, H$_2$ gas was introduced into the PLD chamber during the initial stages of CeO$_2$ growth. Hydrogen is effective in reducing NiO, while having little effect on the CeO$_2$ film. This (001)-oriented CeO$_2$ layer provides an oxide template directly on the metal surface for the subsequent epitaxial growth of additional oxide buffer and HTS layers. The in-plane φ-scans and out-of-plane rocking curves show that the crystallographic texture in the nickel foil is replicated in each of the layers, including the superconducting YBCO film. The in-plane FWHM for all of the layers is ~ 6.8°, indicating excellent epitaxy of the oxide layers with the biaxially textured metal. If the grain-to-grain misorientation angles are uncorrelated with a normal distribution, ~ 90% of the Ni grains have in-plane misorientation angles of 7° or less.

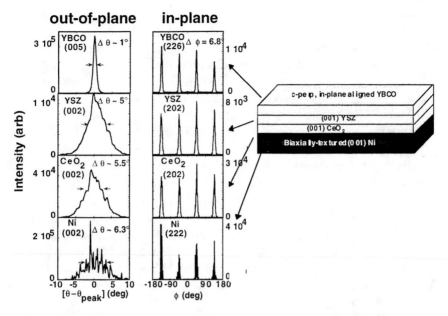

FIGURE 3. X-ray diffraction rocking curves and φ-scans showing the in-plane and out-of-plane texture in a multilayer structure for YBCO growth on rolled-textured nickel. Similar architectures would be required for other applications of textured metals as substrates .

Biaxially-Textured Substrates for Photovoltaics?

With the successful implementation of biaxially-textured substrates for long-length HTS applications, one can begin to consider the applicability of this approach in other thin-film applications, including photovoltaics. Challenges associated with using rolled-textured metal substrates for photovoltaic applications are numerous. The specific rolling-induced texture is highly dependent on material. Identification of an attractive substrate material that possesses a reasonable match in lattice parameter and thermal expansion coefficient, and can be rolled-textured with a useful texture component may prove difficult. In addition, the typical dislocation densities in rolled-textured metals are orders of magnitude larger than that generally required for semiconductor solar cell material. This will require a complex, multilayer buffer layer architecture that not only chemically isolates the semiconductor film from the metal substrate, but also effectively terminates dislocations. With the RABiTS approach, the polycrystalline substrates possess relatively large grain sizes, thus proving attractive for film growth when large grain polycrystalline material is advantageous regardless of the grain boundary misorientation. However, the most fundamental issue that must be addressed in determining the potential benefit of using any biaxially-textured substrate for photovoltaics involves understanding the role of grain boundary misorientation in limiting the solar cell efficiency for photovoltaics. Previous work in this area has primarily focused on Ge grain boundaries (8), with less work reported on Si (9), GaAs (10), and other compound semiconductors. For materials with high sensitivity to low or moderate densities of dislocations, one would anticipate that almost any grain boundary would significantly reduce carrier lifetime and solar cell efficiency. However, there are few studies in which the specific electronic behavior of grain boundaries are determined as the misorientation angle is varied for small angles ($< 10°$), including the role of misorientation on defect passivation. For many photovoltaic materials considered candidates for thin-film applications, the effects of dislocations and grain boundaries on semiconductor properties is less clear. Obviously, a better understanding of how relevant semiconductor properties vary with grain boundary misorientation is needed in evaluating the potential benefits of textured substrates in photovoltaics

ACKNOWLEDGMENTS

This research was sponsored by ORNL, managed by Lockheed Martin Energy Research Corporation, for the U.S. Department of Energy Office of Energy Research and the Office of Energy Efficiency and Renewable Energy, under contract DE-AC05-96OR22464.

REFERENCES

1.   L. L. Kazmerski, in *Polycrystalline and Amorphous Thin Films and Devices*, New York: Academic Press, 1980, ch. 3, pp. 59-133.

2.   Y. Iijima, N. Tanabe, O. Kohno, Y. Ikeno, *Appl. Phys. Lett.* **60**, 769 (1992); R. P. Reade, P. Berdahl, R. E. Russo, S. M. Garrison, *Appl. Phys. Lett.* **61**, 2231 (1992).

3.   M. Fukutomi, S. Aoki, K. Komori, R. Chatterjee, H. Maeda, *Physica C* **219**, 333 (1994).

4.   K. Hasegawa, N. Yoshida, K. Fujino, H. Mukai, K. Hayashi, K. Sato, T. Ohkuma, S. Honjyo, H. Ishii, and T. Hara, in Proceedings of the 1996 International Cryogenics Materials Conference (in press).

7.   N. N. Khoi, W. W. Smeltzer, J. D. Embury, *J. Electrochem. Soc.* **122**, 1495   (1975).

8.   B. Reed, O. A. Weinreich, and H. F. Matere, *Phys. Rev.* **113**, 454 (1959); A. G.   Tweet, *Phys. Rev.* **99**, 1182 (1955); R. K. Mueller, *J. Phys. Chem. Solids* **8**, 157   (1959).

9.   Y. Matukura, *J. Phys. Soc. Japan* **16**, 842 (1961).

10. J. P. Salerno, J. C. C. Fan, R. W. McClelland, P. Vohl, J. G. Mavroides, and C. O. Bozler, *"Electronic Properties of Grain Boundaries in GaAs: A Study of Oriented Bicrystals Prepared By Epitaxial Lateral Overgrowth"*, Technical Report 669,   Lincoln Laboratory, Massachusetts Institute of Technology (1984).

Inexpensive Approach to III-V Epitaxy for Solar Cells

Michael G. Mauk, Bryan W. Feyock, Robert B. Hall,
Kathleen Dugan Cavanaugh, and Jeffrey E. Cotter

AstroPower, Inc. Solar Park, Newark, DE USA 19716-2000
tel: 302-366-0400 e-mail: mauk@astropower.com

Abstract. An approach for low-cost, thin-film polycrystalline GaAs solar cells on large-area silicon-based substrates is described. A proprietary Silicon-Film™ sheet material serves as an inexpensive substrate on which large-grain (> 2 mm) polycrystalline GaAs films can be grown. The GaAs films are grown by a simple close-spaced vapor transport (CSVT) technique. A recrystallized $Ge_{1-x}Si_x$ buffer layer between the GaAs epilayer and Silicon-Film substrate can facilitate growth of the GaAs. A selective mode of growth to reduce thermal stress and lattice mismatch effects, along with a new interconnection scheme using a transparent conducting oxide, is described.

INTRODUCTION AND OVERVIEW

To date, the highest efficiency solar cells have been realized in the GaAs and related alloys: 25.7% (AM1.5G) for a single-junction cell and 30.3% (AM1.5G) for a multijunction cell. GaAs-based solar cells have a proven track record for critical space-power applications in satellites and, in general, GaAs optoelectronic devices are considered very reliable. GaAs also has a highly developed technology base. Further, GaAs films can be grown by variety of processes including hydride and halide chemical vapor deposition, metal organic chemical vapor deposition, molecular beam epitaxy (and related high vacuum evaporative techniques), and liquid-phase epitaxy. As is well known, GaAs is a direct bandgap semiconductor with a strong optical absorption coefficient, and thus, high efficiency solar cells can be achieved in a thin-film configuration.

In view of this, it is perhaps surprising that GaAs is not more prominent as a potential low-cost, high-efficiency thin-film polycrystalline solar cell. On the contrary, efforts in polycrystalline GaAs solar cells crested in the early 1980s, although there appears to be a revival of interest due to some very promising recent results showing high conversion efficiencies in GaAs cells made on polycrystalline Ge. Nevertheless, there are clearly serious problems that have precluded the successful development of high-efficiency GaAs thin film solar cells on low-cost substrates.

We believe a low-cost, high efficiency, thin-film polycrystalline GaAs-based solar cell on an inexpensive substrate is feasible. It is evident, however,

CP404, *Future Generation Photovoltaic Technologies: First NREL Conference*, edited by McConnell
© 1997 The American Institute of Physics 1-56396-704-9/97/$10.00

that the approaches previously developed for II-VI and amorphous silicon solar cells cannot be simply repeated for GaAs. For GaAs solar cells, the most important issues are related to: 1. grain size, 2. lattice mismatch and other defect generating phenomena, 3. thermal expansion mismatch, 4. surface passivation, 5. grain boundary passivation, 6. scale-up to large-area substrates, 7. substrate cost and preparation, and 8. process throughput and other manufacturing issues including environmental and safety considerations such as the use and/or generation of highly toxic gases.

Prior work in this area centered on adapting conventional GaAs epitaxy to thin-film deposition on glass, graphite, or metal sheet substrates. The resulting GaAs films were invariably of low quality due to small (~1 µm) grain size, thermal stress, or film adhesion problems. Modeling and experimental results indicate that minimum grain sizes of 100 to 1000 microns are needed to achieve high efficiencies in a thin-film polycrystalline GaAs solar cell.

Our approach is based on a specially designed low-cost silicon-based substrate with a graded $Ge_{1-x}Si_x$ "buffer" layer. We are adapting Silicon-Film™ material that, for purposes of this work, may be considered a large-grain polycrystalline silicon substrate sheet for heteroepitaxy of GaAs or Ge. This proprietary material is in production for low-cost silicon solar cells (FIGURE 1). As a substrate for GaAs solar cells, certain specifications of the Silicon-Film material, such as purity and minority carrier diffusion length, can be relaxed in the interest of lower cost. In effect, a near-metallurgical-grade silicon sheet material

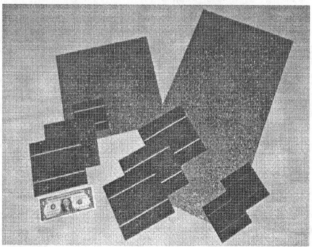

FIGURE 1: Silicon-Film™ substrates and solar cells.

can be used since the substrate serves only as a passive support for the GaAs thin film solar cell. This modified Silicon Film™ substrate is cheaper than quartz, graphite, tungsten, molybdenum, polycrystalline GaAs, or polycrystalline (optical-grade) germanium; and is cost competitive with high-temperature glasses and other metal or polymer substrates. By utilizing a new low-cost silicon-based substrate which is amenable to producing large-grain (> 2 mm) GaAs films, high efficiency devices should be feasible. The use of a silicon substrate is not without problems, however. The chief problems with a silicon substrate are related to thermal expansion mismatch, lattice mismatch, and cross-doping of the GaAs epitaxial layers and silicon substrate. The report of high efficiency (~20%) GaAs-on-silicon solar cells [O'HARA *et al.*] indicates that these problems are manageable.

FIGURE 2 is a schematic of a polycrystalline GaAs solar cell formed on a Silicon-Film™ substrate. The structure includes a $Ge_{1-x}Si_x$ interlayer which functions as a buffer between the silicon-based substrate and the GaAs film. The important feature here is that the grain structure of the germanium and GaAs film(s) replicates that of the substrate. This approach thus provides the large-grain (> 2 mm) material need for high efficiency. For comparison, FIGURE 2b shows a more conventional approach where a seeding layer of polycrystalline Ge is formed by recrystallizing a Ge film deposited on a glass substrate. The Silicon-Film™ substrate approach (**2a**) has two advantages: **1.** large-grain poly GaAs can be realized due to the large-grain Silicon-Film™ substrate, and **2.** the substrate is conductive thus allowing backside metallization for ohmic contact to the base.

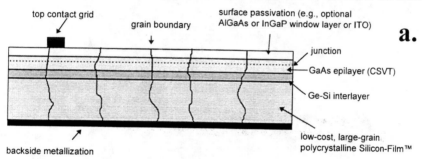

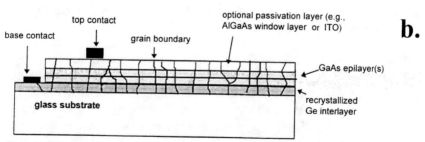

FIGURE 2: Cross-sectional views of polycrystalline GaAs-based thin-film solar cell utilizing: **a.** Silicon-Film™ substrate, and **b.** recrystallized Ge on a glass substrate.

The use of a Ge or Ge-Si compositionally-graded alloy interlayer for GaAs-on-silicon heteroepitaxy has several advantages. Ge can be readily deposited and easily recrystallized, unlike GaAs which decomposes rather than melts, and in contrast to silicon for which the high melting point makes recrystallization on a substrate problematic. Germanium is also less prone to oxidation and therefore provides a clean surface for GaAs epitaxy. Most importantly, Ge is closely lattice matched to GaAs and the Ge interlayer serves as a buffer to bridge the lattice mismatch between GaAs and silicon.

A simple way to form the Ge interlayer is to deposit Ge on the Silicon-Film substrate and melt the Ge layer by radiative heating. Molten Ge will dissolve silicon and on cooling, a $Ge_{1-x}Si_x$ alloy will regrow on the silicon substrate. The $Ge_{1-x}Si_x$ layer will be compositionally graded from pure silicon to nearly-pure Ge at the top surface; i.e., the last portion of molten alloy to solidify is Ge-rich. Ge films can be readily converted to epitaxial $Ge_{1-x}Si_x$ layers (with the same grain structure as the substrate) by a simple heating step. FIGURE 3 is a top-view photomicrograph of a 500-nm thick electron-beam evaporated Ge film recrystallized on a Silicon-Film™ substrate. The recrystallization step consisted of heating to ≈95 °C for 1 hour followed by slow cooling.

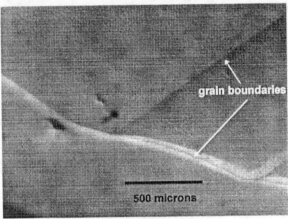

FIGURE 3: Top-view photomicrograph of vacuum-deposited 500-nm thick Ge film on a Silicon-Film™ substrate recrystallized at 800 °C.

The dimensions of Silicon-Film™ substrates can be significantly greater than the diameter of silicon wafers. The present sheet width is nominally 160 mm wide, and several meters long. A sheet width of 320 mm is under development with plans to increase sheet width to 920 mm in the next year. The process is continuous and has potential for very large-area silicon-based displays, detectors, and integrated electronic arrays.

The Silicon-Film™ substrates will be significantly cheaper than large-diameter silicon wafers. Presently, Silicon-Film™ substrates for solar cell

186

applications are being fabricated for \$20/m^2. This is much less than Vycor™ and is comparable to window glass. As an illustration of substrate costs, it is interesting to compare the Silicon-Film™ material with single-crystal silicon wafers and various glass substrates. The rule of thumb for the cost of single crystal silicon is \$1/in^2 or \$1550/m^2. However, historically the price of single crystal silicon has increased with the diameter of the substrate. Rose Associates (Los Altos, Ca) predicts that the price of a prime 300-mm wafer will be \$450 in the 1999-2000 time frame. There is a wide variety of both types and price ranges for glass. However, if high-temperature processing is required for device fabrication, a glass with a higher softening point and close expansion match to GaAs is needed. Corning glass 1737 is often used and is available in float form for approximately \$10/sq. ft. (\$108/m^2) in moderate quantities.

Close-Spaced Vapor Transport of Ge- and GaAs-on-Si

To produce a GaAs-based solar cell on a Silicon-Film substrate we use a close-spaced vapor transport technique (CSVT) to deposit Ge and GaAs. This is a simple epitaxy technique based on reversible chemical reaction between the source material and a transport agent such as HCl or water vapor. The CSVT process is shown schematically in FIGURE 4. A Ge or GaAs source is separated from the substrate by a small gap of several mm. The source and substrate are individually heated using infrared lamps to a temperature of 600 to 800 °C and such that the source is 50 to 100 °C hotter than the substrate. A small amount of a transport agent (e.g., 500 to 2000 ppm$_v$ of water or hydrogen chloride in forming gas) is injected into the ambient. For Ge CSVT, the water vapor reacts with the source to form a volatile germanium oxide, which diffuses to the substrate and is reduced to elemental Ge due the lower substrate temperature.

$$Ge(s) + H_2O(v) \xrightleftharpoons[\text{low T}]{\text{high T}} GeO(v) + H_2(g)$$

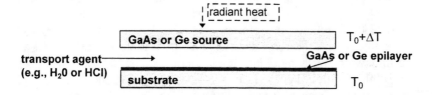

FIGURE 4: Schematic of close-spaced vapor transport (CSVT) of Ge or GaAs.

FIGURE 5 shows a top-view photomicrograph of a 3-micron thick single crystal Ge film grown on a (111) silicon substrate. Some fracture lines are evident due to thermal stress. We are able to achieve growth rates of ~0.5 microns/min using this CSVT process.

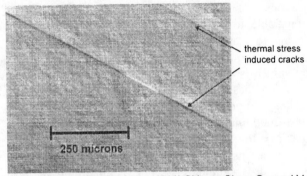

thermal stress induced cracks

250 microns

FIGURE 5: Single-crystal Ge epitaxial film grown on (111) Si by a Close-Spaced Vapor Transport (CSVT) process at 700 °C using water vapor as the transport agent.

The chemical vapor deposition process for GaAs also uses a CSVT technique. The substrate and GaAs source are separated by a small gap of 2 to 10 mm. The substrate and source are individually heated such that the source is 10 to 100 °C hotter than the substrate. A transport agent is introduced into the ambient which reacts with the source to form volatile compounds. For example, in the case of GaAs, water vapor (approx. 2000 ppm_v) in a hydrogen ambient at atmospheric pressure is used as the transport agent. The reaction between the water vapor and GaAs source is

$$GaAs(s) + H_2O(v) \underset{low\ T}{\overset{high\ T}{\rightleftharpoons}} GaO\ (v) + \frac{1}{2}As_2\ (v) + H_2\ (g)$$

The volatile GaO and As diffuse to the silicon substrate where they combine in the reverse of the above reaction (which is favored at the lower temperatures of the substrate) to form GaAs. We have previously applied this CSVT process (TERRANOVA *et al.*, U.S. Patent No. 4,818,278) for heteroepitaxy of GaAs on silicon and InP on silicon. In addition to water vapor, other transport agents are possible including HCl and hydrogen. FIGURE 6 shows a GaAs film grown on polycrystalline germanium wafer at 700 °C with water vapor as the transport agent.

Despite its simplicity, this process produces GaAs-on-silicon films with sufficient quality for optoelectronic devices. We have demonstrated CSVT GaAs-on-silicon films with 300-K mobilities in excess of 3000 cm^2/V-s We also made Zn-diffused light-emitting diodes (LEDs) in CSVT GaAs-on-silicon films with good luminescence efficiency (2 to 5% at room temperature). Schottky-Barrier Au-GaAs-on-Si solar cells cells (with no AR coating, no passivation, and no Ge interlayer) made by CSVT with water vapor as the transport agent exhibited a short-circuit photocurrent of of 7 ma/cm^2 (AM1.5).

FIGURE 7 shows a *selective* mode of GaAs CSVT growth. The silicon substrate is masked with a dielectric film. The mask is patterned by selective etching to expose the silicon substrate. The CSVT epitaxial growth is restricted to the openings of the mask; there is virtually no deposition of GaAs on the mask.

Selective epitaxy reduces the effects of thermal stress. For GaAs-on-silicon heteroepitaxy, YAMAGUCHI measured a 100-fold decrease in thermal stress using selective growth similar to that shown in FIGURE 7.

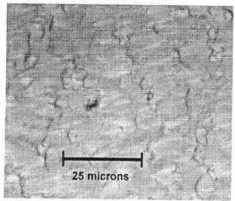

FIGURE 6: GaAs epitaxial film on polycrystalline Ge substrate grown by CSVT at 700 °C using water vapor as the transport agent.

The ability to achieve selective growth on silicon is a consequence of the near-equilibrium growth conditions and may be considered a significant advantage of the CSVT technique for this application.

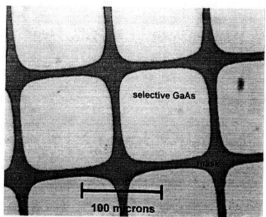

FIGURE 7: Selective CSVT GaAs-on-silicon in window openings on patterned, oxide-masked single-crystal silicon substrate.

The selective GaAs/Ge/Silicon-Film process will lead to a type of solar cell structure shown in FIGURE 8. This device design is based on the selective CSVT epitaxy process described with respect to FIGURE 7. The reduction in

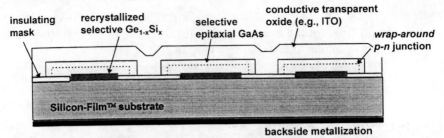

FIGURE 8: GaAs/Ge/Silicon solar cell made by selective heteroepitaxy. This "mosaic" structure provides thermal stress reduction. The selectively-grown solar cell elements are interconnected using a planar overcoating of a conducting, transparent oxide such as indium tin oxide (ITO). A conformal "wrap-around" p-n junction formed by diffusion or epitaxy avoids electrical shorting of the device caused by the conducting oxide overlayer.

thermal stress improves minority carrier lifetime. It also permits thicker layers while avoiding film cracking. By using selective growth, a thicker (> 5 μm) GaAs film can be grown which is desirable since generally the quality of GaAs-on-silicon films improve with thickness. The mask can also function as a diffusion barrier to reduce cross-doping effects.

REFERENCES

M. BÖHM and A.M. BARNETT, "On the Origin of the Shunt Effect in Polycrystalline GaAs *p-n* Junction Devices" *Solar Cells* **20** (1987).

D. COTE *et al.* "Epitaxy of GaAs by the Close-spaced Vapor Transport Technique" *J. Electrochemical Soc.* **133, 9** (1986) 1925.

O. IGARASHI, "Two-Stage Epitaxial Growth of GaP on Si" *Jap J. Appl Phys* **16** (1977) 1863.

W.D. JOHNSTON, JR and W.M. CALLAHAN, "Properties of Polycrystalline AlAs/GaAs on Graphite Heterojunctions for Solar Cell Applications" *J. Elect. Soc.* **125, 6** (1978) 977-983.

B.A. LOMBOS *et al.*, "Thermodynamic Equilibrium Displacement Controlled Epitaxial Growth of GaAs" *J. Crystal Growth* **79** (1986) 455.

M.G. MAUK, S. XU, D. ARENT, and G. BORGHS, "Study of Novel Passivation Techniques for GaAs PN Junctions Solar Cells" *Applied Physics Letters* **54**,3 (16 Jan 89) 213-215.

G. OELGART, G. GRUMM, M. PROCTOR, and F.K. REINHART, "Minority Carrier Recombination in Post-Growth Hydrogenated AlGaAs" *Semiconductor Science and Technology* (1989).

T. NISHINAGA *et al.*, "Epitaxial Lateral Overgrowth of GaAs " *Jap J. Appl Phys* **27** (1988) L964.

T. O'HARA *et al.* "High Eff. GaAs Solar Cells on Si" *IEEE PVSC* (1987).

S.J. PEARTON *et al.*, "Hydrogenation of GaAs on Si" *Appl. Phys. Lett.* **51** (1987).

G. PERRIER *et al.* "Growth of Semiconductors by the Close-Spaced Vapor Transport Technique: A Review" *J. Materials Research* **3, 5** *(1988).*

S.J. TAYLOR *et al.* "Transparent Ohmic Contacts to InP Cells" *1st WCPEC* (1994).

B.-Y. TSAUR *et al*,. "Heteroepi of Vacuum Evapor. Ge Films on Si" *Appl. Phys. Lett.* **38** *(1981).*

B.-Y. TSAUR *et al.*, "Efficient GaAs/Ge/Si Solar cells" *IEEE PVSC* (1982).

G.W. TURNER *et al.* "GaAs Shallow-Homojunction Solar Cells" *IEEE PVSC* (1981).

M. YAMAGUCHI and Y. ITOH, "Efficiency Considerations for Polycrystalline GaAs Thin-Film Solar Cells" *J. Applied Physics* **60, 1** (1986) 413-417.

M. YAMAGUCHI *et al.*, "Analysis of Dislocation Density Reduction in Selective Area Grown GaAs Films on Silicon Substrates" *Applied Physics Letters* **56, 1** (1990) 27-29.

Requirements for a 20%-Efficient Polycrystalline GaAs Solar Cell

Sarah R. Kurtz and Robert McConnell

National Renewable Energy Laboratory, 1617 Cole Blvd., Golden, CO 80401

Abstract. Based on a literature review, we explore the material and performance requirements for high-efficiency, potentially low-cost, GaAs solar cells. The goal is a GaAs solar cell, on a low-cost substrate, having an efficiency greater than 20%. An important issue limiting efficiency for polycrystalline GaAs cells is recombination through deep levels associated with grain boundaries, specifically at the part of a grain boundary that intersects the p-n junction. The effect of this junction recombination on cell efficiency is shown as a function of grain size. We explore the potential impact of grain size, grain-boundary passivation, intragrain defects, impurities, and crystal orientation. The impact of intragrain defects on minority-carrier lifetime and cell efficiency is also discussed. We conclude that two critical parameters for achieving high efficiency are the mitigation of intragrain-defect density and perimeter junction recombination. To achieve over 20% efficiency, dislocation densities need to be reduced to less than $5 \times 10^6/cm^2$ for very large-grain material. Assuming that the intragrain and surface-recombination properties are state-of-the-art, grain sizes of 20-50 μm are needed to reach 20%. Once these conditions are met for GaAs cells fabricated on low-cost substrates, both low cost and high efficiency would be possible.

INTRODUCTION

For years, researchers have worked to develop low-cost and high-efficiency solar cells. Unfortunately, low-cost cells have also had low efficiencies and high-efficiency cells have been expensive. Although silicon costs less per kg than gallium arsenide, the active-layer material requirements for silicon solar cells are significantly greater than for gallium arsenide cells. Even with light trapping, because of the indirect band gap, silicon cells must be at least 20 to 30 μm thick. In contrast, champion GaAs cells have active layers that are about 3 μm thick, and it may be possible with light trapping to use layers less than 1 μm thick. This significant difference in materials requirements implies that the cost of thin-film GaAs cells is potentially similar to the cost of thin-film Si cells (1). Although the cost of organo-metallic chemical vapor deposition (OMCVD) of GaAs cells is currently about $1/cm^2 (2), evaporation (from the elements) and liquid-phase epitaxy (3) of GaAs have the potential for lower costs and high material quality. Also, III-V cells are typically grown in a single process, in contrast with most Si cells, where each of the various layers may require its own process step. Growth of an optimized GaAs cell, with optimal dopant profile, back-surface-field, window, and contacting layers, is accomplished in one process step. Also, the III-V industry is growing rapidly and, because the industry is not as mature as the Si industry, III-V prices have a greater potential for price reduction. A single-junction GaAs cell efficiency of 25.7% (4) has been reported, and multijunction

CP404, *Future Generation Photovoltaic Technologies: First NREL Conference*, edited by McConnell
© 1997 The American Institute of Physics 1-56396-704-9/97/$10.00

III-V cells have passed 30% in efficiency (5, 6). The achievement of these efficiencies with a total materials usage of less than a few micrometers (assuming that the high-efficiency devices can be made on low-cost substrates such as glass or stainless steel) could pave the way for a high-efficiency, low-cost device—exactly what the photovoltaic industry would like to see.

Other thin-film concepts have potentially low costs, but the module efficiencies remain low. Development of a polycrystalline GaAs cell with an efficiency in the 10%-15% range serves little useful purpose unless it shows a clear advantage over existing amorphous silicon, cadmium telluride, and copper indium diselenide technologies. A significant research effort in this area is exciting if efficiencies in the 20% range can be achieved. Thus, this paper analyzes the material requirements that would be needed to reach 20%.

In the 1980s, a significant number of research groups attempted to develop efficient polycrystalline GaAs cells. The best efficiencies were in the range of 9%-12% (7), some of these using approaches that might not be adequately low cost. Among the problems encountered at that time were small grain size, poor intragrain quality, zinc diffusion at grain boundaries (8), and other contamination problems (e.g., when tungsten-coated substrates were used for high-temperature processing). The only recent work on polycrystalline GaAs cells is that at the Research Triangle Institute (9, 10) using polycrystalline Ge substrates. They have reported a confirmed efficiency of 18.2% (9) and an unconfirmed efficiency of over 20% for samples with 1-mm grain size (10, 11). Cast Ge wafers often have grain sizes larger than 1 mm. If the Ge substrates were adequately low in cost, this would already represent a 20% cell with low materials cost. However, the cost of raw Ge is currently high, and it is not clear whether it can be reduced to an acceptable range for use as a substrate for 1-sun, terrestrial solar cells. Thus, it is of interest to investigate other, lower-cost approaches.

A high-efficiency GaAs cell on low-cost substrates through use of a single-crystal template has already been demonstrated. The CLEFT technology has demonstrated an efficiency within about 1% of the wafer GaAs efficiency (12). In fact, theory predicts that the highest GaAs solar cell efficiency can be achieved by placing thin GaAs cells on a reflective substrate (13). Experimentally, "lifted-off" solar cells have shown an increased quantum efficiency and similar crystal quality (compared with before being lifted off) (14). A newer approach that uses conventional single-crystal wafers involves bombarding the surface with hydrogen atoms. The wafer is then bonded to another substrate and annealed. The hydrogen atoms form bubbles under the surface, leading to cleavage of a thin layer (15, 16). The end result is a thin, single-crystal layer on a separate substrate and the original substrate, which is just a little thinner. Some polishing must be done to smooth the new surface. For these "template" approaches, the problem is not achieving high efficiency, but achieving adequate reduction of cost by reusing each GaAs wafer multiple times. At a cost of $1.30/cm^2 for GaAs wafers and assuming a 20% efficiency, a substrate would need to be reused 65 times to keep the wafer cost to $1/peak watt. However, as the cost of GaAs drops, the "template" approach also drops in cost and is a possible way to achieve low-cost modules with efficiencies as high as 30%.

The choice of the best approach for achieving a low-cost GaAs cell with >20% efficiency depends on our understanding of the factors limiting the performance of these devices. The purpose of this paper is to explore the material properties required for a GaAs cell with an efficiency greater than 20%. The basic theory is revisited to show that the recombination current, and therefore, the

192

V_{oc} of the solar cell, is dominated by one specific contribution—the junction recombination at defects. Then, we discuss the requirements on grain size, intragrain properties, impurities, and consistency of crystal orientation.

BACKGROUND - PERIMETER RECOMBINATION

The literature contains several studies (1, 17-19) of the theoretical efficiency of polycrystalline GaAs cells as a function of grain size. The numbers show wide variations primarily because of a lack of understanding of the dark current. There is general agreement that high V_{oc} is more difficult to achieve than high J_{sc}. The V_{oc} is primarily limited by the dark current, I, which we express as (20)

$$I = J_{01}A \, [e^{qV/kT} - 1] + (J_{02B}A + J_{02P}P) \, [e^{qV/2kT} - 1], \tag{1}$$

where A and P are the area and perimeter of the device; q, V, k, and T are the charge of an electron, the bias voltage, Boltzmann's constant, and temperature; and J_{01}, J_{02B}, and J_{02P} are the prefactors for the injection current, the bulk space-charge recombination current, and the perimeter space-charge recombination current. The injection current (first term) increases exponentially with bias with a diode factor of unity. Therefore, when the injection current dominates the dark current, the diode quality factor or ideality factor is unity. In silicon diodes, the achievement of a diode factor of unity (referred to here as kT-type dark current) implies a high-quality diode. However, for GaAs, this is not the case. Champion, one-sun GaAs devices have always been reported to have diode factors of about 2 (2kT-type dark current) at the maximum power point (4, 21). At higher voltages, the dark current may return to kT behavior.

The kT current comes largely from radiative recombination and is proportional to the product of the electron and hole concentrations, np. The np product depends on voltage bias and the intrinsic carrier concentration, n_i, as follows:

$$np = n_i^2 \exp(qV/kT). \tag{2}$$

Studies of the 2kT current (22-26) have shown that the perimeter recombination, J_{02P}, dominates over the bulk recombination, J_{02B}. Given that this limits the performance of even 2-cm x 2-cm single-crystal GaAs solar cells (26), the perimeter 2kT recombination will also dominate the performance of polycrystalline GaAs cells, unless their perimeters (grain boundaries) behave in a qualitatively different way from the free surfaces. Studies of polycrystalline GaAs cells have consistently shown diode factors of 2 or higher (9) and large dark currents, implying that grain-boundary recombination is as problematic as perimeter recombination.

Woodall and Hovel (1) presented an early treatment of the efficiencies of thin-film Si and GaAs cells. Their conclusion was that only 1-µm grain size was needed to achieve a respectable GaAs cell. However, their study is in error because it did not include the 2kT current. In a later study, Lanza and Hovel (17) added a 2kT term and showed that somewhat larger grain sizes were needed. However, they only included the J_{02B} term, rather than the J_{02P} term that is known to dominate, so they still predicted unrealistically high efficiencies for small-grain material. In contrast, Yamaguchi (18) assumed a high perimeter 2kT current and concluded that 1-mm grain size is needed to get to 18% efficiency.

The origin of the 2kT current has been discussed (22, 25, 27); here we give an abbreviated discussion of its origins. Shockley and Read (28) and Hall (29) developed the basic theory for recombination through a defect state. The steady-state recombination rate, R, is given by

$$R = \frac{N_B \quad \sigma_n v_n \sigma_p v_p (np - n_o p_o)}{[\sigma_n v_n (n + n_o) + \sigma_p v_p (p + p_o)]} \tag{3}$$

where N_B is the defect density, σ_n and σ_p represent the capture cross-sections, v_n and v_p are the carrier velocities, and n_o and p_o are the equilibrium electron and hole concentrations. For a given bias voltage, the np product is a constant, and R is maximized when $\sigma_n v_n n = \sigma_p v_p p$, or when the Fermi level is near midgap. This condition is satisfied in part of the space-charge region of the junction of a p-n diode. Sah et al. (30) integrated the recombination across the junction and showed that the total recombination current, J_{02}, is

$$J_{02} = 1/2 \, qN_B \sigma v (np)^{1/2} w_0, \tag{4}$$

where σ and v are now averaged for electrons and holes; w_0 is the effective width over which the recombination takes place:

$$w_0 = \pi kT/qE_0, \tag{5}$$

where E_0 is the electric field at the junction. Eq. (4) shows a dependence on $(np)^{1/2}$, instead of the np dependence found for the kT current. From Eq. (2) and the dependence on $(np)^{1/2}$, we see that the voltage dependence of J_{02} should involve 2kT.

For GaAs homojunctions, the calculated current from Eq. (4) is two orders of magnitude smaller than the observed current (22) because N_B is usually quite small for GaAs. (This may not be the case for heterojunctions, especially if the two materials are lattice mismatched.) However, the free surface is itself a primary source of "defect" states. Thus, recombination at the perimeter of the junction dominates J_{02}. If dislocations thread through the junction, they will also increase J_{02}.

Henry et al. (22) derived an expression for the perimeter recombination current:

$$J_{02P} (V) = qs_o L_s n_i \exp(qV/2kT), \tag{6}$$

where s_o is the surface recombination velocity and L_s is an effective surface diffusion length (of minority carriers diffusing along the surface channel). Henry (22) measured an order of magnitude more recombination current than what he would have predicted from Eqs. (4) and (5) for the perimeter recombination. Other groups (25, 27) have given different presentations of this theory, but agree with the basic form of Eq. (6). However, L_s is determined experimentally; an accurate prediction of L_s requires a numerical calculation. Lundstrom (25) also showed by numerical calculation and experiments that kT-type (in addition to 2kT-type) perimeter current is both expected and observed at high biases. The observed kT perimeter current was always smaller than the 2kT perimeter current. Although the minute details of the mechanism and calculation of the perimeter current have not been clarified, there is a clear consensus that the perimeter 2kT current limits the performance of both single-crystal and polycrystalline GaAs solar cells.

GRAIN SIZE AND GRAIN PASSIVATION

As discussed above, the studies in the literature do not give a unified view of the expected efficiency of polycrystalline GaAs cells as a function of grain size. Our purpose is to estimate the minimum grain size that is necessary for efficiencies near 20%. Of course, one can place a bound by assuming that the grain passivation is perfect and then grain size becomes unimportant. However, the more useful bound is based on today's practical limit on surface passivation.

Many passivation techniques have been investigated over the years for polycrystalline and single-crystal GaAs. For grain-boundary passivation in polycrystalline GaAs, some success has been obtained with H (31). Hydrogen passivation of GaAs on Si has reduced the diode reverse leakage current (32). Ruthenium was successful in passivating Schottky diodes, but not p-n junctions. Attempts with thermal oxidation, water-vapor oxidation, anodic oxidation, hydrogen plasma, and nitrogen plasma were unsuccessful in improving performance of polycrystalline homojunctions (7).

For single-crystal, free surfaces, various sulfur treatments have been successful at temporarily or permanently passivating the surface. Se has shown similar effects (33). Nitridization of GaAs surfaces has also been shown to reduce the diode leakage current (34). High-band-gap layers like $Al_xGa_{1-x}As$ or $Ga_{0.5}In_{0.5}P$ are routinely used to passivate GaAs layers (as with a back-surface field or window layer) and <1800 cm/s surface recombination velocities have been achieved with n^+ layers on n-type GaAs (35). In contrast, band-gap narrowing for highly doped p-type GaAs prevents it from being a useful passivating layer. ZnSe has also been reported to passivate GaAs (36). Although, the oxidized GaAs surface is known to have a high recombination velocity, in the last few years some Al-containing III-V materials have been successfully passivated by a moisture-induced oxide layer (37, 38).

Assume Grain-Boundary Passivation is Similar to Free-Surface Passivation

We assume that the best passivation reported for single-crystal GaAs may be used as an estimate of the best passivation that we can obtain today for grain boundaries in polycrystalline GaAs. This assumption is not likely to be quantitatively accurate, but should be qualitatively correct because the states that are problematic (unpaired electrons on Ga and As) are similar. However, if oxidation of grain-boundary surfaces can be avoided, there could be a significant difference between them and oxidized free surfaces. Also, a free surface may be easier to passivate than the polycrystalline grain boundaries, which require diffusion of the passivant down grain boundaries. Because the recombination is primarily taking place in the junction region at the perimeter, it is unclear whether a dopant-dependent passivation scheme will work. On the other hand, because the grain-boundary recombination is dominated by the recombination at the junction, it may not be necessary to passivate the entire surface of the grain to see improved solar-cell properties. Thus, the assumption that grain-boundary passivation is similar to free-surface recombination should be taken as a starting point.

Select J_{02P} Value Measured on a Solar Cell

An ultra-low interfacial recombination velocity of <1.5 cm/s has been reported for the $Ga_{0.5}In_{0.5}P/GaAs$ heterointerface (39). However, this was obtained for undoped GaAs with a carrier concentration of 2-5 x 10^{14} cm^{-3}. Optimal GaAs cell efficiencies require higher doping levels that are known to increase the interfacial recombination velocity (40). Thus, data obtained from passivated champion solar cells are more relevant to polycrystalline GaAs solar cells. Also, by using data measured directly on solar cells, we avoid the problem of determining L_s in Eq. 6. Sulfur treatment with Na_2S (20), $(NH_4)_2S$ (20), or As_2S_3 (41) has been shown to decrease the J_{02} dark current of solar cells and to decrease the surface recombination velocity at a free surface (42, 43). When written in the form of a perimeter current, the lowest values that have been obtained are 2.79 x 10^{-13} A/cm for S-passivated (20) solar cells and 3.75 x 10^{-13} A/cm (4) or 8.8 x 10^{-13} A/cm (21) for unpassivated solar cells (also see ref. (43)).

Calculate Efficiency Versus Grain Size

Using

$$J = J_{01}e^{qV/kT} + J_{02P}P\, e^{qV/2kT}/A - J_{sc}, \qquad (7)$$

with J_{sc} of (21) 27.75 mA/cm^2, I-V curves were calculated for various perimeter/area ratios, and the resulting efficiencies were calculated and plotted in Fig. 1. It is very important to note that this trivial calculation has ignored changes of J_{sc} and J_{01} with grain size. For the curve labeled "best reported free surface," a value of 2.79 x 10^{-13} A/cm was used for J_{02P}, as measured for a S-passivated (20) cell. The physical significance of this curve can be understood as follows: If one takes tiny square single-crystal cells, of size specified by the value on the x axis, and assembles many of these into a larger device in such a way as to not introduce any series resistance or loss of current, one can expect to obtain the efficiency plotted in Fig. 1. The top curve assumes that the quality of the tiny single-crystal cells is equivalent to today's best-passivated GaAs solar cells. The 2.79 x 10^{-13} A/cm perimeter current (20) was not permanent. However, it is only slightly smaller than 3.75 x 10^{-13} A/cm, which was obtained for an untreated cell (4). A value of 8.8 x 10^{-13} A/cm has been reported in two different references (20, 21) and represents a typical "good" untreated cell. In all cases, a J_{01} of 1.2 x 10^{-19} A/cm^2 (21) was used and no series-resistance losses were included.

Yamaguchi used a value for the recombination current that reflected what had frequently been achieved for polycrystalline GaAs, rather than what *could* be achieved. More recent data (4, 9, 10) have already shown that his curves are pessimistic compared with what *can* be achieved (see data in Fig. 1 that lie above Yamaguchi's curve).

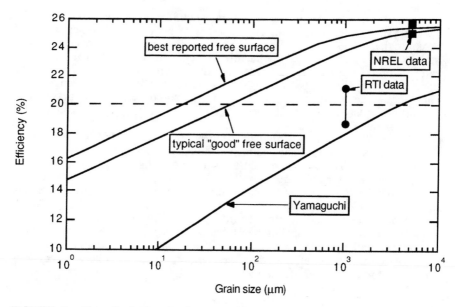

FIGURE 1. The effect of perimeter recombination current, J_{02P}, on the expected efficiency of polycrystalline GaAs cells as a function of grain size. The values assume single-crystal quality (long diffusion lengths), current collection, and J_{01} current independent of grain size and may be overly optimistic, as discussed in the text. Yamaguchi's curve and the NREL data were taken from refs. (4, 18). The top and bottom (unconfirmed) RTI data points are for devices with areas of 0.25 cm^2 and 4 cm^2, respectively (10).

The relative importance of the perimeter junction recombination is shown by comparing the slopes of the new curves with the slope of Yamaguchi's curve in Fig. 1. Yamaguchi included the effects of reduced J_{sc} as well as the effect of the perimeter recombination. The similarity of the two slopes shows that the perimeter junction recombination is the dominant effect and that the J_{sc} loss is a lesser effect.

Factors Affecting Uncertainty in Calculation

It is impossible to predict exactly how well grain boundaries can be passivated, but next, we discuss some of the issues involved. The Na$_2$S and (NH$_4$)$_2$S passivation (20) that was used as a basis for the top curve of Fig. 1 is not permanent. Application of As$_2$S$_3$ (41, 43) has been reported to lead to permanent passivation, but the only report of the use of As$_2$S$_3$ for GaAs solar cells showed higher dark currents for both the unpassivated and passivated cells (41). No one has attempted to use As$_2$S$_3$ (41) passivation on polycrystalline GaAs cells, so it is not known whether it can be made to diffuse along the grain boundaries without adverse effects to the rest of the cell; but it is clearly something that should be tried. Little is known regarding the passivation of GaAs grain boundaries by adjacent grains. For very-low-angle grain boundaries or when symmetry (e.g. twin crystals) allows adjacent crystal faces to bond to each other, it may be possible that some grain boundaries are relatively benign. However, to date, no studies show that GaAs grain boundaries have significantly lower recombination rates than free surfaces. Thus, we assume here that passivation of grain boundaries is similar to passivating free surfaces. Finally, whereas single-crystal GaAs cells can be mass produced without much concern about control of the crystalline quality, it may be very difficult to scale up the production of polycrystalline GaAs cells.

The small perimeter recombination current that was used for the top curve in Fig. 1 is a record value that one cannot count on duplicating even for single-crystal devices. Thus, the lower curve is a more realistic curve of what we can expect to achieve for tiny single-crystal GaAs cells. If we assume that this can be translated directly to polycrystalline materials, then a 50-μm-size grain is required to reach a 20% efficiency, ignoring other losses. Additional losses of J_{sc} and fill factor were neglected, which could realistically increase the grain-size requirement. Even then, this assumes that the crystal quality is as perfect as can be obtained with epitaxial growth on carefully prepared GaAs wafers. Yamaguchi also ignored intragrain effects (discussed below), which are likely to reduce the efficiency for all but the most perfect crystals.

Implications of Necessary Grain Size on Growth Technique

Even if we take the optimistic value of 20 μm predicted by Fig. 1 as the grain size with which we might hope to achieve 20%, this puts some limits on the types of processes that we could use. "Normal" grain growth usually results in crystals with grain size similar to the thickness of the layer (44). We can estimate the approximate GaAs thickness by assuming an efficiency of 20%, a projected GaAs wafer cost of \$1.30/cm^2, and a GaAs wafer thickness of 300 μm. If the material cost is limited to \$1/peak watt, then we can afford to use a GaAs layer that is about 4-5 μm thick. Obviously, this is a guideline, not a definite limit. Nevertheless, to keep the cost in the necessary range and to reach 20% efficiency, normal grain growth will be insufficient, and more sophisticated growth techniques need to be used. When a thin film is annealed, the crystallites with the lowest energies (those with low index surfaces) consume adjacent crystallites that have higher energy orientations (44). This is called secondary grain growth, and it typically results in crystal sizes an order of magnitude larger than the film thickness. However, the secondary grain growth is most strongly driven for very thin films (on order of 100 nm or less), so the achieved grain size is typically on the order of 1 μm. In general, Ge tends to form large crystals much more readily than GaAs. Because high-efficiency GaAs cells can be grown on Ge, it may be easier to achieve high-efficiency polycrystalline GaAs cells by first fabricating a large-grain, but thin film of Ge. Annealing of Ge films on oxides has demonstrated 1-μm grain growth with some controlled orientation (45, 46).

A myriad of other approaches have been proposed for growing large-grain thin films. Solid-phase crystal growth from crystallites seeded through a shadow mask (so that the grain size is controlled by the spacing of the holes in the shadow mask) (47, 48) has achieved grain sizes of 30 μm. The SENTAXY approach also uses artificial nucleation sites, but grows the crystals from the vapor (49). We have begun looking at a different approach whereby small grains of GaAs are nucleated at widely spaced intervals on a glass substrate. These can then be grown laterally using selective epitaxy and controlled growth parameters. So far, we have grown what appear to be single crystals as large as 20 μm. Theoretically, one should be able to achieve very large-grain material with a consistent orientation. A comprehensive summary of growth of oriented crystals on amorphous substrates can be found in reference (50).

It seems probable that a technique for growing adequately large crystals can be found. However, these crystals also need to be of the highest quality.

INTRAGRAIN DEFECTS

Even in large-grain, single-crystal material, there is ample evidence that dislocations and other defects can limit the performance of GaAs. We do not attempt to survey all of the possible intragrain defects. The most extensive set of data in the literature on this subject is associated with growth of GaAs on Si. These represent essentially single-crystal layers with very high intragrain defect densities. The study of Vernon et al. (51) shows that a dislocation density of < 5 x 10^6/cm^2 is needed to achieve an efficiency of about 20%. To study a wider range of dislocation densities, they used a lattice-mismatched GaAsP layer to controllably introduce dislocations in the solar cells. As the dislocation density increased to 10^6/cm^2, the V_{oc} and J_{sc} showed very little variation, but the fill factor (FF) decreased. At a dislocation density of 10^7/cm^2, the V_{oc}, J_{sc}, and FF were reduced from their best values by about 14%, 5%, and 9%, respectively. This demonstrates the relative sensitivity of V_{oc} to crystallographic defects. The minority-carrier lifetimes for double heterostructures grown under similar conditions to the solar cells have also been correlated with the dislocation density (52). Fig. 2 is constructed from the efficiency versus dislocation density data from ref. (51) and the lifetime versus dislocation density data (for similar samples) from ref. (52). Although this curve reflects only one data set, it can serve as a guide. Using Fig. 2, we note that a minority-carrier lifetime of > 10 ns is needed to achieve 20% efficiency.

IMPURITIES

The importance of purity of semiconductors has been known for many years. Table 1 lists the energy levels associated with common impurities. The elements are grouped as n-type and p-type and by whether they are shallow or deep levels.

Oxygen is not a major problem for OMCVD growth of GaAs because the high growth temperature and reducing conditions can prevent incorporation of oxygen. However, oxygen is strongly incorporated if Al is added to the layers, and GaAs readily oxidizes in air, so in a more general sense it is considered a "problem" contaminant. Its uptake by grain boundaries of films sitting in air has not been carefully studied. Carbon is a common contaminant. At the 10^{17}/cm^3 level, it often incorporates only as a p-type dopant and can be a useful dopant for the base. The dopants Sn, C, Ge, and Si can be either n-type or p-type, but most commonly they are found to be n-type in GaAs, except for C, which is usually p-type. Carbon and silicon may be chosen as the best p- and n-type dopants, respectively, because they diffuse at a negligible rate compared with the other dopants. It is useful to keep all unintentional dopants (C, Si, Ge, Sn, S, Se, Te, Li, Be, Mg, Zn, and Cd) at the $10^{16}/cm^3$ level, although $10^{17}/cm^3$ may not have large effects on the solar-cell performance as long as the final device has an appropriately doped (53) base (~1 x $10^{17}cm^{-3}$) and emitter (~2 x $10^{18}cm^{-3}$).

Small amounts (<1%) of group III and group V elements such as Al, In, P, and Sb change the lattice constant and band gap, but should have minimal effect on optoelectronic properties. Cu and Ni are known to form stable arsenides and need to be avoided. Cr and Fe are often used to make GaAs semi-insulating, so these should be avoided as well. We observe that growth of GaAs on bare soda-lime glass at 650°C results in a Na level of $10^{17}/cm^3$ in the GaAs layer (as measured

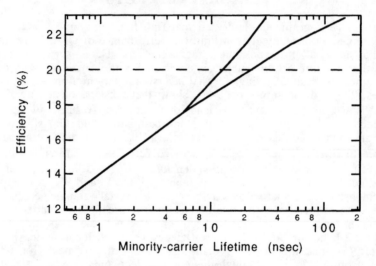

FIGURE 2. Observed efficiency (51) as a function of measured minority-carrier lifetime (52) for samples with varying dislocation densities. The lifetimes were measured on double heterostructures made on companion devices to the solar cells. The two branches of the curve at higher lifetimes were taken from the two curves in ref. (52). These data reflect observed values for one data set and will vary for other samples.

TABLE 1. Ionization energy levels associated with common impurities in GaAs (54-56). Multiple levels are separated by commas. Hyphens denote the range of reported values.

Element	From conduction band (meV)	Element	From valence band (meV)	Element	From valence band (meV)
S	6.1	Li	23,50	Co	160,560
Se	5.9	Mg	28-30	Cu	140,240,440
Te	5.8-30	Be	28-30	Cr	790
Sn	6.0	Sn	171	Mn	900-950
C	6	C	19-27	Fe	380,520
Ge	6.1	Ge	30-40	Ca	160
Si	5.8	Si	30-35	Ni	210-350,420
		Zn	24-31	Au	90
O	400,750	Cd	21-35	Ag	110

by Secondary Ion Mass Spectroscopy). The literature has very little data about Na in GaAs (57), but Na may behave similarly to Li, i.e., be a shallow dopant (54). A previous study of GaAs grown on Ge-coated NaCl implies that Na is not necessarily a serious problem (58). Atomic hydrogen often passivates defects (31, 32) and is beneficial if it stays in place. Impurity complexes can greatly change the tolerable impurity levels, and segregation of impurities to grain boundaries can have both beneficial and detrimental effects.

CRYSTAL ORIENTATION/TEXTURE EFFECTS

If one desires to grow >20%-efficient polycrystalline GaAs on low-cost substrates, it may be useful to control the texture (orientation) of the crystals. We will use the convention here that "texture" refers to the plane of the crystal that is parallel to the substrate surface. This is also sometimes referred to as "single-axis" texture. A sample that has most of its grains oriented so that the (100) plane is parallel to the substrate's surface is referred to as having strong (100) texture. "Two-axis" texture refers to crystals that have a preferential in-plane alignment as well. The orientation of the surface is much more important for the growth of GaAs than it is for the growth of Si because of the lower symmetry.

A variation in the orientation of the growth surface has strong effects on the incorporation of dopants, intrinsic defects, and other impurities. A simple rule for understanding this is that dopants or impurities sitting on group III sites incorporate more strongly onto surfaces with A-type steps, whereas dopants sitting on the group V sites incorporate in larger numbers at B-type steps (60). More complex models are necessary to understand all of the data. The effect of the crystal orientation is highly dependent on growth conditions (growth temperature, group V pressure, etc.). A study of S, Zn, Si, Te, and Sn incorporation showed variations of 1 to 2 orders of magnitude as the crystal orientation was varied (59). Kondo et al. (60) reported similar variations in the incorporation of a number of elements as a function of crystal orientation and presented a summary table describing the incorporation of Mg, Zn, Cd, Fe, S, Se, Te, O, Si, C, and Sn in GaAs. Sometimes dramatic effects are seen. For example, Si doping causes n-type conduction on (111)B GaAs, but p-type on (111)A GaAs (61). Effects like this can allow one-step growth of lateral p-n junctions by growing on a patterned substrate. Extrapolating this concept to the uncontrolled texture of a polycrystalline GaAs layer, it is clear that undesirable results may be obtained. Despite these large variations, some dopants, under some growth conditions, incorporate much more uniformly. Also, GaAs solar cells are usually insensitive to doping level variations of a factor of 2 or 3 in the emitter and up to an order of magnitude in the base. Thus, although it is very important for researchers to be aware that the doping levels are changing across the films, highly textured films (i.e., films for which all of the crystals have the same exposed surface) are not necessary to fabricate appropriately doped GaAs solar cells.

A second result of non-uniform orientation is a variation in growth rate. Experiments and simulations have shown how, under certain conditions, growth rates may vary by 1 to 2 orders of magnitude as a function of substrate orientation (62). Venkatasubramanian et al. (9) reported that tunneling between the doped emitter and doped base may become large in GaAs solar cells grown on randomly oriented polycrystalline Ge substrates, presumably because of variable growth rates. They found that the problem could be mitigated by increasing the thickness of an undoped ("spacer") layer between the emitter and base (9). However, the grain size in their study was close to 1 mm, whereas growth rate variations are a much more severe problem when the grain size is comparable to the Ga adatom surface diffusion length (typically on the order of micrometers).

Two-axis texture can produce low-angle grain boundaries. Majority-carrier transport across a single GaAs-GaAs grain boundary has been measured to be closer to ohmic when the grain-boundary angle is small (63, 64). This is useful for reducing the emitter sheet resistance, thus simplifying the grid structure that is needed. The effect of grain-boundary angle on minority-carrier lifetime has not been well documented, but there is some reason to believe that low-angle grain boundaries would cause less minority-carrier recombination.

Various techniques for growing large-grain polycrystalline GaAs require control of the orientation of the grains. For example, lateral growth similar to that used in the CLEFT technique requires careful control of the crystal orientation. In this way, uniform texture of the GaAs films can have an indirect benefit to the quality of the final device (by helping to increase the grain size).

Venkatasubramanian has reported (unconfirmed) an efficiency over 20% for a GaAs solar cell grown on a polycrystalline Ge substrate with randomly oriented, 1-mm grains. Thus, we can conclude that no preferential texture is required to grow a 20% polycrystalline cell.

Ion-beam-assisted deposition has demonstrated the ability to control the two-axis texture and has been very successful in growing superconductor films with properties similar to those of single-crystal superconducting films (65). Rolled and annealed metal foils also have a controlled texture and large (30-100-μm) grain sizes, which can serve as a template for growth of textured, large-grain material (66) assuming that adequate intragrain properties can be achieved.

Our conclusion regarding the requirement for two-axis texture is that there is inadequate data available to assess the need/advantages of in-plane alignment. However, one-axis texture is likely to be very useful for uniform doping and growth rate.

CONCLUSIONS

The large space-charge recombination current observed for polycrystalline GaAs solar cells originates from recombination through defect states at grain boundaries and through defect states in the space-charge region at intragrain defects. The space-charge recombination current is the most important loss mechanism observed for polycrystalline GaAs solar cells, and it must be reduced to achieve high efficiencies. The two most common causes of it are small grain size (and the associated grain-boundary recombination in the p-n junction) and intragrain defects that penetrate the p-n junction. The acceptable grain size is highly dependent on the passivation of the grains, but extrapolating from today's best single-crystal GaAs cells, a grain size of 20-50 μm is needed to reach 20%. If data in the literature can be used to estimate future performance, then the threading dislocation density must be reduced to less than $5 \times 10^6/cm^2$ to achieve an efficiency over 20%. Uniform crystallographic orientation is of less importance than the inter- and intragrain properties, but may be useful toward growing more uniform layers. Developing a low-cost, high-efficiency, 1-sun GaAs cell depends on identifying a growth technique that can satisfy these needs and/or a new method for passivating GaAs grain boundaries and other defects.

ACKNOWLEDGMENTS

We would like to thank B. Reedy for the SIMS measurements, C. Kramer for sample preparation, and R. Venkatasubramanian for pointing out the confusion regarding differences in "quality" diode factors for GaAs and Si. We also thank E. Yablonovitch, M. Wanlass, D. Friedman, H. Atwater, J. Olson, R. Noufi, K. Ramanathan, and J. Bernard for useful conversations. This work was completed under Contract No. DE-AC36-83CH10093.

REFERENCES

The reference list is not meant to be comprehensive. We chose for reasons of relevance and convenience and recognize that we have omitted many references of very high quality.

1. Woodall, J. M. and Hovel, H. J., "Outlooks for GaAs terrestrial photovoltaics," *J. Vac. Sci. Technol.* **12**, 1000-1009 (1975).
2. Emcore, Aixtron, *personal communication*.
3. Mauk, M., "Inexpensive Approach to III-V Epitaxy," *this conference* (1997).
4. Kurtz, S. R., Olson, J. M., and Kibbler, A., "High efficiency GaAs solar cells using GaInP$_2$ window layers," in *Proc. of the 21st IEEE Photovoltaic Specialists Conference*, 1990, 138-40.
5. Friedman, D. J., Kurtz, S. R., Bertness, K. A., Kibbler, A. E., Kramer, C., Olson, J. M., King, D. L., Hansen, B. R., and Snyder, J. K., "30.2% efficient GaInP/GaAs monolithic two-terminal tandem concentrator cell," *Progress in Photovoltaics: Research and Applications* **3**, 47-50 (1995).
6. Takamoto, T., Ikeda, E., Kurita, H., and Ohmori, M., "Over 30% efficient InGaP/GaAs tandem solar cells," *Appl. Phys. Lett.* **70**, 381-383 (1997).
7. Chu, S., *Thin-Film Gallium Arsenide Solar Cell Research Final Report*, SERI Subcontract XL-4-04018-1(1985).
8. Nelson, N. J., *Thin Films of Gallium Arsenide on Low-Cost Substrates* Report # E(04-3)-1288 for U.S. Energy Research and Development Administration (1977).
9. Venkatasubramanian, R., O'Quinn, B. C., Hills, J. S., Sharps, P. R., Timmons, M. L., Hutchby, J. A., Field, H., Ahrenkiel, R., and Keyes, B., "18.2%(AM1.5) efficient GaAs solar cell on optical-grade polycrystalline Ge substrate," in *Proceedings of the 25th IEEE Photovoltaic Specialists Conference*, 1996, 31-36.
10. Venkatasubramanian, R., O'Quinn, B. and Siivola, E., "High-efficiency GaAs solar cells on mm and sub-mm grain-size polycrystalline Ge substrates," in *NREL/SNL Photovoltaics Program Review:* Lakewood, CO, 1996, A.I.P. **394**.
11. Venkatasubramanian, *private communication (subcontractor report)*.
12. Gale, R. P., McClelland, R. W., King, B. D., and Gormley, J. V., "High-efficiency thin-film AlGaAs-GaAs double heterostructue solar cells," in *Proceedings of the 20th IEEE Photovoltaic Specialists Conference*, 1988, 446-450.
13. Lush, G. and Lundstrom, M., "Thin film approaches for high-efficiency III-V cells," *Solar Cells* **30**, 337-344 (1991).
14. Lush, G. B., Patkar, M. P., Young, M. P., Melloch, M. R., Lundstrom, M. S., Vernon, S. M., Gagnon, E. D., Geoffroy, L. M. and Sanfacon, M. M., "Thin-film GaAs solar cells by epitaxial lift-off," in *Proc. of the 23rd IEEE Photovoltaic Specialists Conference*, 1993, 1343-1346.
15. Bruel, M., "Application of hydrogen ion beams to silicon on insulator material technology," *Nucl. Instr. Meth. Phys. Res. B* **108**, 313-319 (1996).
16. Bruel, M., "Silicon on insulator material technology," *Elec. Lett.* **31**, 1201-1202 (1995).
17. Lanza, C. and Hovel, H. J., "Efficiency calculations for thin-film polycrystalline semiconductor p-n junction solar cells," *IEEE Trans. on Elec. Dev.* **ED-27**, 2085-8 (1980).

18. Yamaguchi, M. and Itoh, Y., "Efficiency considerations for polycrystalline GaAs thin-film solar cells," *J. Appl. Phys.* **60,** 413 (1986).
19. Mohammad, S. N., Sobhan, M. A. and Qutubuddin, S., "The influence of grain boundaries on the performance efficiency of polycrystalline gallium arsenide solar cells," *Solid-State Elec.* **32,** 827-834 (1989).
20. Carpenter, M. S., Melloch, M. R., Lundstrom, M. S. and Tobin, S. P., "Effects of Na_2S and $(NH_4)_2S$ edge passivation treatments on the dark current-voltage characteristics of GaAs *pn* diodes," *Appl. Phys. Lett.* **52,** 2157-2159 (1988).
21. Tobin, S. P., Vernon, S. M., Bajgar, C., Geoffroy, L. M., Keavney, C. J., Sanfacon, M. M. and Haven, V. E., "Device processing and analysis of high efficiency GaAs cells," *Solar Cells* **24,** 103-115 (1988).
22. Henry, C. H., Logan, R. A. and Merritt, F. R., "The effect of surface recombination on current in $Al_xGa_{1-x}As$ heterojunctions," *J. Appl. Phys.* **49,** 3530-3542 (1978).
23. de Lyon, T. J., Casey, H. C., Timmons, M. L., Hutchby, J. A. and Dietrich, D. H., "Dominance of surface recombination current in planar, Be-implanted GaAs p-n junctions prepared by rapid thermal annealing," *Appl. Phys. Lett.* **50,** 1903-1905 (1987).
24. Stellwag, T. B., Melloch, M. R., Lundstrom, M. S., Carpenter, M. S. and Pierret, R. F., "Orientation-dependent perimeter recombination in GaAs diodes," *Appl. Phys. Lett.* **56,** 1658-1660 (1990).
25. Lundstrom, M. S., Melloch, M. R., Pierret, R. F., Carpenter, M. S., Chuang, H. L., Dodd, P. E., Keshavarzi, A., Kausmeier-Brown, M. E., Lush, G. B. and Stellwag, T. B., *Basic studies of III-V high efficiency cell components,* Report # NREL/TP-451-4850 (1990).
26. DeMoulin, P., Tobin, S. P., Lundstrom, M. S., Carpenter, M. S. and Melloch, M. R., "Influence of perimeter recombination on high-efficiency GaAs p/n heteroface solar cells," *IEEE Electron Device Letters* **9,** 368-370 (1988).
27. Fossum, J. G. and Lindholm, F. A., "Theory of grain-boundary and intragrain recombination currents in polysilicon p-n-junction solar cells," *IEEE Trans. on Electron Dev.* **ED-27,** 692-700 (1980).
28. Shockley, W. and Read, W. T., *Phys. Rev.* **87,** 835 (1952).
29. Hall, R. N., *Phys. Rev.* **87,** 387 (1952).
30. Sah, C. T., Noyce, R. N. and Shockley, W., *Proc. IRE* **45,** 1228 (1957).
31. Pearton, S. J. and Tavendale, A. J., "Hydrogen passivation of grain boundaries in polycrystalline gallium arsenide," *J. Appl. Phys.* **54,** 1154-1155 (1983).
32. Pearton, S. J., Wu, C. S., Stavola, M., Ren, F., Lopata, J., Dautremont-Smith, W. C., Vernon, S. M. and Haven, V. E., "Hydrogenation of GaAs on Si: effects on diode reverse leakage current," *Appl. Phys. Lett.* **51,** 496-498 (1987).
33. Sandroff, C. J., Hegde, M. S., Farrow, L. A., Bhat, R., Harbison, J. P. and Chang, C. C., "Enhanced electronic properties of GaAs surfaces chemically passivated by selenium reactions," *J. Appl. Phys.* **67,** 586-588 (1990).
34. Pearton, S. J., Haller, E. E. and Elliot, A. G., "Nitridization of gallium arsenide surfaces: effects on diode leakage currents," *Appl. Phys. Lett.* **44,** 684-686 (1984).
35. Wolford, D. J., Gilliland, G. D., Kuech, T. F., Smith, L. M., Martinsen, J., Bradley, J. A., Tsang, C. F., Venkatasubramanian, R., Ghandi, S. K. and Hjalmarson, H. P., "Intrinsic recombination and interface characterization in "surface-free" GaAs structures," *J. Vac. Sci. Technol.* **B9,** 2369-2376 (1991).
36. Ghandhi, S. K., Tyagi, S. and Venkatasubramanian, R., "Improved photoluminescence of GaAs in ZnSe/GaAs heterojunctions grown by organometallic epitaxy," *Appl. Phys. Lett.* **53,** 1308-1310 (1988).
37. Dupuis, R. D., "III-V compound semiconductor native oxides - the newest of the semiconductor device materials," *Compound Semiconductor* 32-34 (1997).
38. Dallesasse, J. M., Holonyak, N., Sugg, A. R., Richard, T. A. and El-Zein, N., *Appl. Phys. Lett.* **57,** 2844 (1990).
39. Olson, J. M., Ahrenkiel, R. K., Dunlavy, D. J., Keyes, B. and Kibbler, A. E., "Ultralow recombination velocity at $Ga_{0.5}In_{0.5}P$/GaAs heterointerfaces," *Appl. Phys. Lett.* **55,** 1208 (1989).
40. Aspnes, D. E., "Recombination at semiconductor surfaces and interfaces," *Surface Science* **132,** 406-421 (1983).
41. Stellwag, T. B., Dodd, P. E., Carpenter, M. S., Lundstrom, M. S., Pierret, R. F., Melloch, M. R., Yablonovitch, E. and Gmitter, T. J., "Effects of perimeter recombination on GaAs-based solar cells," in *Proc.of the 21st IEEE Photovoltaic Specialists Conference,* 1990, 442-447.
42. Yablonovitch, E., Sandroff, C. J., Bhat, R. and Gmitter, T., "Nearly ideal electronic properties

204

of sulfide coated GaAs surfaces," *Appl. Phys. Lett.* **51**, 439-441 (1987).

43. In Yablonovitch, E., Gmitter, T. J. and Bagley, B. G., "As$_2$S$_3$/GaAs, a new amorphous/crystalline heterojunction for the III-V semiconductors," *Appl. Phys. Lett.* **57**, 2241-2243 (1990) a lower surface-recombination current is reported, but the structure contains more AlGaAs than GaAs, casting doubt as to whether this number can be duplicated on conventional GaAs structures.
44. Thompson, C. V., "Grain growth in polycrystalline silicon films," in *Proceedings of the Mat. Res. Soc. Symp. Proc.* **106**, 1988, 115-125.
45. Yonehara, T., Smith, H. I., Palmer, J. E. and Thompson, C. V., "Surface-energy-driven graphoepitaxy in ultrathin films of Ge," in *Extended Abstracts of the 16th (1984 International) Conference on Solid State Devices and Materials*, 1984, 515-518.
46. Yonehara, T., Thompson, C. V. and Smith, H. I., "Abnormal grain growth in ulta-thin films of Ge on insulator," in *Mat. Res. Soc. Symp. Proc.* **25**, 1984, 517.
47. Yang, M. and Atwater, H. A., "Controlled grain size and location in Ge thin films on silicon dioxide by low temperature selective solid phase crystallization," *Mat. Res. Soc. Symp. Proc.* **403**, 113-118 (1996).
48. Yang, C. M. and Atwater, H. A., "Selective solid phase crystallization for control of grain size and location in Ge thin films on silicon dioxide," *Appl. Phys. Lett.* **68**, 3392-3394 (1996).
49. Kumomi, H., Yonehara, T., Nishigaki, Y. and Sato, N., "Selective nucleation based epitaxy (SENTAXY)," *Appl. Surf. Sci.* **41/42**, 638-642 (1989).
50. Givargizov, E, *Oriented Crystallization on Amorphous Substrates*, New York: Plenum, 1991.
51. Vernon, S. M. and Tobin, S. P., "Experimental study of solar cell performance versus dislocation density," in *Proceedings of the 21st IEEE Photovoltaic Specialists Conference*, 1990, 211-216.
52. Ahrenkiel, R. K., Al-Jassim, M. M., Keyes, B., Dunlavy, D., Jones, K. M., Vernon, S. M. and Dixon, T. M., "Minority carrier lifetime of GaAs on silicon," *J. Electrochem. Soc.* **137**, 996-1000 (1990).
53. Lundstrom, M. S., "Device physics of crystalline solar cells," *Solar Cells* **24**, 91-102 (1988).
54. Milnes, A. G., *Deep Impurities in Semiconductors*, New York: John Wiley & Sons, 1973.
55. Ghandhi, S. K., *VLSI Fabrication Principles*, New York: John Wiley &Sons, 1983.
56. Watts, R. K., *Point Defects in Crystals*, New York: John Wiley & Sons, 1977.
57. Stojic, M., Kostic, D. and Stosic, B., "The behaviour of sodium in Ge, Si, and GaAs," *Physica* **138B**, 125-128 (1986).
58. Shuskus, A. and Cowher, M., "Fabrication of monocrystalline GaAs solar cells utilizing sacrificial NaCl substrates," in *Proceedings of the Advanced High Efficiency Concepts Subcontractors Review Meeting*, 1983, 89-94.
59. Bhat, R., Caneau, C., Zah, C. E., Koza, M. A., Bonner, W. A., Hwang, D. M., Schwarz, S. A., Menocal, S. G. and Favire, F. G., "Orientation dependence of S, Zn, Si, Te, and Sn Doping in OMCVD growth of InP and GaAs - application to DH lasers and lateral p-n junction arrays grown on non-planar substrates," *J. Cryst. Growth* **107**, 772-778 (1991).
60. Kondo, M., Anayama, C., Okada, N., Sekiguchi, H., Domen, K. and Tanahashi, T., "Crystal-lographic orientation dependence of impurity incorporation into III-V compound semicon-ductors grown by metalorganic vapor phase epitaxy," *J. Appl. Phys.* **76**, 914-927 (1994).
61. Pavesi, L., Piazza, F., Henini, M. and Harrison, I., "Orientation dependence of the Si doping of GaAs grown by molecular beam epitaxy," *Semicond. Sci. Technol*. **8**, 167-171 (1993).
62. Jones, S. H., Salinas, L. S., Jones, J. R. and Mayer, K., "Crystallographic orientation dependence of the growth rate for GaAs low pressure organometallic vapor phase epitaxy," *J. Electron. Mater.* **24**, 5-14 (1995).
63. Salerno, J. P., McClelland, R. W., Fan, J. C. C., Vohl, P. and Bozler, C. O., "Growth and characterization of oriented GaAs bicrystal layers," *Proc. of the 16th IEEE Photovoltaic Specialists Conference*, 1982, 1299-1303.
64. Kish, F. A., Vanderwater, D. A., Peanasky, M. J., Ludowise, M. J., Hummel, S. G. and Rosner, S. J., "Low-resistance ohmic conduction across compound semiconductor wafer-bonded interfaces," *Appl. Phys. Lett.* **67**, 2060-2062 (1995).
65. Norton, D. P., Goyal, A., Budai, J. D., Christen, D. K., Kroeger, D. M., Specht, E. D., He, Q., Saffian, B., Paranthaman, M., Klabunde, C. E., Lee, D. F., Sales, B. C. and List, F. A., "Epitaxial YBa$_2$Cu$_3$O$_7$ on biaxially textured nickel (001): An approach to superconducting tapes with high critical current density," *Science* **274**, 755-757 (1996).
66. Wu, X. D., Foltyn, S. R., Arendt, P., Townsend, J., Adams, C., Campbell, I. H., Tiwari, P., Coulter, Y. and Peterson, D. E., "High current YBa$_2$Cu$_3$O$_{7-\delta}$ thick films on flexible nickel substrates with textured buffer layers," *Appl Phys. Lett.* **65**, 1961-1963 (1994).

Thin Self-Supporting Polycrystalline Silicon Substrates - A New Role for EFG Cylinders

J. P. Kalejs

ASE Americas, Inc., 4 Suburban Park Drive, Billerica, MA 01821

Abstract Large scale cost effective crystalline silicon utilization in the photovoltaic industry is dependent on the availability of a low cost method to produce a crystal in a wafer form suitable for subsequent device processing and encapsulation. Very thin (1-10 microns) devices typically require a substrate for support and handling. This paper describes an approach capable of production of 50-100 μm thick crystalline silicon "thick films" without the aid of a substrate, i.e., in a self-supporting form, using melt growth at much higher rates and at a potentially lower cost than is currently being practiced. This approach uses a modification of a well established method of Edge-defined Film-fed Growth (EFG) which today is already being used to produce polycrystalline silicon wafers on a commercial scale.

INTRODUCTION

Ribbon growth produces a wafer substrate which can keep material utilization and silicon cost to a minimum. Ribbon is grown down to thicknesses as low as 100 μm. In contrast, conventional CZ boule growth and ingot casting both require costly slicing and polishing and waste over 50% of the material in kerf. Wafer thickness limits appear to be about 200 microns. In the other limit of current practice, crystalline silicon substrates are produced as very thin polycrystalline films by CVD or LPE deposition. These methods are slow and more costly compared to ribbon growth; further, the device manufacturer now must cope with mostly undesirable effects arising from the foreign substrate material These range from impurities, to imposed microcrystallinity, to thermal incompatibility and added cost of a new material. Films up to 10 to 20 μm in thickness are available from these methods, but issues of quality and substrate compatibility remain as barriers to commercialization.

I discuss here a novel form of a self-supporting substrate material and its production method as a candidate for a low cost crystalline silicon substrate intermediate to the cases discussed above. This material is obtained by growing a thin, hollow silicon cylinder, with wall thicknesses of 50 to 100 μm, by the Edge-defined Film-Fed Growth (EFG) technique. EFG is used currently to produce octagon tubes, from which 10 cm x 10 cm wafers of 300 μm average thickness are obtained.

CP404, *Future Generation Photovoltaic Technologies: First NREL Conference*, edited by McConnell
© 1997 The American Institute of Physics 1-56396-704-9/97/$10.00

These form the material base for module manufacture at ASE Americas on a commercial scale. The production method of low wafers and cost solar cells based on thin crystalline silicon cylinders are the subject of this paper.

ROTATING EFG CYLINDER TECHNOLOGY

The EFG process for growth of silicon wafers has undergone many scaleups in the 25 years that it has been under development (1). Productivity has increased by more than an order of magnitude in this time. Wafer production costs have decreased proportionately. The current configuration of EFG photovoltaic wafer technology, the octagon, consists of a hollow eight-sided tube, where each side or facet has a width of 10 cm. Other growth parameters are given in Table 1. A large diameter cylinder configuration is proposed here as the next generation of EFG technology which would further raise the throughput and decrease the cost of the 21st century EFG product. The productivity parameters of this concept are compared within the historic progression of EFG technology development in Table 1.

Table 1. Historic operating parameters for EFG wafer production and projected output from an advanced cylindrical EFG format.

Year	1980	1995	2010
Crystal Format			
Shape	5 cm wide ribbon	Eight 10 cm wide ribbons (octagon)	1 m dia. cylinder
Length	75 m	4.6 m	2 m
Thickness	400 µm	300 µm	75 µm
Growth speed	1.75 cm/min	1.75 cm/min	3 cm/min
Wafers and Cells			
Wafers/tube	-	360	2
Wafers/module	432	216	1
(wafer area)	$(50 \times 100 mm^2)$	$(100 \times 100 mm^2)$	$(1570 \times 2000 mm^2)$
Cell efficiency	10%	14%	20%
Substrate Production			
Power	5 W/h	115 W/h	1000 W/h
Hours/module (power)	42 (220 W)	2.6 (300 W)	0.6 (600 W)

The gain of nearly an order of magnitude in productivity with the proposed new EFG technology is evident in Table 1. This also introduces the novel concept of utilizing a single wafer in the form of a large area sheet of about 1.5 x 2 m in the construction of one module. A large reduction in the time needed to grow enough substrate material is also evident - down to less than 20 minutes for the current 300 W module. This is accomplished by the increase in effective perimeter gained with the large diameter cylinder, the projected higher cell efficiency, and a higher crystal pull rate. The rotation permitted by the cylindrical configuration allows reduction of

Figure 1. Conceptual schematic of EFG wafer production from a crystal in the shape of a 1 m diameter hollow thin-walled cylinder.

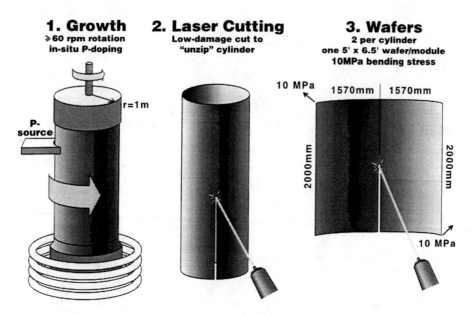

1. Growth
≥60 rpm rotation
in-situ P-doping

2. Laser Cutting
Low-damage cut to
"unzip" cylinder

3. Wafers
2 per cylinder
one 5' x 6.5' wafer/module
10MPa bending stress

After growth of a cylinder which already has the required junction on one sheet surface, the wafer next needs to go through processing providing bulk and surface passivation, metallization and lamination. A possible processing sequence for integrating the wafer into a laminate is shown in Figures 2 and 3. The detailed steps needed will depend on the geometry of the wafer cut from the cylinder. One option is to process intact the total grown length of 2 meters of the "unzipped" cylinder by metallizing and encapsulating it in a semi-continuous belt-to-belt transfer process. If other size wafers are desired, smaller areas, e.g., ranging from 10 cm x 10 cm to 40 cm x 40 cm, could be cut and processed individually.

A long term cost goal of the order of $1/W for a crystalline silicon photovoltaic module product requires high cell efficiencies. Efficiencies of 20% are feasible for thin polycrystalline silicon substrates produced from high quality EFG cylinders. Well known methods that produce 20% efficient crystalline silicon cells and above need to be applied to these thin substrates to reach this goal. These include: bulk passivation, light trapping and reduction of recombination at surfaces and under metallization. While the path to 20% EFG cells on thin wafers does not involve new or unproven processes, it does requires novel engineering and manufacturing approaches. Schemes depend in detail on the cell configuration required to meet the target 20% cell efficiency.

Two solar cell design concepts which can support the efficiency goal are the emitter wrap-though (EWT) cell (5) and the point contact cell (6). The goal in

Figure 2. Front (a) and back(b) laminate component formation sequence for large area EFG wafer and module production concept.

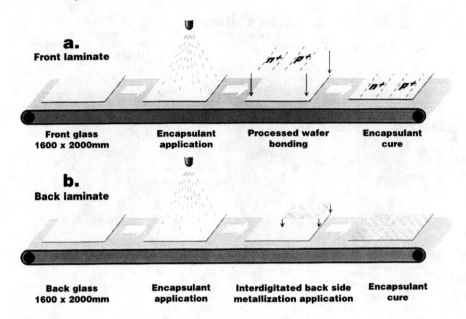

a.
Front laminate

| Front glass
1600 x 2000mm | Encapsulant
application | Processed wafer
bonding | Encapsulant
cure |

b.
Back laminate

| Back glass
1600 x 2000mm | Encapsulant
application | Interdigitated back side
metallization application | Encapsulant
cure |

Figure 3. Integration sequence for front and back laminate components for large area EFG wafer and module production concept.

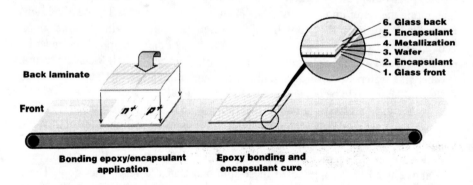

Back laminate

Front

6. Glass back
5. Encapsulant
4. Metallization
3. Wafer
2. Encapsulant
1. Glass front

| Bonding epoxy/encapsulant
application | Epoxy bonding and
encapsulant cure |

adopting either one of these cell structures is the same: they allow all metallization to be removed to the back plane of the cell and have promise of single step interconnection and lamination. The EWT cell configuration incorporates holes drilled through the cell to bring metallization from the front to the back of the cell. This removes the busbar shadowing and assists in reducing series resistance. The

the wall thickness nonuniformity, hence growth of a crystal with a much reduced wall thickness.

EFG cylinders have been grown with rotation at various wall thicknesses down to 100 μm (2). Thickness uniformity increases as the speed of rotation is raised. It has been demonstrated that growth at up to 100 rpm rotation can produce cylinders with uniformity of better than 10 μm at a thickness of 100 μm. Growth lengths were up to a meter and tube diameters were 15 cm. Extension of this method to grow thinner and larger diameter cylinders is straightforward and does not require any new technology. The cylindrical symmetry results in a reduced thermal stress and higher as-grown material quality than in ribbon (3). This factor allows increase of another important growth parameter contributing to throughput - the growth speed. Dislocation densities are low, and the minority carrier diffusion lengths in the cylindrical EFG material are higher than in ribbon. The cylindrical shape allows the crystal to be grown without the aid of a substrate, as the cylinder is self supporting as it emerges from the crystal growth furnace.

21ST CENTURY EFG TECHNOLOGY - ONE WAFER PER MODULE

The novel form of a thin crystalline silicon substrate material grown in the form of a large cylindrical sheet presents a provocative challenge for device manufacturers. Laser cutting methods exist to either "unzip" the cylinder by making a single line cut along its length and then opening it up into a large area sheet, or to cut it into arbitrary area squares or rectangular strips. This is shown conceptually in Figure 1. It has been shown that the best lasers are either copper vapor or excimer, which have low cutting damage. They produce a wafer edge strength which supports the stress needed to flatten the cylinder without fracture. For example, to flatten a cylindrical section of 10 cm chord length with a radius r=100 cm requires a stress of the order of 10 MPa, compared to typical fracture strengths for laser cut wafers of 60 MPa. A new opportunity now arises - to build a large area module using a *single wafer* of the order of 1.5 x 2 m area.

In order to aid the processing engineer, some solar cell processing can be incorporated into the EFG crystal growth process. This reduces the number of handling and processing steps required to make a module. For example, *in situ* p-n junction formation in the furnace during crystal growth has been demonstrated, and used to form a viable junction for solar cell fabrication (4). The diffusion source, a solid phosphorus-doped oxide, is placed so that it dopes only on one surface of the tube (P-source in Fig. 1), and the cylinder emerges from the furnace with an appropriate depth junction of 0.2-0.3 microns. This reduces the number of subsequent handling steps in cell fabrication, as well as eliminating the need for a junction isolation step.

211

point contact cell has been developed for high concentrator applications, where it has already shown to be capable of efficiencies of 27%. The back side point contact cell has achieved a record efficiency of over 22% at one sun illumination. In the point contact cell design, alternating n^+ and p^+ regions are formed all on the back surface of the wafer. The front (illuminated) side of the cell consists only of an oxidized and textured/antireflection coated surface. An interdigitated back-side metalllization configuration handles all the current collection. This design is particularly suited to the large wafer concept of one wafer per 3 m^2 module, since each side of the wafer can be processed separately before it is integrated with front and back encapsulant and protective glass layers. A lamination process which uses encapsulant curing temperatures in the same range as those for interconnection is also desirable, so that the fully processed wafer can be handled efficiently and gently in the lamination step. These wafer, metallization and lamination configurations can utilize low cost processing schemes which use belt furnaces with high throughput.

As single cell area increases, a choice also has to be made on module voltage characteristics. Leaving the wafer as a single large solar cell, in the one extreme, restricts the module voltage to 0.6-0.7 V in the case of crystalline silicon. To service conventional applications requiring voltages up to several hundred volts, many modules need to be series connected. The wafer size can be progressively scaled down using laser cutting to increase the number of individual solar cells within the module, and so raise module voltage.

CONCLUDING REMARKS

This paper proposes a novel wafer production method which offers the potential for producing very thin (50-100 μm) crystalline silicon wafers in the form of large area sheets which are self-supporting, i.e., do not require a substrate to produce. This method would use EFG to grow a large diameter hollow cylinder directly from the melt. High speed rotation of the cylinder, up to 100 rpm, is used to achieve thickness uniformity at the very thin wall thicknesses required for high efficiency cell structures of 20%. A unique feature of this approach is that the cylinder can lead directly to a large area single wafer, e.g., of 1.5 m x 2 m area, and the concept of producing a module consisting of one wafer or cell. Large diameter hollow cylinder growth increases crystalline silicon wafer production rates by an order of magnitude over conventional rates being practiced in manufacture of commercial photovoltaic products today.

Processing of the wafer into a module is simplified by the ability, with the EFG method, to grow in a p-n junction during this large area wafer production process. Simple additional cell and module processing concepts are envisioned to take advantage of this large wafer format. These favor single back plane metallization and single-step metallization and lamination, and have the potential of reducing module manufacturing costs based on crystalline silicon wafers to the $1/W level.

REFERENCES

1. M.J. Kardauskas, M.D. Rosenblum, B.H. Mckintosh and J.P. Kalejs, "The Coming of Age of a New PV Wafer Technology - Some Aspects of EFG Polycrystalline Silicon Sheet Manufacture", *Twenty-Fifth IEEE PVSC*, 383-88 (1994).

2. L.E. Eriss, R.W. Stormont, T. Surek and A.S. Taylor, "The Growth of Silicon Tubes by the EFG Process " *J. Crystal Growth* **50**, 200-205(1980).

3. J.P. Kalejs, A.A. Menna , R.W. Stormont and J.W. Hutchinson, "Stress in Thin Hollow Cylinders Grown by the Edge-defined Film-Fed Growth Technique", *J. Crystal Growth* **104**, 14-19(1990).

4. B.R. Bathey, M.C. Cretella and A.S. Taylor, "Method of Fabricating Solar Cells", U.S. Patent No. 5,156,978(1992).

5. J.M. Gee and T.F. Ciszek, "The Crystalline-Silicon Photovoltaic R&D Project at NREL and Sandia", paper presented at the NREL/SNL Program Review Meeting, 18-22 November, 1996, Lakewood CO (in press).

6. R.M. Swanson, "Point Contact Solar Cells; Modelling and Experiment" *Solar Cells* **7**, 85-118(1986).

INNOVATIVE CONCEPTS

Thermophotovoltaic Principles, Potential, and Problems

Timothy J. Coutts

National Renewable Energy Laboratory
1617 Cole Blvd., Golden, CO 80401

Abstract

This paper discusses the attractions of thermophotovoltaics (TPV) and reviews the potential performance of single-junction devices. It then offers speculations regarding the next generations of TPV devices, including tandem cells and devices consisting of polycrystalline thin films.

Introduction

Thermophotovoltaics (TPV) is a re-emerging topic that has been growing significantly in national interest during the last 5 years. The subject was actively studied as early as the middle 1960s, at which time Pierre Aigrain (1) suggested converting infrared (IR) radiation with a photovoltaic (PV) converter to produce electricity. Other authors during this time discussed the use of TPV for both military (2) and nonmilitary (3) applications. The subject was vigorously researched until the early 1980s, when funding diminished. One of the reasons for this was the unavailability of high-quality PV converters with optical and electronic characteristics that were well-matched to typical IR spectra. The earliest configurations generally consisted of a silicon or germanium PV converter used in conjunction with a selective radiator consisting of a rare-earth oxide (4). The performance of the systems was inadequate for various reasons, which are inappropriate for discussion here.

TPV has successfully exploited the high-efficiency solar cell program of the U.S. Department of Energy (DOE) and, in particular, the work performed on high efficiency converters based on III-V semiconductors. This has resulted in rapid progress in the field and there are now several high-quality PV converters available for systems use. In the opinion of the author, this is the main reason for the resurgence of interest in TPV as a means of generating electricity.

At present, most of the funding is military in origin, the details of which are inappropriate here. The Defense Advanced Research Project agency (DARPA) is also funding TPV projects and the Army Research Office also appears to be moving enthusiastically into the field. As will be seen, there are many attractions to the military for using TPV systems, and we may expect the short-term economic thrust to derive from military applications. The estimated total U.S. funding in 1997 for

CP404, *Future Generation Photovoltaic Technologies: First NREL Conference*, edited by McConnell
© 1997 The American Institute of Physics 1-56396-704-9/97/$10.00

TPV research and development is in the range of $20-30 million per year. It is stressed, however, that this is a "best-guess," which is not necessarily reliable.

In addition to the attractions of TPV to military agencies, there are many nonmilitary applications for TPV systems. The market for these is probably on the order of $500 million per year for well-developed, reliable systems. However, most recent work has been narrowly focused on the PV converters rather than on complete systems, and this is presently one of the weaknesses of the field. There is a lack of a significant body of work on the modeling and realization of actual systems.

A typical TPV system will consist of a fuel and a means of burning it, a radiant surface heated by the fuel, a semiconductor PV converter, a means of recirculating unusable sub-bandgap radiation, and a cooling and heat recuperation system. Although various individuals have experience in one or more of these components, there is very limited experience on constructing and characterizing a complete system. The exceptions are probably the military organizations involved, from which little information is available, and small companies such as JX Crystals, Inc. (5) and others of similar size. Consequently, the most pressing short-term need is for systems experience and characterization.

Attractions and Applications of TPV

There are many attractions of TPV systems, but it must be realized that these are to be regarded as potential attractions. Again, the attractions are based on idealized modeling; very few organizations have the experience of complete systems.

High Power Densities

As we will show later, depending on the precise details of the system design and operating conditions, power densities in the range of 1 to 5 W cm^{-2} of converter area may be expected. However, the modeling performed to date has been highly idealized and generally assumes that there are no parasitic losses associated with the conversion. Nevertheless, prototype systems (5) appear to have demonstrated electrical power density outputs in the range of 1-2 W cm^{-2}. Clearly, higher radiator temperatures would yield higher electrical power densities, all other things being equal. However, higher operating temperatures imply a higher cell temperature, more radiator and system durability problems, and an increasing need for an efficient heat recuperation system.

Quietness

Like PV, TPV conversion involves no moving parts and it can be expected to be essentially silent. This has obvious strategic advantages for the military, but there

218

are also nonmilitary applications for which quietness may be attractive (e.g., recreational vehicles). Many such vehicles are located on campsites; an owner who has paid upwards of $100,000 for his vehicle may not be averse to paying a premium for electricity that is quiet compared with the competition (mainly diesel generation). This application alone is expected to be particularly lucrative.

Low Maintenance

Because of the absence of moving parts, maintenance requirements are expected to be minimal. However, there is no long-term experience in operating radiant surfaces at temperatures of approximately 1500 K, as are needed for high power density outputs.

Low NO_x Emissions

This potential advantage depends sensitively on the design and operating conditions of the burner. The higher the operating temperature, the higher the power density output, but the NO_x emissions increase dramatically with higher burner temperatures. Consequently, it will be necessary to optimize the radiator operating temperature to achieve high power densities and yet ensure low NO_x emissions.

Cogeneration

A TPV system must include a heat recovery system, which is part of the cell and system cooling, and may be used to preheat the incoming fuel before the burner. This will lead to more efficient use of the fuel and a higher flame temperature. An alternative to this mode is to use the heat extracted from the system as a means of providing local or community heating. Community heating systems are already in widespread use in Scandinavia, where TPV is being seriously considered as a means of cogenerating heat and electricity (6).

Versatility

In principle, TPV may be fueled by almost any combustible material or even a radioactive source. However, a specific burner is likely to be optimal only for the specific fuel for which it has been designed. Nevertheless, for a well-designed system, one may expect that the actual burners may easily be interchanged so that different fuels can be used (for example, natural gas, propane, or diesel). Again, to the knowledge of the author, there is no experience of this.

Potential Applications

Potential applications can broadly be divided into two categories—military and nonmilitary. Although there is some duplication between these, their applications are distinct. A market survey was recently conducted by NREL (7). The implication of the survey results is that the market sizes, for the applications identified as appropriate, are potentially several billion dollars per year. However, it must be assumed that TPV systems could only command (at least in the early stages) perhaps 5% of this. Hence, if the market size for small capacity generators (up to 2 kW) is $2 billion per year and TPV generators command 5% of this, market size may approach $100 million per year. Building a successful TPV industry requires a substantial amount of consumer education (i.e. the public, the various military organizations, and potential federal funding agencies). There is presently not a plan in place to conduct such widespread education, although NREL is developing ideas on what a TPV national plan might contain. We shall now discuss some of the possible applications.

Military

i   Vehicles

It is possible that in certain military vehicles the large-scale use of TPV systems could be made. The attraction, of course, would be their quietness. The fact that supply and demand would also be in phase and that the power would be available quickly are additional attractions not provided by other electricity-generating schemes.

ii   Battery Charging

The U.S. military is a substantial customer for batteries, most of which must be recharged in either base or relatively front-line positions. The quietness of TPV systems and their ability to be scaled to the desired application are attractive features.

iii   Man-Portable Power

The modern soldier, and certainly the soldier of the future, is expected to be carrying an increasing amount of electronic equipment requiring substantial amounts of battery, or some alternative, power. The soldier is expected to operate as a quasi-independent entity and such power will be essential to his/her survival. At a workshop at Duke University in June 1996 (8), various specialists in the field addressed which of the applications being considered by the U.S. Army could best be met by TPV systems. Several more than those itemized above were considered, but battery charging was the most appealing to the specialists present because they concluded it was the most feasible in the short-term. The specialists were not overly confident about the development of man-portable power supplies because of the need to accommodate ruggedness, safety to the solder, minimal heat signature, and the ability to work in arbitrary orientations. The U.S. military typically places large orders for batteries and for battery chargers on a spasmodic basis. Consequently, the military forces are potentially significant, if erratic, customers.

Nonmilitary

The NREL marketing report (7) identified a number of nonmilitary applications that could amount to substantial markets. It was concluded that the sum total of these would be at least equivalent to that of the military market.

i   Recreational vehicles.

As mentioned earlier, recreational vehicles are expensive and the increment required to purchase a quiet, low-maintenance, low-polluting generation system may not be a significant deterrent to the owner, even if the power is relatively costly. The competition at the moment is provided mainly by diesel generators or PV modules. Both of these have a number of implicit disadvantages, including noise, low reliability, erratic insolation, or high pollution.

ii   Uninterruptible Power Supplies (UPS)

UPS, or auxiliary power supplies, constitute a rapidly expanding market as the adoption of computers for massive storage of data becomes increasingly wide-spread internationally. Many developing countries are not at all technologically backward and make significant use of digital computers. However, many of these countries do not have a reliable infrastructure and, in particular, lack reliable grid electricity. Many of them have scheduled brown- outs or black-outs on almost a daily basis, and allowance must be made for these in using electronic computers. Consequently, the NREL market survey suggested that there may be significant markets in this area for TPV electricity generators.

iii   Remote Homes

There are many hundreds of thousands of remote homes in North America and the rest of the world. Many of these are out of reach of grid electricity and are in low insolation areas. Where it is available, grid-line extensions are excessively expensive. Consequently, many such homes could use TPV electricity generators for their domestic needs.

The above examples of the type of markets that may be available to TPV generators is certainly not intended to be an exhaustive list. For example, under the category of recreation vehicles, one could also include yachts. TPV generators may be very attractive to yacht owners, whereas PV panels may be unappealing because of the unreliability of the solar radiation, and potential corrosion problems. Consequently, we believe that the real market size for well-developed TPV systems is probably placed conservatively at several hundreds of millions of dollars per year.

Principles of Operation and Potential Performance

TPV generation of electricity uses an IR source of radiation from a radiator at a temperature in the range of 1300 K to 1700 K. This is probably the feasible operating range; temperatures below the minimum of the range are likely to lead to electricity power density outputs that are too low, whereas temperatures above the top of the range may present long term issues of durability and system cooling. Two basic configurations have been envisioned: systems based on selective radiators and those based on broad-band radiators.

Selective Radiators

The rare-earth oxides radiate in a relatively narrow band of wavelengths, the peak of which is determined by their electronic structure. The rare-earth oxides (the chemical series starting with cerium and finishing with ytterbium) behave in this manner because their valence electrons are screened by outer electrons. Consequently, their atoms (or ions) are prevented from interacting with other nearby atoms. Hence, bands do not form and the absorption and emission spectra are more like those of ions in a gas than of ions in a solid. It has typically been argued that the emissivity of the materials off-band is low, while that in-band approaches unity. Consequently, the system envisioned would be based on a selective radiator with its emission band tuned to correspond to the band-edge of the semiconductor converter. Provided the short-wave and long-wave radiation received by the semiconductor converter is minimal, the only wavelengths received correspond to the near-band-edge wavelengths. These are expected to be converted very efficiently, provided that the selective radiator behaves in the nearly-ideal manner described. In practice, materials like ytterbia and erbia emit in bands corresponding to the bandgap of silicon and germanium (or GaSb or $In_{0.53}Ga_{0.47}As$), respectively. A high-temperature ytterbia radiator, at a temperature of approximately 2000 K, operating in conjunction with a silicon solar cell, is expected to function efficiently. The attraction of this is that low-cost cells are, in principle, already available. An erbia selective radiator could operate effectively at a lower temperature in conjunction with a lower bandgap semiconductor converter.

Unfortunately, the selective radiators do not behave in the ideal manner described and there is always significant off-band emissivity. For wavelengths greater than 3 μm, the emissivities of these materials rises rapidly towards unity and, if an efficient system is to be constructed, a means must be devised for reflecting the long-wavelength photons back to the radiator to conserve energy and maximize system efficiency. In the short wavelengths, the higher energy photons are absorbed by the cell and generate excess minority charge, but they have more energy than needed to do this. The additional energy is dissipated as heat in the cells. Ideally, therefore, a means of returning both the longer and shorter wavelength photons to the selective radiator is required. This may be achieved in a variety of ways that have been discussed at the two conferences on TPV organized by NREL (9, 10).

Broad-Band Radiators

This approach uses a radiator that behaves according to the Planck radiation law. The radiated spectrum extends across a wide wavelength range and, for a radiator temperature of, say, 1500 K, the peak in the radiated spectrum occurs at about 2.5 μm, or an energy of about 0.5 eV. This required a low bandgap converter, for which many options are possible. These will be discussed later. The long-wave tail of the Planck spectrum extends to at least 10 μm (at which the radiated flux is still about 12% of the maximum.) All the photons between 2.5 μm and 10 μm must be recirculated as described for the selective radiator systems. Thus, photon recirculation is common to both selective and broad-band approaches.

Potential Performance

The TPV converter modeling performed to date has been highly idealized and makes a number of sweeping assumptions that are most unrealistic. For example, most modeling has assumed that all sub-bandgap photons can be returned to the radiator, that there are no parasitic losses due to series resistance or various optical losses, and that there is no heating of the cell beyond, perhaps, 300 K. More realistic modeling needs to be performed. It has generally been assumed that the short circuit current density can be obtained simply by assuming that every photon with energy greater than that of the bandgap generates an electron/hole pair, which is collected and delivered to the external circuit. Most modeling, and certainly that considered here, has been concerned with broad-band radiators, which are either black or gray. The reverse saturation current density has been obtained using an approximation originally devised by Nell et al. (11) and subsequently used by Wanlass et al. (12). Having obtained the short-circuit current density and the reverse saturation current density, it is then possible to obtain the open circuit voltage.

The fill-factor of the cell was calculated by Wanlass et al., using an approximation originally devised by Green (13) that appears to be accurate to within 1%, down to bandgaps of approximately 0.3 eV. Consequently, the modeling has made successful use of this calculation because all the semiconductors involved have bandgaps greater than 0.3 eV.

Figure 1 shows the power density output as a function of bandgap, with the radiator temperature being treated parametrically. Radiator temperatures in the range of

1200-2000 K are considered in the model and, as can be seen, for the intermediate range of 1600 K, power density outputs of almost 6 W cm^{-2} can be expected. If one assumes that this is a factor of 2-3 times too optimistic, then we would speculate that outputs of 2-3 W cm^{-2} seems possible. These calculations assume that the converter temperature remains at 300 K. The bandgap at which the power density reaches its maximum is relatively independent of radiator temperature.

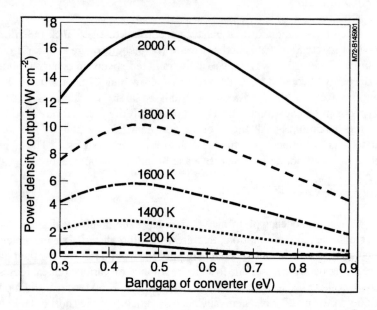

Figure 1: Variation of the modeled power density output of TPV converters as a function of bandgap: the radiator temperature being treated parametrically. The converter temperature was taken as 300 K, and the external quantum efficiency was taken as 0.95 above the bandgap of the semiconductor.

If, possibly more realistically, we allow for a significant increase in temperature of the converter, the optimum bandgaps will shift to higher values to accommodate the rapid rise in magnitude of the reverse saturation current density. Thus, at a device temperature of 350 K, we may expect the optimum bandgaps to be in the range of 0.6-0.65 eV.

First-Generation TPV Semiconductors

Figure 2 shows the well-known diagram of energy gap vs. lattice parameter for various III-V semiconducting compounds and alloys. As was indicated in Fig. 1, the optimum bandgap for most conceivable radiator temperatures is in the range 0.5-0.7 eV. By drawing an abscissa from the y axis of Fig. 2, we note that the first tie line to be intersected is that between InAs and GaAs. The individual points on the diagram represent 10% compositional increments. The point of intersection with the InAs-GaAs tie-line has a lattice parameter that does not correspond to that of any of the commonly available binary compounds. The options are to grow the device on either an InP or InAs substrate, but in both cases the ternary alloy would be mismatched to the binary compound substrate. Nevertheless, this is the approach that has been used in the last few years by groups from NREL (14), the Research Triangle Institute (15), the Spire Corporation (16), and NASA-Lewis Research Center (17).

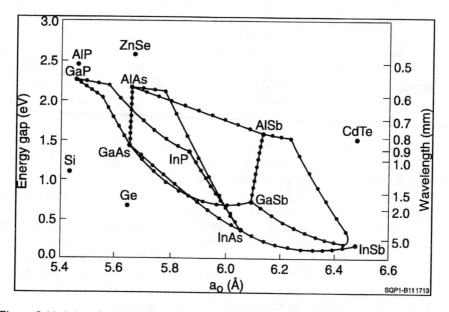

Figure 2: Variation of energy gap and lattice constant for various III-V compounds and alloys.

An alternative would be to utilize GaSb, either as a substrate material or as the active device. This approach has been used by Fraas, et al. (5) and discussed at this conference (18). To date, the latter is the only operational III-V-based converter to have been reduced to practice. A further alternative is the quaternary alloy $Ga_xIn_{1-x}As_ySb_{1-y}$. As can be seen from Fig. 2, it is possible to grow this alloy lattice-matched on either an InAs or a GaSb substrate. Much work has been performed on this system by the technical community developing mid-range IR lasers. However, to date, there does not appear to be convincing evidence that the additional complexity introduced by the fourth element is justified by the relatively minimal change in device performance beyond that of lattice-mismatched InGaAs. However, the system has an inherent advantage. If the device layers are grown on a GaSb substrate, then, in principle, the substrate is transparent to all radiation not absorbed by the active layers and a back surface reflector of long-wave photons may be realized. To date, however, this raises an additional difficulty that semi-insulating GaSb substrates are not available and free carrier absorption would be problematic. This is not the case with the lattice mismatched InGaAs on an InP substrate. Semi-insulating InP is readily available. In a paper at this meeting, Fraas et al. (18) discuss the economics of GaSb in relation to InP, and make a convincing argument for it.

Next-Generation TPV Converters

Although commercially available devices based on GaSb are the first available in the field, GaSb substrates are not ideal for more advanced devices. The present generation of devices are fabricated by diffusion into a p-type GaSb substrate to form the junction. Not only does the absence of a semi-insulating GaSb substrate make photon recirculation using a back surface reflector difficult, if not impossible, it also excludes the possibility of more advanced tandem TPV converters. In this section, we shall speculate on a few of the many possibilities available for tandem converters. None of these has yet been reduced to practice.

TPV tandem converter modeling proceeds in precisely the same way as for solar cells. The approach is as described earlier with the exception that one calculates the short circuit current densities generated by both the top and the bottom cell of the tandem pair. The smaller of these is then selected as the short circuit current density generated by the device. Implicitly, therefore, it is assumed that the two sub-cells are connected in series. Following this, the steps proceed as for a single junction device. The individual open-circuit voltages and fill-factors are calculated; the final step is to add the power densities generated by the individual cells. These are added to give the total power density output of the tandem structure. The modeled performance necessarily assumes ideal behavior of both devices, an approximation that is almost certainly untrue in practice.

Normally, one represents the performance of tandem cells with a diagram showing contours of equal power density output (W cm^{-2}) with the vertical and horizontal axes being the bandgaps of the top and bottom cells, respectively. At NREL, a code has been written in Mathematica©, into which the bandgaps of the top and bottom cell, the temperature of the radiant source, and the temperature of the device may be input as parameters. Fig. 3 shows the output of such a calculation for a radiator at a temperature of 1500 K. As with single junction cell modeling, the procedures are excessively idealized and need to be refined because of the gross assumptions made. They assume, for example, that all long-wave photons with an energy less than that of the lower cell bandgap are returned to the radiator, that both cells have a "box-car" quantum efficiency of 90%, and that there are no parasitic losses. Bearing this in mind, we see from Fig. 3 that the maximum power density outputs predicted for a tandem pair are in excess of 5 W cm^{-2}. This would occur for a top cell with a bandgap 0.54-0.58 eV and a bottom cell with a bandgap of 0.36-0.42 eV. If one allows for realistic factors such as cell heating, parasitic losses, optical and electrical parasitic losses, then these values change significantly.

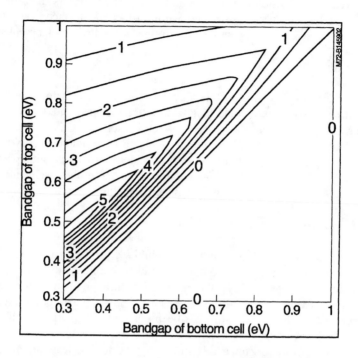

Figure 3: Modeled power density output of tandem cells. The numbers in the body of the figure are measured in Watts cm^{-2}. The model assumed a radiator with an emissivity of unity at a temperature of 1500 K, a uniform external quantum efficiency of 95%, and a converter temperature of 300 K.

Note that a single-junction cell with an optimized bandgap of ~0.55 eV would generate a power density output of 3-4 W cm^{-2}, so the added complexity of going to a tandem converter would yield approximately 25% more power. In addition, the use of a lower bandgap cell beneath a top cell of about 0.53 eV, would decrease the current density by a factor of two and would increase the voltage. The reduction of current density would lead to a significant reduction in I^{2}R losses which, in turn, would permit fewer grid fingers to be used and a reduction in shadowing losses.

In$_x$Ga$_{1-x}$As

Tandem devices could be constructed from lattice-mismatched InGaAs grown on InP, but the problem is that both cells would be under biaxial compression. The bandgap of the bottom cell would be close to that of InAs, and the possibility arises, perhaps, of growing the tandem structure with the bottom cell of the binary compound and only the upper InGaAs cell in compression. In principle, the compressive stress in the upper cell should be significantly less than for devices grown on an InP substrate. However, we are, once again, faced with the near impossibility of avoiding free carrier absorption in the low bandgap substrate.

$Ga_xIn_{1-x}As_ySb_{1-y}$

The quaternary alloy, GaInAsSb, is an attractive proposition for a tandem cell because the entire structure could be grown lattice-matched on a GaSb substrate. However, bearing in mind the absence of semi-insulating GaSb substrates, free carrier absorption in the substrate seems likely to be problematic. Note that it should also be possible to fabricate entire tandem structures with passivating layers of a higher bandgap. This could consist of InAsP in the case of the InGaAs alloys or GaAlAsSb in the case of the quaternary alloys.

Ordered Ga_2AsSb

This alloy is one of the III-V family that exhibits crystallographic ordering. As discussed by Horner et al. (19), ordering could be exploited to make a multiple junction device using precisely the same alloy throughout the structure. An ordered alloy exhibits very different electrical and optical properties to the disordered alloys. Extensive work has been performed by Gomyo et al. (20) on the alloy, but very little work has been performed on Ga_2AsSb. However, calculations by Wei and Zunger (21) indicate that the bandgap suppression that occurs when the arsenic antimony are ordered on alternating (111) planes of the anion sublattice is much larger than with Ga_2InP. They calculated that the fully-ordered alloy would have a bandgap of 0.38 eV, whereas the fully disordered alloy should have a bandgap of 0.83 eV. Horner et al. argued that it may be possible to grow ordered Ga_2AsSb lattice-matched on a semi-insulating InP substrate by controlling the growth parameters. Following this, a less-ordered alloy could be grown with a bandgap chosen to be appropriate to that of the radiant surface.

Assuming, more realistically, that the alloy could be grown with 80% ordering, the bandgap of the bottom cell may be reduced to, perhaps, 0.5 eV with reasonable expectation. This would mean that for a radiant temperature of 1500 K, the bandgap of the top cell required to achieve current matching would be approximately 0.64 eV. This line of reasoning can be extended to a triple-junction cell, in which the bandgaps of both the middle and top cells are dictated by the temperature of the radiant surface. In all cases, 80% ordering of the bottom cell was assumed giving a bandgap of 0.5 eV.

Although this is an appealing prospect because the alloy composition remains constant throughout the structure (only the degree of ordering is changed), the power density outputs predicted are disappointing and certainly not worth the added complexity. The reason for this is the bottom cell bandgap of 0.5 eV is too large; nearly 100% is required for this approach to be potentially useful. Consequently, of the tandem structures hypothesized so far, this is by far the most speculative and dependent on major research breakthroughs.

Thin-Film Converters

Although some calculations on the economics of TPV conversion have been performed by Fraas et al. (18) showing that the cost of GaSb substrates is likely only to be a minimal fraction of the total system costs, no extensive experience is available from which to draw firm conclusions. In DOE's Solar Cell program, approaches based on high efficiency converters using single crystal III-V alloys, as well as thin-film converters of lower cost, have been developed. For TPV converters, only production experience will confirm whether Fraas' economic calculations are realistic. Consequently, there does not seem to be any a priori reason to neglect low-cost thin film compounds and alloys.

A significant amount of work was done in the early 1980s on polycrystalline III-V solar cells (see, for example, Yamaguchi and Itoh, [22]) but the grain boundaries were generally found to be problematic. Further consideration of polycrystalline GaAs, and the issue of grain boundaries, was presented at this conference (23). These materials were made by inherently low-cost growth processes, which may be amenable to production scale-up. The II-VI family of compounds and alloys may prove more attractive as thin-film converters, because of the possibly less-damaging effect of grain boundaries. Fig. 4 shows the diagram of lattice parameter vs. bandgap for the II-VI compounds and alloys and this immediately raises a number of attractive possibilities, none of which, to the knowledge of the author, is presently being investigated. The abscissa at 0.55 eV first crosses the tie-line between PbS and ZnS. The former can be doped both n- and p-type and has been extensively investigated for photoconducting applications. Consequently, it may prove profitable to investigate adding a small amount of zinc to the PbS to increase the bandgap.

Continuing the abscissa horizontally shows that there are several more alloys that may be worthy of investigation. These would also include the material HgCdTe, which has been extensively investigated for IR detectors. The disadvantage with this material is the relatively high vapor pressure of the mercury, which may damage long-term reliability.

Consequently, there are various II-VI compounds and alloys worthy of investigation that could be manufactured relatively cheaply. Grain boundaries in solar cells usually have an injurious effect, unless the device is operated under optical concentration. Wanlass et al. (24) demonstrated that heavily defected InP grown on a GaAs substrate exhibited high efficiencies when operated under optical concentration.

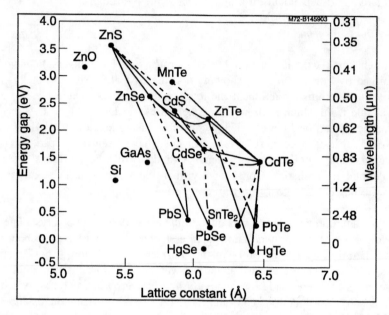

Figure 4: Variation of the bandgap and lattice constant for various II-VI compounds and alloys.

It was shown that the efficiency of the device increased more rapidly than for a homoepitaxially-grown device for concentrations of up to 10 suns, after which the rate of increase of efficiency was similar for both devices (although the absolute value was less than that of the homoepitaxial device). The main point is that for a polycrystalline device, which in some ways resembles a defected heteroepitaxial device, perhaps the same phenomena occurs. If this is the case, polycrystalline II-VI semiconductors may operate well under the intense fluxes prevalent in TPV systems.

Conclusions

In this paper we have briefly reviewed the potential promise of TPV and indicated the range of materials currently under investigation. The paper has concentrated on the semiconductor converters themselves rather than the other components in a TPV system, and it cannot be regarded as a complete review of TPV. However, we have indicated that there are several options available for first-generation TPV devices that may be perfectly adequate to enable systems to be manufactured. We have also speculated on next-generation TPV devices based on both tandem structures and polycrystalline II-VI semiconductors. So far as we are aware, no work is presently being done on these systems, and there are many opportunities for original research.

Many fundamental questions remain unanswered in TPV, which make realistic modeling of the converters difficult, if not impossible. These shortcomings are increased in number when one examines the behavior of complete systems. For example, the equilibrium flux distribution in the optical cavity, when photon recirculation is included, is presently unknown. We know nothing about the fundamental recombination processes in these low bandgap materials and, although it has been speculated that Auger processes may be important, to the knowledge of the author this has not been confirmed. There is also little information about the effect of imperfections, such as dislocations or grain boundaries at very high flux densities. In the case of mismatched devices, it is almost certain that the strain has an effect on the band structure of the semiconductor. Consequently, the effective mass, the intrinsic carrier concentration, and the reverse saturation current density may change in an unknown manner.

These are but a few of the many fundamental questions pertaining to the semiconductor converter that have not been answered. There are many more questions when one considers complete systems. For the future of TPV, perhaps the most important question of all is "where is the required funding going to come from"? At present, no single entity is funding all of the subtopics within TPV that warrant funding, and it is possible that the subject is being hindered by this absence.

References

1. Aigrain, P., "Thermophotovoltaic Conversion of Radiant Energy; Unpublished lecture series at MIT," (1956).

2. Kttl, E., "Thermophotovoltaic Energy Conversion," presented at the 20th Power Sources Conference, Red Bank, New Jersey, (PSC Publications and Communications, Red Bank, New Jersey), (1966), pp. 178-182.

3. Shapiro, S.J., "Thermophotovoltaic Spectral Analysis of a TPV System," presented at the 21st Annual Power Sources Conference, Red Bank, New Jersey, (PSC Publication, Red Bank, New Jersey), edited by Daniel, A. F., (1967), pp. 138-142.

4. Guazzoni, G.E., "High temperature spectral emittance of oxides of erbiium, samarium, neodymium and ytterbium.," Applied Spectroscopy, **26** (1), 60-65 (1972).

5. Fraas, L., Ballantyne, R., Samaras, J., and Seal, M., "Electric Power Production using New GaSb Photovoltaic Cells with Extended Infrared Response," presented at the First NREL Conference on Thermophotovoltaic Generation of Electricity, Copper Mountain, Colorado, (American Institute of Physics, New York), edited by Coutts, T. J. and Benner, J. P., (1995), pp. 44-53.

6. Broman, L., and Marks, J., "Co-Generation of Electricity and Heat from Combustion of Wood Powder Utilizing Thermophotovoltaic Conversion," presented at the The First NREL Conference on Thermophotovoltaic Generation of Electricity, Copper Mountain, Colorado, (American Institute of Physics, New York), edited by Coutts, T.J. and Benner, J.P., (1994), pp. 133-138.

7. Johnson, S., Final Report No. RAK-5-15377, National Renewable Energy Laboratory, Golden, Colorado 80401, (January 7, 1996).

8. Rose, M.F., *Prospector VIII- Workshop on Thermophotovoltaics for the Army*, Vol. **1**, (Auburn University, New Orleans, Louisiana), (1996).

9. Coutts, T.J., and Benner, J.P., *Proceedings of the First NREL Conference on Thermophotovoltaic Generation of Electricity*, AIP Conference Series, Vol. **321**, (American Institute of Physics, New York), (1994).

10. Benner, J.P., Coutts, T.J., and Ginley, D.S., *Proceedings of the Second NREL Conference on Thermophotovoltaic Generation of Electricity*, AIP Conference Series, Vol. **358**, (American Institute of Physics, New York), (1995).

11. Nell, M.E., and Barnett, A.M., "The Spectral p-n Junction Model for Tandem Solar-Cell Design," IEEE Transactions in Electron Devices, **ED-34** (2), 257-266 (1987).

12. Wanlass, M.W., Coutts, T.J., Ward, J.S., Emery, K.A., Gessert, T.A., and Osterwald, C.O., "Advanced High-Efficiency Concentrator Tandem Solar Cells," presented at the Twenty Second IEEE Photovoltaic Specialists Conference, Las Vegas, Nevada, (IEEE, Piscataway, New Jersey), (1991), pp. 38-45.

13. Green, M.E., *Solar Cells: Operating Principles, Technology, and Systems Applications*, (University of New South Wales, Kensington, New South Wales, Australia), (1986).

14. Ward, J.S., Wanlass, M.W., Wu, X., and Coutts, T J., "Issues in the Design of High-Performance Contacts to GaInAs TPV Converters," presented at the First NREL Conference on Thermophotovoltaic Generation of Electricity, Copper Mountain, Colorado, (American Institute of Physics, New York), edited by Coutts, T. J. and Benner, J. P., (1994), pp. 404-411.

15. Sharps, P.R., Timmons, M.L., Venkatasubramian, R., Hills, J.S., O'Quinn, B.O., Hutchby, J.A., Iles, P.A., and Chu, C.L., "Thermal Photovoltaic Cells," presented at the First NREL Conference on Thermophotovoltaic Generation of Electricity, Copper Mountain, Colorado, (American Institute of Physics, New York), edited by Coutts, T.J. and Benner, J.P., (1994), pp. 194-201.

16. Wojtczuk, S., Gagnon, E., Geoffroy, L, and Parodos, T., "In$_x$Ga$_{1-x}$As Thermophotovoltaic Cell Performance vs. Bandgap," presented at the First NREL Conference in Thermophotovoltaic Generation of Electricity, Copper Mountain, Colorado, (American Institute of Physics, New York), edited by Coutts, T. J. and Benner, J. P., (1994), pp. 177-187.

17. Wilt, D. M., Fatemi, N. S., Hoffman, R. W., Jenkins, P. P., Scheiman, D., Lowe, R., and Landis, G. A., "InGaAs PV Device Development for TPV Power Systems," presented at the First NREL Conference on Thermophotovoltaic Generation of Electricity, Copper Mountain, Colorado, (American Institute of Physics, New York), edited by Coutts, T. J. and Benner, J. P., (1994), pp. 210-220.

18. Fraas, L., "Low Cost Processing III-V Processing," presented at the First NREL Conference on Future Generation Photovoltaic Technologies, Denver, Colorado, (American Institute of Physics, New York), edited by McConnell, R. D., (1997), to be published.

19. Horner, G., Coutts, T. J., and Wanlass, M. W., "Novel TPV Converters Based on Ga$_2$AsSb," presented at the First NREL Conference on Thermophotovoltaic Generation of Electricity, Copper Mountain, Colorado, (American Institute of Physics, New York), edited by Coutts, T. J. and Benner, J. P., (1994), pp. 145-151.

20. Gomyo, A., Suzuki, T., Kobayashi, K., Kawata, S., Hino, I., and Yuasa, T., "Evidence for the Existence of An Ordered State in Ga$_{0.5}$In$_{0.5}$P Grown by Metalorganic Vapor Phase Epitaxy and its Relation to Band-Gap Energy," Applied Physics Letters, **50** (11), 673-675 (1987).

21. Wei, S.-H., and Zunger, A., "Band-Gap Narrowing in Ordered and Disordered Semiconductor Alloys," Applied Physics Letters, **56** (7), 662-664 (1990).

22. Yamaguchi, M., and Itoh, Y., "Efficiency Considerations for Polycrystalline GaAs Thin-Film Solar Cells," Journal of Applied Physics, **60** (2), 413-417 (1986).

23. Kurtz, S. R., and McConnell, R., "Requirements for a 20%-Efficient Polycrystalline GaAs Solar Cell," presented at the First NREL Conference on Future Generation Photovoltaic Technologies, Denver, Colorado, (American Institute of Physics, New York), edited by McConnell, R., (1997), to be published.

24. Wanlass, M. W., Coutts, T. J., Ward, J. S., and Emery, K. A., "High-Efficiency Heteroepitaxial InP Solar Cells," presented at the Twenty-Second IEEE Photovoltaic Specialists Conference, Las Vegas, Nevada, (IEEE), (1991), pp. 159-165.

233

Photovoltaic Cells Based on Vacuum Deposited Molecular Organic Thin Films

V. Bulović[*], P.E. Burrows[*], D.Z. Garbuzov[†], M.E. Thompson[‡]
A.G. Tsekoun[†], and S.R. Forrest[*]

[*]Department of Electrical Engineering, Center for Photonics and Optoelectronic Materials
Princeton University, Princeton, NJ 08544

[†]Sarnoff Corporation, Princeton, NJ 08542

[‡]Department of Chemistry, University of Southern California, Los Angeles, CA 90089

Abstract. We present an overview of the metal/organic - Schottky-type, and organic/organic heterojunction photovoltaic (PV) cells grown by vacuum deposition of molecular organic thin films. Recent developments leading to multilayer organic PVs are highlighted. The use of organic phosphors for enhancement of the UV response of conventional PV technologies is also discussed.

The need to develop renewable energy sources has stimulated new approaches for production of efficient low-cost photovoltaic cells. Photovoltaics (PVs) based on organic materials such as molecular organics, conjugated polymers, and liquid crystals are emerging as an alternative technology [1] to more conventional approaches based on inorganic semiconductors such as silicon and gallium arsenide. Photosensitivity of organic molecules together with their high optical absorption coefficient and compatibility with vacuum deposition, gives promise for realizing large area, thin-film PV cells that can be produced at a modest cost. Furthermore, the demonstrated ability to deposit organic materials on flexible or shaped substrates may eventually lead to the development of lightweight and conformable PVs. At present, organic thin film PVs are not commercially competitive due to their low power conversion efficiencies. However, the immense variability in the molecular composition and layer structure as well as advances in vacuum growth techniques assure the continued improvement in the development of PV cells based on molecular organics.

Crystalline organic semiconductors such as the phthalocyanines (Pc's), perylenes and other relatively low atomic weight polyacenes have been long recognized as having the greatest potential for use in solar cell structures due to their ability to be controllably deposited via conventional vacuum techniques, their relatively high purity, and high mobilities (hence relatively low series

CP404, *Future Generation Photovoltaic Technologies: First NREL Conference*, edited by McConnell
© 1997 The American Institute of Physics 1-56396-704-9/97/$10.00

resistance) [2]. However, in spite of considerable research having been pursued in investigating both organic p-n junction and Schottky barrier crystalline organic photovoltaic (PV) cells over the past 20 years, there has yet to be a demonstration of such a cell whose characteristics are adequate for even the most undemanding of solar conversion applications [1]. Their poor performance can be ascribed principally to the following causes: (i) Due to the intrinsic nature of photoconductivity in crystalline organic materials, where free electron-hole pairs are generated in a *second order* process following absorption and exciton generation, the efficiencies realized to date are very low (typically <1% to AM0 - AM2 illumination). (ii) Due to the limited π-orbital overlap between adjacent molecules in an organic crystalline stack, and due to the presence of numerous defects in the crystalline order in the deposited materials, the free carrier mobilities in many organics are very low (typically $<10^{-3}$ cm^2/V-s). This leads to high film resistance, and hence low power conversion efficiency. (iii) The exciton diffusion length (L_D) in organic thin films is typically considerably shorter than the optical absorption length, $1/\alpha$, where α is the absorption coefficient. Hence, the exciton is often photogenerated far from an interface where dissociation into free carriers can occur, leading to exciton recombination and a concomitant loss in quantum yield. (iv) A final consideration is one of environmental stability of the organic thin film materials, and in particular the stability of the intramolecular bonds to prolonged exposure to ultraviolet radiation.

While these problems still prevent the realization of fully organic, low cost PV cells, significant advances in the growth and processing technology of organic materials worldwide suggest that these devices may eventually find practical application for low cost solar energy conversion. Indeed, recent experiments indicate that there are novel and effective means for growing nearly perfect, multiple quantum well crystalline structures which can significantly increase the quantum efficiency and carrier lifetime in organic materials. Extending the photoactive region width by inserting multiple junctions, with a combined width $\sim 1/\alpha$, has been explored recently due to the uniformity control inherent in organic molecular beam deposition growth technique [3]. Furthermore, materials which are grown with a high degree of stacking order have enhanced π-orbital overlap, which results in increased carrier mobility and concomitant decrease in bulk layer resistance.

To date, two basic thin film molecular organic solar cell structures have been demonstrated: the metal/organic Schottky-type cell, and the organic/organic heterojunction bilayer cell employing two ohmic metal contacts. The Schottky cell consists of a light absorbing organic sandwiched between two metal contacts, at least one of which is rectifying. The rectifying contact typically occurs between a p-conducting organic material and a low work function metal such as Al or Mg, or n-conducting films and a high work function metal such as Au. Photogeneration occurs near the rectifying contact where the photogenerated

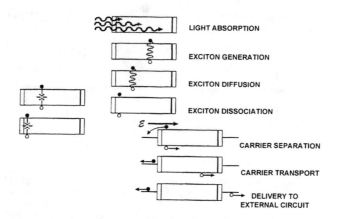

FIGURE 1. Photocurrent generation process in a Schottky-type organic solar cell

charge is separated by the contact built-in field. The photogeneration process is illustrated in Fig. 1. The absorbed solar energy first results in the highly efficient generation of excitons in the film bulk [4] (described by the process $S_o + h\nu \rightarrow S_o^*$, where S_o and S_o^* denote the molecular ground and excited states, respectively). Due to the concentration gradient, these excitons diffuse to a contact, impurity, interface, or other inhomogenity within the structure (denoted M), at which point they dissociate into free carrier pairs (via $S_o^* + M \rightarrow e + h$), or alternatively suffer recombination. Although the ionization process is not fully understood [5], it is often attributed to exciton dissociation in the presence of the built-in electric field surrounding the crystal defect or an interface. If this dissociation occurs in an otherwise neutral region of the film, the free electron-hole pairs generated from exciton ionization are still localized in vicinity of the defect where they recombine without being collected in the external circuit. Hence, to significantly increase η two conditions must be met: (i) the film must be relatively free of random defects which generate local electric fields, and (ii) the exciton and free carrier diffusion lengths must be sufficiently long such that these particles can migrate to regions of the film where the built-in electric field from an adjacent junction is high enough to separate the free electron-hole pairs prior to recombination.

In contrast, in the organic/organic heterojunction cell, photogeneration occurs at the organic/organic interface. In this case, absorption can occur in either of the two organic films, doubling the width of the photoactive region. Exciton dissociation at the organic/organic interface generates carriers that are separated by the two contacting organic materials, with electrons in the material with a higher electron affinity and holes in material with a lower ionization potential [6]. The built-in potential at this junction, determined by the size of the HOMO-LUMO gap energy offset [7], can aid this process. Charge separation is followed

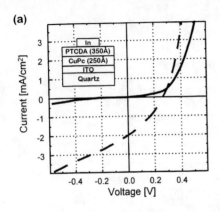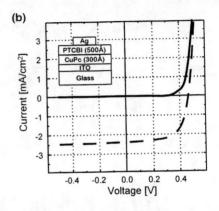

FIGURE 2. (a) PTCDA/CuPc heterojunction solar cell with V_{OC} = 260mV, I_{SC} = 2.0mA/cm^2, ff = 0.35, η = 1.80%, (b) CuPc/PTCBI solar cell with V_{OC} = 450mV, I_{SC} = 2.3 mA/cm^2, ff = 0.65, η = 0.95%

by carrier transport to the contacts and delivery of power to an external load. Each step of this photogeneration process can have low probability, contributing to low power conversion efficiencies of organic PVs. To date, the highest power conversion efficiency of η = 1.8% has been reported for a 3,4,9,10-perylinetetracarboxylic dianhydride/copper phthalocyanine (PTCDA/CuPc) heterojunction cell [7] under an illumination of 10mW/cm^2. Here the power conversion efficiency is defined as η = ff V_{OC} I_{SC} / P_{INC}, where ff is the forward-biased fill factor, V_{OC} is the open circuit voltage, I_{SC} is the short circuit current, and P_{INC} is the incident optical power. At the higher incident power of 75 mW/cm^2 (equivalent to AM2 illumination), a maximum power conversion efficiency η = 0.95% was reported for a CuPc/3,4,9,10-perylenetetracarboxylic bis-benzimidazole (CuPc/PTCBI) bilayer cell [6]. One problem with organic-based PVs is their high contact and series resistance which leads to space-charge build-up with increasing incident power. Indeed, it was shown [8] that there is a roughly two-fold decrease in efficiency resulting from space-charge effects as the power density increases from 10 to 75 mW/cm^2, suggesting that the result for the CuPc/PTCBI and PTCDA/CuPc cells are nominally equal. The I-V characteristics of these p-P heterojunction cells [7], as well as their schematic cross sections are shown in Fig. 2. Of particular note is the low series resistance of the latter structure, as indicated by the abrupt increase in current at V > V_{OC}. This leads to the high fill factor, although the author observes that the ff characteristics degrade after prolonged exposure of the cell to air [6].

It has been suggested that increased order within the film also results in an increased excitonic and carrier diffusion length, which ultimately should produce a higher carrier mobility along the stacks. These effects have been extensively

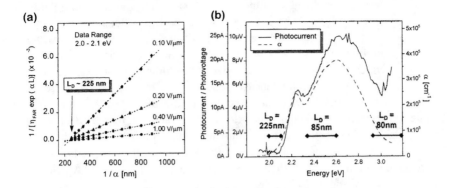

FIGURE 3. (a) Liner fit to inverse quantum yield $1/[\eta_{FAR}\exp(\alpha d)]$ at different applied fields versus inverse absorption coefficient $(1/\alpha)$ for E = 1.99 to 2.10 eV region yields diffusion lengths of $L_D = 225 \pm 15$ nm. (b) Similar fits for E = 2.36 to 2.60 eV and E = 2.92 to 3.20 eV range yield $L_D = 88 \pm 6$ nm and $L_D = 79 \pm 7$ nm, respectively.

studied [9,10] in the model compound, PTCDA, which is well-known to have extremely close π-π stacking with long-range order. Assuming that all exciton dissociation (and subsequent free carrier generation) occurs at contacts in a metal/organic/metal cell, Ghosh and Feng [11] have shown that the photocurrent density is governed by exciton diffusion, $J \propto -D_{EX}\, dn(x)/dx$. Here, D_{EX} is the exciton diffusion constant, while $n(x)$ is the exciton density as a function of distance with respect to the contact on which the light is incident (i.e. the "near" contact), and is proportional to the optical flux $\phi(x)$, given by $\phi(x) = \phi_o \exp(-\alpha x)$ (ϕ_o is the flux at the near contact). From this analysis, it can be shown that the quantum yield due to exciton generation primarily at the near and far interfaces depends only on $\alpha(\lambda)$ and $L_D(\lambda)$. Plots of $1/\eta_{NEAR}$ or $1/[\eta_{FAR}\exp(\alpha d)]$ (film thickness, d) as a function of $1/\alpha$, yield a straight line with an intercept on the $1/\alpha$ axis providing the exciton diffusion length. Results of this analysis for the low energy absorption tail of PTCDA films grown by OMBD [10] are shown in Fig. 3a. A summary of exciton diffusion lengths in different energy ranges (Fig. 3b) reveals that at short wavelengths where Frenkel excitons are generated [12], the diffusion length is found to be $L_D = 88 \pm 6$ nm, consistent with independent studies of Karl, et al. [9]. However, the diffusion length of the CT exciton [12] at low energies (~2.0 eV) is $L_D = 225 \pm 15$ nm which to our knowledge is the longest published diffusion length for an exciton in an organic thin film. This clearly suggests the need for extended stacking order to achieve long diffusion lengths, which in turn contribute to increased quantum yields.

Material purity also strongly affects the organic PV quantum efficiency, as recently reported in a detailed study of DM-PTCDI/H$_2$Pc heterojunction cells and

PVs based on similar phthalocyanine, perylene, and porphyrinic compounds [8]. For example, for ZnPc/DM-PTCDI cells the power efficiency for sublimed materials is $\eta = 0.29\%$ and decreases by the factor of four (to $\eta = 0.07\%$) for materials which undergo solvent washing. Impurities which act as ionic trapping centers, or which disrupt the stacking order of the films ultimately increase series resistance, cause charging (and hence space-charge build-up) and decrease the mobility of the charge carriers and excitons in the film. Furthermore, trapped charge can shift the Fermi-energy positions within the film, thereby decreasing (or increasing) the built-in voltage, ultimately determining the width of the photoactive region. The implication of the above results is that the continued improvement in η is linked to improvements in material quality which govern exciton diffusion lengths and carrier mobilities.

Recent demonstrations of organic PV cells using various organic thin films in both Schottky and organic/organic heterojunction geometries have typically yielded $\eta < 1\%$. It is, therefore, apparent that relying just on the improvement in material quality is not likely to generate PVs that perform at levels which will make them attractive for use in practical solar energy conversion applications. Rather, improved materials should be used in conjunction with innovate device concepts. For example, increasing the built-in potential (and hence increasing V_{OC}) has been attempted using several different approaches, the most notable being the insertion of additional organic layers [13], or even fabricating series-connected "tandem" PV cells in a single stack [14]. These methods have proven successful in increasing V_{OC} which extends the effective width of the photoactive region, although they generally also result in an increase in series resistance (with its concomitant decrease in ff), thereby limiting the overall power efficiency of PVs to values somewhat less than those reported for the best, single junction cells.

A more promising concept is of an organic multilayer stack optimized for the overlap between the optical, excitonic and carrier distributions. The increased number of interfaces provides extra exciton dissociation sites which in turn result in photocurrent increase. This was observed for VOPc/PTCDA multilayer structures where photocurrent yield increased eight fold as compared to a single heterostructure cell [3]. The benefit of multiple interfaces, however, is diminished by an increase in the series resistance of the stack. Optimizing the tradeoff between these two effects will provide the most likely means for achieving the long sought-after goal of low cost power generation via the use of efficient thin film molecular organic PV cells.

Another promising use of organic thin films in PVs is as downconverting phosphors layered on the surface of conventional solar cells. It is well known that the efficiency of conventional Si photodiodes and solar cells decreases rapidly in the near-UV range, while numerous organic thin films exhibit a strong UV absorption followed by energy downconversion and re-radiation. The re-radiated

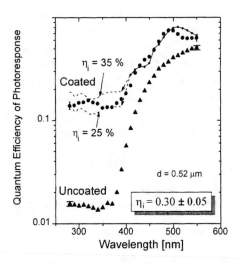

FIGURE 4. Photoresponse quantum efficiency of an uncoated silicon photodiode (triangles), and the same diode coated with 520 nm of Alq$_3$ (circles). Calculated internal luminescence efficiency is 30 ± 5 %.

light can be more readily detected by the Si solar cell than the original UV radiation, thus, a down-converting organic thin film on top of a conventional solar cell effectively increases the solar cell efficiency in the UV range. The downconversion is due to the Frank-Condon shift in organic molecules which red-shifts the luminescence with respect to absorption, rendering the organic films transparent to their own radiation. Organic thin films transparent in the visible can then also serve the secondary function of an antireflection coating.

Experimental and theoretical studies of organic fluorescent collectors for Si solar cells and organic coated UV-enhanced Si photodetectors have been pursued for almost twenty years [15,16,17,18]. In the initial studies thick (2-5 µm) layers of laser dye molecules in sol-gel or in plastic matrices were used, and wavelength down conversion (from 400 nm to 510 nm) with efficiency exceeding 90% was demonstrated [16]. Later, deposition of "monocompositional" solid films of coronene [17] and metachrome [18] on CCD sensor surfaces resulted in the commercial production of UV-enhanced cameras. It appears, however, that further studies of successful utilization of such wavelength down-converters combined with Si photodiodes are limited to recent work utilizing aluminum tris-(8-hydroxy quinoline) (Alq$_3$) [19,20], material used in numerous organic light emitting device structures. Alq$_3$ has strong (>10^5 cm^{-1}) absorption for λ < 370 nm, photoluminescence peak at λ = 530 nm (where silicon PV cell efficiency is very high), and has a thin film photoluminescence efficiency of 30%. An Alq$_3$ film deposited directly on top of a conventional Si solar cell increases the

UV collection by an order of magnitude (Fig. 4). A similar organic coating with a 100% PL conversion efficiency would increase the overall solar cell efficiency by as much as 11% under AM0 illumination. Hence, it is probable that new organic materials will lead to the realization of UV-enhanced Si photodiodes and solar cell energy converters with efficiencies high enough for commercial applications.

We would like to thank NREL, AFOSR, NSF MRSEC program and Universal Display Corp. for their generous support of this research.

REFERENCES

1. Wöhrle, D., and Meissner, D., *Adv. Mater. 3, 129 (1991)*.
2. References in Table I are examples of vacuum deposited thin film solar cells.
3. Arbour, C., Armstrong, N.R., Brina, R., Collins, G., Danziger, J., Dodelet, J.-P., Lee, P., Nebesny, K.W., Pankow, J., and Waite, S., *Mol. Cryst. Liq. Cryst., 183, 307 (1990)*.
4. Chaiken, R.F., Kearns, D.R., *J. Chem. Phys, 45, 3966 (1966)*.
5. Bounds, P.J., Siebrand, W. *Chem. Phys. Lett, 75, 414 (1980)*.
6. Tang, C.W., *Appl. Phys. Lett. 48, 183 (1986)*.
7. Forrest, S.R., Leu, L.Y., So, F.F., Yoon, W.Y., *J. Appl. Phys. 66, 5908 (1989)*.
8. Wohrle, D, Kreienhoop, L., Schnurpfeil, G., Elbe, J., Tennigkeit, B., Hiller, S.,Schlettwein, D., *J. Mater. Chem., 5, 1819 (1995)*.
9. Karl, N., Baner, A., Holzäpfel, J., Marktanner, J., Möbus, M., Stölzle, F., *Mol. Cryst. Liq. Cryst., 252, 243 (1994)*.
10. Bulović, V., and Forrest, S. R., Che*m. Phys. 210, 13 (1996)*.
11. Ghosh, A.K., and Feng, T.J., *J. Appl. Phys. 49, 5982 (1978)*.
12. Bulović, V., Burrows, P. E., Forrest, S. R., Cronin, J. A., and Thompson, M. E., *Chem. Phys. 210, 1(1996)*.
13. Hiramoto, M., Fukusumi, H., Yokoyama, M., *Appl. Phys. Lett. 61, 2580 (1992)*.
14. Hiramoto, M., Suezaki, M., Yokoyama, M., *Chem. Lett. 327 (1990)*.
15. Goetzberger, A., and Greubel, W., *Appl. Phys. 14, 123 (1977)*.
16. Viehmann, W., *SPIE Measurements of Optical Radiation, 196, 90 (1979)*.
17. Blouke, M.M., Cowens, M.W., Hall, J.E., Westphal, J.A., and Christensen, A.B., *Appl. Opt. 19, 3318 (1980)*.
18. Sims, G., Griffin, F., and Lesser, M., *SPIE Optical Sensors and Electronic Photography, 1071 (1989)*.
19. Garbuzov, D.Z., Bulović, V., Burrows, P.E., and Forrest, S.R., *Chem. Phys. Lett., 249, 433, (1996)*.
20. Garbuzov, D.Z, Forrest, S.R., Tsekoun, A.G., Burrows, P.E., Bulović, V., and Thompson, M.E., *J. Appl. Phys. 80 , 4644 (1996)*.

Thin-Film Filament-Based Solar Cells and Modules

J.R. Tuttle, E.D. Cole, *T.A. Berens, *J. Alleman, and *J. Keane

DayStar Technologies, 789 Clarkson St. Suite 1105, Denver, CO 80218
National Renewable Energy Laboratory, 1617 Cole Blvd., Golden CO 80401 USA
(ph) 303-837-9657, (fax) 303-837-8190, e-mail: DayStarTec@aol.com

Abstract. This concept paper describes a patented, novel photovoltaic (PV) technology that is capable of achieving near-term commercialization and profitability based upon design features that maximize product performance while minimizing initial and future manufacturing costs. DayStar Technologies plans to exploit these features and introduce a product to the market based upon these differential positions. The technology combines the demonstrated performance and reliability of existing thin-film PV product with a cell and module geometry that cuts material usage by a factor of 5, and enhances performance and manufacturability relative to standard flat-plate designs. The target product introduction price is $1.50/Watt-peak (Wp). This is approximately one-half the cost of the presently available PV product. Additional features include: increased efficiency through low-level concentration, no scribe or grid loss, simple series interconnect, high voltage, light weight, high-throughput manufacturing, large area immediate demonstration, flexibility, modularity.

INTRODUCTION

The deployment of PV technologies continues to grow annually with 1996 module sales of 91 MW (≈$360M), a 13% growth with respect to 1995 sales. Within that sales volume, 82.9% was product based upon single and poly-crystalline Silicon (c-Si), 15.8% upon amorphous Silicon (a-Si), and only 1.3% upon polycrystalline thin-film (PTF) technologies [1]. Even with this high annual growth, module and system prices remain levelized at $4 and $8 per Watt-peak range, respectively, representing either a stabilization of manufacturing costs, or increases in margin. Photovoltaic Insiders Report [2] recently reported on a projected 28.5 MW of production capacity for all thin-film technologies within the next two years, compared to a total of 1.2 MW in 1995. This new infusion of thin-films into the overall mix marks the initiation of a new era in PV. The degree to which this new era is explosive in growth will depend on the degree to which there are continued technology advances in both the performance and cost arenas.

CP404, *Future Generation Photovoltaic Technologies: First NREL Conference*, edited by McConnell
© 1997 The American Institute of Physics 1-56396-704-9/97/$10.00

Solar cells and modules fabricated from polycrystalline $Cu(In,Ga)Se_2$ (CIGS)-based thin films are strong candidates for high performance and low cost [3]. The high performance criterion is satisfied by total-area conversion efficiencies near 18% [4] for the CIGS technology. The low-cost criterion is satisfied for most thin-film technologies through low materials usage, monolithic integration, and low manufacturing costs. Unfortunately, the latter has not been completely realized due to cost-performance tradeoffs and the high capitalization costs associated with pre-economies of scale/volume (EOSV) production levels. Secondly, significant development effort remains to address basic issues such as large-area film deposition and product yield in a manufacturing environment.

In this contribution, we describe a new, thin-film based PV technology developed by DayStar Technologies that realizes low manufacturing and initial capitalization costs while maintaining the high performance of laboratory processes. It is also uniquely designed to incorporate existing thin-film based PV product fabricated on compatible substrate materials. DayStar Technologies (formerly CoGen Solar) was created in March 1996 and began development work on this new PV technology in Sept. 1996. Unlike monolithic flat-plate designs where there exists a 2-4% absolute difference in performance between localized cell and overall module performance, DayStar's unique design guarantees nearly a even correlation between what is realized in small area and what is realized in large area. The design incorporates low-level concentration (2-10x) and a flexible interconnect structure. The additional cost of tracking for the 5-10x product is more than accounted for by the added performance and low-cost of the module product. We will additionally describe a cell manufacturing process that is simple and obviates the requirement for glass substrates, vacuum breaks, mechanical or laser scribing, or wet processing. The result is a set of PV products for large-scale Utility and/or small-scale rural electrification applications.

DESIGN STRATEGY

One way to effect an optimal design for a PV technology is to examine the successes and challenges of existing technologies. In Table 1, they are summarized. The common themes in the challenges are a high cell/module performance differential (all), high capital equipment costs (all), high materials usage and costs (c-Si), large-area process uniformity (PTF), and low module performance (a-Si, PTF). In the area of successes, easy translation to large areas (c-Si) with associated uniformity, low materials usage (a-Si, PTF), laborless cell interconnect (a-Si, PTF), low energy payback (a-Si, PTF), potential for mechanical flexibility (a-Si), continuos in-line processing (a-Si, PTF), and processing that can coat-tail on existing technologies (all).

We present here the basics of a PV technology that has the potential for high performance at the cell and module level, easy translation to large areas, low material usage and high yield, low capital costs, continuous in-line processing, mechanical flexibility, automated cell interconnect, low-energy payback, simple processing that does not require large area uniformity, and processing technologies that are accomplished in other industries.

Table 1   Successes and challenges facing existing flat-plate PV technologies.

Technology	Successes	Challenges
c-Si	High Performance Easy to translate to large areas Reliable Existing Semi ind. Predictable mat'l	Batch Processing High (⇑) capital $ ⇑ mat'l usage and $ ⇑ energy payback Haz. processing Cell / module perf Δ Labor $
a-Si	Low (⇓) mat'l usage Large-area process Monolithic integ'n Existing TFT ind. Building apps. ⇓ energy payback In-line processing Simple mat'ls chem.	⇓ performance Complexity / stability ⇑ capital $ Haz. processing Acceptability Cell / module perf Δ Labor $
PTF	⇓ mat'l usage Large-area process Monolithic integ'n ⇓ energy payback ⇑ perf. potential In-line processing	Complex mat'l chem Process uniformity Processing temps ⇑ capital $ Acceptability Cell / module perf Δ Labor $

CELL AND MODULE DESIGN

In Fig. 1, the first generation patented design for DayStar's *SunSparc* cell and interconnected module are shown [5]. The design incorporates a 1-mm dia. stainless-steel (SS) filament as the substrate upon which the solar cell material is deposited. SS filament is flexible, inexpensive, strong, and lightweight. With existing filament-based technology markets in place, the equipment systems required for handling and film deposition presently exist. The absorber surface is curved and forms an ideal surface for illumination by the refracted light.

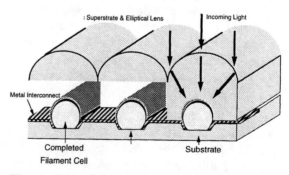

Fig. 1   Schematic of 1st generation filament cell and module technology design.

DayStar has presently evolved this design into a superior technology that further simplifies product development efforts. A patent is presently being filed. This design replaces the non-planar surface required for refractive optics with a planar surface and reflective optics. The total material requirements have been reduced by a factor of two. Solid cylindrical filaments have been replaced by planar ribbon filaments to facilitate cell fabrication and interconnection. The cell and module manufacturing processes have been decoupled, allowing for the incorporation of existing thin-film PV product fabricated on a suitable substrate material and appropriately modified for DayStar's design.

Upon completion of the filament cell fabrication process, the individual filament cells are interconnected via a conducting interconnect plane where each cell is oriented such that both the front and back cell contacts are unilluminated. This eliminates losses associated with illuminated "dead" interconnect regions that exist in standard flat-plate designs. The interconnected cells are assembled into the superstrate at the focus of a linear optical element. Concentration levels from 3:1 to 20:1 are easily achievable in this design. Because this module is built from smaller cells, its size is not limited by deposition equipment but by filament length and numbers. Superstrate thickness is only 4-mm for 5:1 concentration, making this module lightweight and potentially flexible. The superstrate also provides environmental encapsulation for the cells. With these design components, a level of cell performance translates into an equivalent level of module performance.

FILAMENT CELL DEPOSITION PROCESS

Thin-film CIGS-based solar cell fabrication processes have been developed at the National Renewable Energy Laboratory (NREL), under contract to the United States Department of Energy (USDOE), by one of the principles of DayStar while employed at NREL. The Midwest Research Institute (MRI), operator of NREL, retains intellectual property rights to the absorber fabrication technologies. DayStar has negotiated a non-exclusive licensing agreement to utilize the absorber fabrication method as a component of the manufacturing process. The process has produced world-record 1-sun (17.7% total-area) and 20-sun (17.7% total-area) flat-plate device performance (Fig. 2), the latter representing a 2.9% absolute improvement relative to the 1-sun control.

For this cell and module development effort, we have chosen two processing schemes. The first is to employ the high-efficiency cell processing to demonstrate the efficacy of the cell and module design. In this approach, 4-mm dia. soda-lime glass rod is used facilitate manual handling of the cell materials. At present, we have successfully fabricated a 9.6% 1-sun cell and a 7.6% 42.0-cm^2 interconnected mini-module (Fig. 3). The primary loss mechanism when using 4-mm dia.

substrates is series resistance (Rs) in the transparent conducting oxide (TCO). We have modeled this loss as a function of filament diameter and find that this component decreases nearly 80% for smaller dia. filament. This results in cell performance greater than 12% at 1-sun. We are presently in the process of fabricating prototypes utilizing the present module design and will report on the results in future publications.

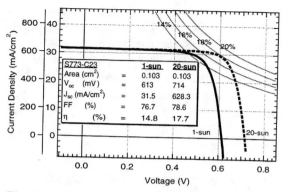

Fig. 2 I-V curves for champion 20-sun device.

For the manufacturing development effort, we are investigating a far simpler cell process. Our product cost-analysis and module design suggest that a fabrication process leading to a 10% performance level is adequate to support

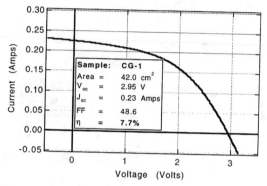

Fig. 3 I-V curve of a mini-module consisting of five interconnected, 4-mm filament solar cells.

the introduction of DayStar PV modules at prices below $1.50/Wp. We have previously initiated this work in cooperation with NREL [4,6].

The next-generation CIGS-based thin-film device will ideally have the following characteristics. The back contact and substrate combination will offer superior reproducibility to the present Mo/SLG system through controlled introduction of required impurities (e.g. Na, O). This will expand the list of potential substrates and back-contact metals to those that may be more optimally suited to CIGS thin-film processing. The absorber will be fabricated in a manner that minimizes in-situ process control and high-temperature processing. This will drastically reduce the cost of manufacturing equipment. Finally, the heterojunction partner will be formed in-situ with the absorber to relieve the necessity for a vacuum break and a CBD process. This will improve reliability and throughput and will reduce cost.

A generic flowchart illustrating what this cell fabrication process will look like is shown in Fig. 4. The three unconventional components of this process scheme are the use of SS for a substrate, sputtering to deliver the CIGS precursor material, and a non-CdS / in-situ junction formed without a break in vacuum. Each of these components are presently under investigation. The best results to date for each of these process perturbations individually are shown in Table 2. The intention is to put the three together into a simple, continuous, scalable, manufacturing process at a performance level in excess of 10%.

Table 2 Best cell performance for alternative cell processes and components.

Cell Process Component	Best Cell η
SS Substrate	13.2%
Sputtered CIGS precursor	9.1%
In-situ CIGS junction	12.7%

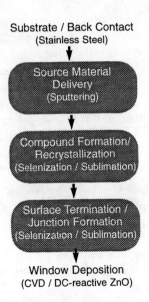

Fig. 4 Generic CIGS-based cell manufacturing process.

MANUFACTURING PERSPECTIVE

In addition to the design features and performance capability of this technology, associated manufacturing ease with greatly reduced capital outlay are beneficial results of the unique cell and module geometry. In Fig. 5, DayStar's manufacturing perspective is schematically illustrated. The cell and module manufacturing processes have been separated and simplified. Although DayStar plans on

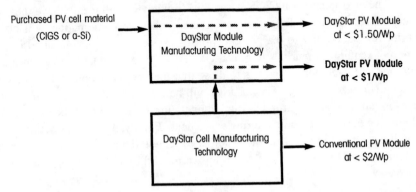

Fig. 5 Schematic illustrating relationship between cell and module manufacturing technologies

248

developing an internal cell manufacturing capability, it is not part of the critical path for Pilot-Phase development. The module manufacturing technology can take existing thin-film PV materials and down/up leverage the cost/power. The result is a cost-competitive product with a small capital investment in module integration hardware. Cell development efforts will progress in parallel with the module technology. Thin-film filament coating techniques are well established using a form of magnetron sputtering (both DC and AC), chemical vapor deposition (CVD), electrodeposition, etc. Filament coating systems have advantages over planar deposition in uniformity, materials utilization, and throughput (a factor of 3-10). This advantage, coupled with low-level concentration, provide a throughput improvement factor of 6-100 as compared to planar technologies. System sizes and capital equipment costs are similarly reduced with projections indicating EOSV attained at just under 3 MWp/yr production levels.

DayStar is designing and developing components for a small scale pilot manufacturing system that will produce a unique PV product from a packaging perspective. The product will be a series of interconnected filament cells up to 2m wide. The module will then be cut into appropriate lengths, external contact posts applied and environmentally sealed around the edges. Each of these processes are well established in other technology areas and simple to apply.

TECHNOLOGY DEVELOPMENT STRATEGY

DayStar is in the process of creating its business model, prototyping and evolving the cell and module design and developing "blueprints" for manufacturing systems. It is our goal to demonstrate the capabilities of the module technology by implementing existing high performance processes as well as those intended for the manufacturing environment. Filament coating technology is being examined for applicability to NREL cell fabrication processes. Absorber and junction processes are being refined in a manner consistent with the unique requirements of the cell geometry. Although CIGS is presently the material system of choice, the geometry is also applicable to a-Si and substrated CdTe. At this time, there is no published data on a-Si cell performance under any degree of concentration, and CdTe cells in the substrate configuration are significantly inferior to those in the superstrate configuration. Balance of system (BOS) components uniquely designed for DayStar's *SunSparc* module technology are concurrently under development. In short, capabilities in process development, materials, manufacturing component adaptation, BOS, and appropriate business models are converging for this breakthrough technology.

CONCLUDING REMARKS

In this paper, we have described the basic aspects of a new PV cell and module design that will be mated with low-cost, high-performance thin-film CIGS or other solar cell technology to produce a PV module product capable of realizing low cost and high performance. Commercial "off-the-shelf" technology components are consistent with manufacturing that is decoupled from the module size. Low cost products are projected in the very near term for production levels of only 3 MWp/yr.

ACKNOWLEDGEMENTS

The authors wish to thank J. Dolan, A. Duda, and S. Ward from NREL for technical assistance. This work was performed under a Cooperative Research and Development Agreement (CRADA) between DayStar Technologies and NREL under Contract No. DE-AC02-83CH10093 to the U.S. Department of Energy.

REFERENCES

1. Photovoltaic Insider;s Report, **16** (2), (Feb. 1997).

2. Photovoltaic Insider;s Report, **15** (10), (October 1996).

3. K. Zweibel, H.S. Ullal, B.G. von Roedern, R. Noufi, T.J. Coutts, M.M. Al-Jassim, *Proceedings 23rd IEEE Photovoltaic Specialists Conference*, Louisville, KY, 1993, pp. 379-388.

4. John R. Tuttle, T.A. Berens, J. Keane, K.R. Ramanathan, J. Granata, R.N. Bhattacharya, H. Wiesner, M.A. Contreras, and R. Noufi, *Proceedings 25th IEEE Photovoltaic Specialists Conference*, May 13-17, 1996, Crystal City, VA, pp. 797-800.

5. Eric D. Cole, United States Patent # 5,437,736, "Semiconductor Fiber Solar Cells and Modules", Aug. 1, 1995.

6. J.R. Tuttle, T.A. Berens, S.E. Asher, M.A. Contreras, K.R. Ramanathan, A.L. Tennant,R. Bhattacharya, J. Keane, and R. Noufi, *Proceedings 13th European Photovoltaic Solar Energy Conference*, Nice, France, Oct. 23-27, 1996, pp. 2131-2134.

Photovoltaic Application of Resonance Light Absorption and Rectification

Guang H. Lin[*] and John O'M. Bockris[†]

[*]Solarex, 826 Newtown-Yardley Road, Newtown, PA 18940
[†]Chemistry Department, Texas A&M University, College Station, TX 77843

ABSTRACT. In this paper, we suggest a new concept, direct energy transfer from light to dc electricity by means of resonance light absorption and rectification. The experimental validation of resonance light absorption by a fabricated subnanostructure and high frequency rectification in the range corresponding to that of visible light is reported. The subnanostructure consists of a parallel dipole antenna array. A resonance peak signal of short circuit current was observed. This peak signal appeared only at the polarization of incident light parallel to dipole element. The short circuit peak signal was also observed at a photon energy less than that of the optical bandgap. The peak angle exhibited the predicted relation with the wavelength of the incident light. The possible photovoltaic applications of the experimental observations are discussed.

INTRODUCTION

There are two major intrinsic energy losses in conventional photovoltaic devices. An incident photon, with the energy less than the optical bandgap, would not be absorbed by the solar cell material. On the other hand, an incident photon, with energy greater than the optical bandgap, creates only one electron-hole pair.[1] The energy of the electron-hole pair is at least equal to the optical bandgap. The excess part of the photon energy is eventually transferred into useless heat.

Direct energy transfer from light to dc electricity by means of resonance light absorption and rectification could eliminate the two energy losses. In this concept, subnanostructures, in either dipole or monopole forms, are used to absorb the energy of high frequency electromagnetic (EM) radiation - light, resonantly. High frequency

---

[1]The chance of one photon creates two or more electron-hole pairs is very low.

CP404, *Future Generation Photovoltaic Technologies: First NREL Conference*, edited by McConnell
© 1997 The American Institute of Physics 1-56396-704-9/97/$10.00

diodes are used to rectify the EM wave directly into dc electric current.

A nano- or subnanostructure relating to the conversion of EM radiation to electricity has been gradually developed. Fletcher, et al., suggested an EM wave energy converter (1). This converter has an antenna energy absorber joint with a half- wave rectifier. An EM wave absorber and a rectifier, working in the microwave region, have been constructed and tested. Marks (2) suggested a high efficiency device for the direct conversion of solar light to electric power by using a similar structure as that of Fletcher. No experiment was done to support his idea. Thus, two separate principles, resonance photon absorption by antennae and electric rectification at the high frequency corresponding to that of light radiation, have to be validated by experiments. The first experimental observation of a functioning antenna structure, working in the far infrared region, was made by Japanese scientists, Yasuoko et. al. (3). They studied thin film antenna structures at 10.6 μm wavelength. An antenna, used for light absorption, was coupled with a Schottky type diode for rectification. A photo voltage output signal was observed. Recently, Canham (4) fabricated a randomly distributed nanostructure on a silicon surface by means of an electrochemical method. The wavelength of the peak photoluminescence signal was measured as a function of the porosity of the silicon. This experiment supplied indirect evidence of antenna absorption in the wavelength of visible light. To the best of our knowledge, no direct experimental observation of resonance absorption by an antenna structure and coupled rectification have been reported in the literature at the frequencies corresponding to that of visible light.

In this paper, the first experimental observations of resonance light absorption by a fabricated subnanostructure and high frequency rectification, in the range corresponding to that of visible light, are reported. The results provide a solid validation to both resonance light absorption and coupled electric rectification in the frequency region of visible light. The principles developed in this paper could be used to fabricate high efficiency solar cells.

EXPERIMENTAL

Solar light radiation has a wavelength spectrum range from about 290 nm to 2.3 μm with random polarization. Two characteristics, viz, wavelength distribution and random polarization, proffer difficulties to the solar cell designs. Therefore, for research aimed at validation of resonance absorption and rectification, radiation with single wavelength and a specific polarization was used.

Subnanostructure Design

There are three key problems in the cell design. One is that the effective

252

length of the fabricated structure equals $n\lambda / 2$ (n is an integer, 1, 2, 3...) for the resonance photon absorption. The second is the thickness of the metal strip. Generally, the thicker the metal layer, the more light that is absorbed. However, thicker layers may reverse the electric field in the metal. Thus, the designed thickness of the metal layer was in the range of 2000 Å to 4000 Å. The third key problem is that associated with very high frequency rectification. The frequency of the absorbed light is around 10^{14} to 10^{15} Hz. Such high frequencies require a very thin n layer. The designed thicknesses of the n layer are 100 Å, 200 Å and 300 Å, respectively.

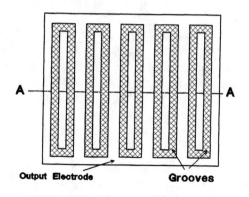

FIGURE 1. Top view of designed subnanostructure.

A dipole antenna array structure was selected for fabricating the test cell. The top view of the designed antenna array is shown in Fig. 1. The array consists of two sets of antenna elements. The elements in the first set are joined together to form an output electrode. The elements in the second set are isolated from each other, and connected to the back electrode through the p-n junction. The width of the metal strips is about 1 μm. The distance between two metal strip is also 1 μm. The length of the metal strip is 4.8 mm, and the size of the antenna array is 4.8 mm × 4.8 mm. The side view of the designed cell is shown in Fig. 2. Both the top and bottom electrodes are made using the same metal, either aluminum or gold.

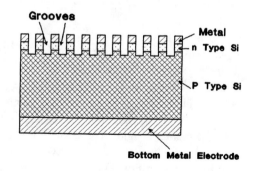

FIGURE 2.   Side view of the designed structure (A-A cross section of figure 1).

Subnanostructure Fabrication

A boron doped p-type single-crystal silicon wafer was selected to be the substrate of the test cell. The resistivity of the p-type silicon is about 1 - 10 ohm-cm. The thickness of the wafer is about 14 - 16 mil. The front surface is mirror polished. A phosphorus-doped n layer was formed on the polished side of the p-type silicon by using two methods, ion implantation or plasma enhanced chemical vapor deposition.

The sample cells were fabricated by means of a lithography technique. The processes include lithography, developing, front metal coating, lift-off, silicon etching, and back electrode deposition. The experimental details, with regard to the substrate and fabrication, have been published (5).

Testing

The schematic experimental arrangement for the cell testing setup is shown in Fig. 3. Three light sources were used in the experiments. One is an Ar ion laser (Coherent, Model I-100-2), that can supply five lines with different frequencies. The second is a He-Ne laser (Hughes 4000) with a wavelength of 632.8 nm. The third is a monochromatic light provided by a 1000 W tungsten-halogen lamp with a 600 lines/mm grating. A long pass optical filter

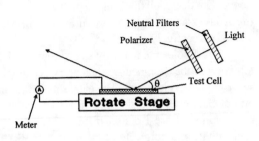

FIGURE 3. Schematic experiment arrangement.

was used to cut off the second harmonic oscillation generated from the grating. Polarized light, obtained by means of a polarizer (Oriel, polarizing beam splitter cube, 26165) is then incident on the test cell, which is mounted on a rotating stage to vary the incident angle. The test cell was connected in such a way that the top metal layer, which surrounds the carved grooves and isolated metal strips, is one electrode (shown in Fig. 1) and the bottom metal layer (shown in Fig. 2) is the another.

RESULTS AND DISCUSSION

Photo response of the cell was observed as the light was incident on the test cell. The short circuit current as a function of the incident angle (as defined in Fig. 3) of the light at different wavelength and different polarization was measured and analyzed.

The short circuit current as a function of the incident angle of the He-Ne laser light with the parallel polarization (the electric field of the light is parallel to the direction of the antenna array) is shown in Fig. 4.   A strong resonance output peak above the background photovoltaic signal was observed at the incident angle of 31.5°.   Since the effective length of an antenna is proportional to $sin^2\theta$, where θ is the incident angle, the resonance absorption condition is

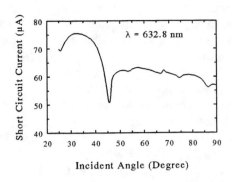

FIGURE 4.   Short circuit as a function of incident angle at the parallel polarization.

$$d = \frac{\lambda}{2\sin^2\theta}$$

(1)

where d is the length of the antenna, λ is the wavelength of the incident light. By using Eq. 1 and the wavelength of the He-Ne laser, 632.8 nm, the calculated antenna length is 1.16 μm.

How may one explain this strong peak?   Does it come from antenna resonance absorption or from the photovoltaic pn junction?

A classical calculation shows that the light absorption at the base of the p-n junction silicon surface decreases as the incident angle increases for light with a polarization parallel to the plane of the incidence. However, no single peak behavior is expected. Thus, simple photovoltaic phenomenon cannot explain the peak signal. On the other hand, if the signal does come from the resonance absorption, this signal must have the following two characteristics. One is that the peak angle changes with the frequency of the applied light, as indicated by Eq. 1. The second is that the peak is observed only when the incident light has a polarization parallel to the direction of the antenna array. Two sets of experiments were designed to find out if these two characteristics of resonance absorption are observed.

The first experiment is to change the light polarization. Figure 5 shows the short circuit current as a function of the incident angle of the He-Ne laser light with the polarization perpendicular to the direction of antenna array. No peak behavior was observed. Thus, the output signal with a maximum peak value was observed only with the light polarization parallel to the direction of the antenna elements, which provides a solid validation that the peak signal does come from the fabricated antenna.

The angle dependence of the short circuit current at perpendicular polarization can be explained by means of geometric optics. The calculated results are in qualitative agreement with the experimental measurements.

The second experiment is designed to observe the short circuit current peak angle at different light frequencies. Six laser lines (five lines from the Ar ion laser, one from He-Ne laser) were used for the experiments. It was observed that the resonance angle at the peak short circuit current became smaller as the frequency of the incident light was increased. Figure 6 shows $sin^2\theta$ as a function of wavelength of the incident light, θ is the incident angle. A linear relation was observed as predicted by Eq. 1. The antenna length calculated from the slope of the line in Fig. 6, according to Eq. 1, is 1.07 μm, consistent with the direct measurement from electron microscopy, 1.04 μm. The calculated antenna length from diffraction measurement is 1.05 μm.[2]

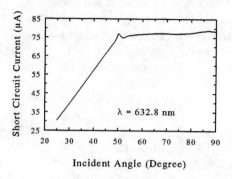

FIGURE 5.   Short circuit current as a function of incident angle at perpendicular polarization.

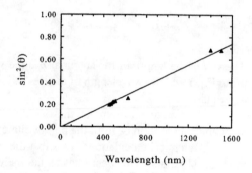

FIGURE 6.   Peak angle as a function of wavelength of the incident light.

Both these two experiments prove that the peak signal came from resonance absorption and rectification from the fabricated subnanostructure. In order to give further evidence that the peak signal does not arise from a photovoltaic source, a third experiment was designed and tested.

Monochromatic light with wavelengths of 1400 nm and 1500 nm was used to measure the short circuit currents, respectively. The photon energies of the applied lights are about 0.88 and 0.83 eV, respectively, well below the bandgap energy of the silicon, 1.1 eV. However, a peak signal was observed as the incident

---

[2] Diffraction pattern of the sample cell formed by He-Ne laser light consists of 6 spots, corresponding to m = ± 0.5, ± 1.5, and ± 2.5. However, this will not cause a single maximum output of short circuit current. The antenna length can be obtained by measuring the angle of these 6 spots. The calculated length is 1.05 μm

angle was changed for both the wavelengths. These peaks cannot be explained by the creation of electron-hole pair under sub-band gap light illumination. Further, the angle-wavelength behavior of the two experiments, also shown in Fig. 6, is consistent with Eq. 1. As shown in Fig. 6, all the experiment points (5 Ar laser lights, one He-Ne laser, and two monochromatic lights) are located on a straight line, predicted by Eq. 1. Thus, this peak corresponds to the resonance light absorption and rectification.

The next question relates to the possibility of a nonlinear effect. Does the peak response come from two photon absorption? To answer this question, a linear test was performed. A set of neutral filters was used to variate the incident laser light intensity into the test cell. The peak signal of the short circuit current as a function of light intensity was measured. A linear relation was observed. Thus, a nonlinear two photon effect is not the explanation of the peak output of the short circuit current signal.

The last question is about the rectification. Does the electrical signal come from a thermal emf effect? Two experimental observation were used to answer this question. As we pointed in the experimental section, the cell is mounted on a rotating stage. The temperature of the back metal electrode remains constant. When light is incident on the top electrode, the temperature difference between the front and back electrodes would first increase, then approach a stable value. If the signal came from a thermal emf effect, the short circuit current should present a transient process, and increase from zero to a maximum stable value after the light is incident upon the device. No such phenomenon was observed. The signal was stable and corresponded to the light incident onto the surface of the device. The second experimental observation applied a heat gun. Hot air (about 80 °C) blew onto the front part of the device with and without light incident, the observed changes of short circuit current were about 0.5 µA, less than 1 % of the rectification signal. Thus, a thermal emf effect is not the reason for the observed electrical signals.

We should point out that the mechanisms for both resonance light absorption and rectification by subnanostructures are not well understood. For example, in Fig. 4, one sees a well defined valley and other fine structures in the plot of current vs. incident angle. Do these come from resonance reflection or high order multipole moment effects? Both experimental and theoretical work should be performed to research this new phenomenon.

Resonance light absorption and rectification can be applied to photovoltaic technology. A test cell, which responds to white solar light with random polarization, was designed and fabricated by means of lithography technique. Initial experiments indicated an enhancement of quantum efficiencies in both high and low frequency regions (6).

Other than using micro-electronic techniques, there are many low cost ways to fabricate resonance nanostructures for photovoltaic applications. Synthesizing optically sensitive polymers with a dimension of a half wavelength of light is one of

257

them. The spacing of large optically sensitive molecules with a distance of half wavelength can also make resonance structures. It should point out that the performance of an existing photovoltaic solar cell would be improved by forming resonance structures on the surface. These structures could be constructed by low cost surface post-treatment methods: e.g., electrochemical treatments to produce porous surfaces, pulsed ion implantations, or proton bombardments with a suitable mask.

CONCLUSIONS

A resonance peak output signal was observed from the fabricated subnanostructure. This peak signal only appeared when the incident light polarization is parallel to the direction of the dipole antenna elements. The peak angle varied with the incident light wavelength according to Eq. 1. The intensity of the peak signal varied linearly with the incident light. The peak signal was also observed at photon energies smaller than the bandgap energy of silicon. It follows that a resonance light absorption and rectification in the visible light region has been experimentally observed. The experimental observations open an avenue to the fabrication of high efficiency solar cells.

ACKNOWLEDGMENT

This work was supported by a U.S. Government Agency. We thank Dr. G. Frazier in Texas Instruments for helpful discussion; Drs. C. Farrell and J. Speidall in IBM, Drs. R. Papania and M. Hanes in Westinghouse, Mr. V. Swanson and Dr. R. Atkins in the Solid State Institute, Texas A&M University, are to be thanked for help in making the micro structure. We thank Mr. R. Abdu for the help in making the test cell, and Dr. M. Z. He for the help in depositing the n type thin films and in coating the aluminum back electrode.

REFERENCES

1. Fletcher, J. C., and Balley, R. L., US patent No. 3760257 (1973).
2. Marks, A. M., US Patent No. 4445050 (1984).
3. Yasuoko, Y., *The 10th International Conference on Infrared and Millimeter waves* ,(IEEE Press, 1985, p. 27.
4. Canham, L. T., *Appl. Phys. Lett.*, **57**, 1046, (1990).
5. Lin, G.H., Abdu, R., and Bockris, ,J. O'M., *J. Appl. Phys.* **80,** 565 (1996).
6. Lin, G.H., and Bockris, J. O'M., "Antenna Absorption of Visible Light and the Production of Electricity", Final Report, U.S. Government Agency, (1994).

Field Effect Solar Cell

H.Fujioka*, M.Oshima*, C.Hu**, G.Collins[†], M.Sumiya[††],
M.Maruyama[††], M.Kawasaki[††], and H.Koinuma[††],

*Department of Applied Chemistry, the University of Tokyo,Hongo, Tokyo, 113 Japan
**Department of Electrical Engineering, U.C.Berkeley, Berkeley, CA 94720
[†]Department of Electrical Engineering, Colorad State University, CO 80523
[††] Ceramics Materials and Structures Laboratory ,Tokyo Institute of Technology,
Yokohama, 4259 Japan

Abstract. We have investigated a p-i-n a-Si:H solar cell structure which can eliminate detrimental effect of a transparent electrode and a heavily doped window layer using a two-dimensional device simulator. The cell utilizes an inversion layer induced by field effect instead of the heavily doped window layer while maintaining p-i-n junction locally to keep the built-in potential high and stable. Device simulation has revealed that the conversion efficiency of p-i-n a-Si:H solar cells can be improved by 30% with the use of this cell structure. This improvement is mainly due to the increase in the photo-currents, which can be explained by the increased quantum efficiency for light with short wavelength.

INTRODUCTION

An a-Si:H p-i-n solar cell has attracted much attention since the report on the cell with conversion efficiency of 5.5% in 1977(1). Its conversion efficiency, however, has remained rather low when it is compared with that of crystalline Si solar cells. This low conversion efficiency can be attributed mainly to the low built-in potential and the poor electrical properties of a-Si:H such as the low carrier mobility and the low carrier lifetime. The low built-in potential, which is due to existence of the tail states in its forbidden gap, causes lower open circuit voltage (Voc) than we expect from its bandgap. The poor electrical properties enhance photo-carrier recombination and lead to small photo-currents. The other reason for the low conversion efficiency is the large absorption of photons in the heavily doped p-type window layer, which stems from the large absorption coefficient of a-Si:H. It is known that the photo-generated carriers in the window layer recombine quickly and do not contribute to photo-current(2). To solve this problem on the window layer, use of p-type hydrogenated amorphous silicon-carbide (a-SiC:H) has been proposed(3). The a-SiC:H layer not only reduces light absorption in the p-type

CP404, *Future Generation Photovoltaic Technologies: First NREL Conference*, edited by McConnell
© 1997 The American Institute of Physics 1-56396-704-9/97/$10.00

layer but also acts as a barrier against back injection of electrons from the intrinsic layer. Indeed a-SiC:H window layer improved conversion efficiency but quantum efficiency for light with short wavelength is still much lower than unity(2). Recently, it was also pointed out that the junction between transparent conducting oxide (TCO) and the p-type a-Si:H window layer has serious detrimental effect on the conversion efficiency(4-5) because the junction causes negative electric field in the p-type layer. These analyses made us reach to an idea that the conversion efficiency can be improved if the area covered with TCO and p-type a-Si:H can be reduced while maintaining the basic p-i-n junction properties. This idea can be realized by the use of an inversion layer formed by field effect instead of the heavily doped window layer while maintaining p-i-n junction locally. In fact, a similar type of cell structure was already examined with crystalline Si, but an a-Si:H solar cell with this structure has never been investigated, to the best of our knowledge. In this paper, we will discuss advantages of the a-Si:H field effect solar cells.

DEVICE STRUCTURES

A Schematic cross-section of the field effect solar cell is shown in Fig.1. As a reference, a conventional p-i-n solar cell is also shown in this figure. In the field effect cell structure, the p-type window layer, which could cause degradation in photo-currents, does not exist except for the region just under the front contact metal. The purpose of the p-type layer under the front metal of the field effect cell is to assure the large and stable built-in voltage. In practice, this structure can be achieved by etching of the p-type layer using the front contact metal as a mask. The electric field across the insulator to form the surface inversion layer can be achieved by several methods, which include the use of fixed charge in the insulator(6) and the use of ferroelectric materials. In this simulation, we assumed for simplicity that the structure has a transparent electrode with a negative applied voltage on a thin insulator to induce field effect, but any method should lead to similar results as far as the electric field in the insulator is large enough. An additional advantage of this structure is the fact that we can choose the work function of the front contact metal which offer best results. In this paper, however, we assumed ITO for the front contact metal to better comparison with the conventional cell. The front contact metal dependence on cell performance will be discussed elsewhere.

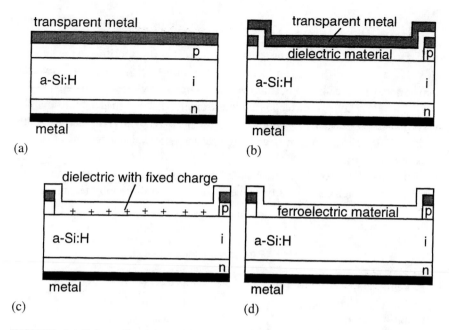

FIGURE 1. Schematic cross-sections of (a) a conventional solar cell and the field effect solar cells with (b) a dielectric material, (c) a dielectric material which includes fixed charge, and (d) a ferroelectric material.

SIMULATION

Table 1 summarizes the electrical properties and parameters for hydrogenated amorphous Si, which were used in this simulation. Most parameters except for the mobility gap were taken from Chatterjee's paper(4). Surface recombination velocity at the ITO/a-Si:H interface was assumed to be 10^7cm/s, which is close to thermal velocities of carriers. Although reflectivity at the backside metal was assumed to be 0.8, reflection at the front surface was not taken into account to concentrate on intrinsic performance of the cells. For simulation of defective materials such as amorphous Si, it is important to model deep levels in its forbidden gap appropriately. We assumed distribution of donor-like levels in the lower half of the forbidden gap and acceptor-like deep levels in the upper half as shown in Figure 2. This distribution is almost identical to that assumed in the reference 4. The donor-like states become positive when they ionize, while the acceptor-like states become negative. For p-type and n-type layers, higher deep level concentration

$(6 \times 10^{18} \text{cm}^{-3})$ than that for the intrinsic layer was assumed because heavy doping causes formation of extra defects. No additional trap was assumed at the insulator/a-Si:H interface because the field effect cell is not sensitive to the interface trap due to the high electric field near the interface(6). Photo-currents of conventional p-i-n cells can be improved theoretically by reducing the thickness of p-type window layer(4), which is obvious from the discussion in the Introduction. We assumed 200Å thick p-type layer because the p-type layer thinner than 200Å is difficult to deposit uniformly and often causes serious decrease in Voc(7). The thickness of intrinsic layer for both samples was assumed to be 5000Å. The width and spacing of the comb front contact were assumed to be 1μm and 30μm, respectively, which can be easily defined using photo-lithography technique.

TABLE 1. Parameters for this simulation

mobility gap	1.7eV
mobility (electron)	$25 \text{ cm}^2\text{V}^{-1}\text{sec}^{-1}$
mobility (hole)	$6 \text{ cm}^2\text{V}^{-1}\text{sec}^{-1}$
charged capture cross-section	$1 \times 10^{-14} \text{cm}^2$
neutral capture cross-section	$1 \times 10^{-16} \text{cm}^2$
barrier height of ITO/p-a-Si:H contact	1.15eV
donor tail energy	46meV
acceptor tail energy	27meV
thickness of p-type / n-type layer	200Å / 200Å
doping concentration in p-type layer	$6 \times 10^{18} \text{cm}^{-3}$
doping concentration in n-type layer	$6 \times 10^{18} \text{cm}^{-3}$
photon flux	AM-1.5 100mWcm^{-2}
absorption coefficient	experimental results
i-layer thickness	5000Å
gap states density in p-type and n-type layer	$6 \times 10^{18} \text{cm}^{-3}\text{eV}^{-1}$
effective density of states at conduction/valence band edge	$2 \times 10^{20} \text{cm}^{-3}$ / $2 \times 10^{20} \text{cm}^{-3}$

Since the field effect cell involves carrier motion in both lateral and vertical directions to the surface, it is essential to analyze this cell using a two dimensional simulator. We used TMA's MEDICI two-dimensional device simulator. The simulation were based on the simultaneous solution of the hole- and electron-continuity equations and Poisson's equation. Poisson equation for this material is given by:

$$\varepsilon \nabla^2 \psi = -q \left(p - n + N_D - N_A + \int_{E_v}^{E_c} N t_D(E) f(E) dE - \int_{E_v}^{E_c} N t_A(E) f(E) dE \right) \tag{1}$$

where n and p are the electron and hole concentrations, φ is the electrostatic potential, q is the electron charge, ε is the permittivity of a-Si:H, N_D and N_A are the ionized donor and acceptor concentrations, and $Nt_D(E)$ and $Nt_A(E)$ are the donor-like and acceptor-like trap concentrations. Although most research groups used T=0 approximation to determine the ionized trap concentration, we used accurate Shockley-Read-Hall expression to obtain the trap occupation function f(E):

$$f(E) = \frac{n\sigma_n v_{th} Nt(E) + \sigma_p v_{th} Nt(E) n_{ie} \exp(-E/kT)}{\sigma_n v_{th} Nt(E)\{n + n_{ie} \exp(E/kT)\} + \sigma_p v_{th} Nt(E)\{p + n_{ie} \exp(E/kT)\}} \tag{2}$$

where σ_n and σ_p are the capture cross-section for electrons and holes, k is the Boltzmann constant, v_{th} is the thermal velocity of carriers, T is the temperature, and n_{ie} is the intrinsic carrier concentration of a-Si:H. Continuity equations for electrons and holes are given by:

$$\frac{\partial n}{\partial t} = \frac{1}{q} \vec{\nabla} \cdot \vec{J}_n - U \tag{3}$$

$$\frac{\partial p}{\partial t} = \frac{1}{q} \vec{\nabla} \cdot \vec{J}_p - U \tag{4}$$

where $\vec{J}_n$ and $\vec{J}_p$ are the electron and hole currents. The recombination rate for this material can be calculated using the Shockley-Read-Hall expression:

$$U = \int_{E_v}^{E_c} \frac{pn - n_{ie}^2}{\dfrac{n + n_{ie} \exp(E/kT)}{\sigma_p v_{th} Nt(E)} + \dfrac{p + n_{ie} \exp(E/kT)}{\sigma_n v_{th} Nt(E)}} dE \tag{5}$$

The electron and hole currents are given by the following expressions:

$$\vec{J}_n = -q\mu_n n \vec{\nabla} \phi_n \tag{6}$$

$$\vec{J}_p = q\mu_p p \vec{\nabla} \phi_p \tag{7}$$

263

where μ_n and μ_p are the electron and hole mobilities and ϕ_n and ϕ_p are the quasi-Fermi potentials for electrons and holes, respectively.

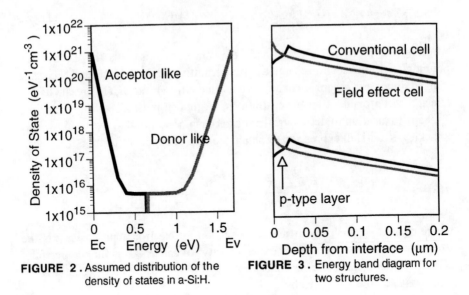

FIGURE 2. Assumed distribution of the density of states in a-Si:H.

FIGURE 3. Energy band diagram for two structures.

RESULTS AND DISCUSSIONS

Our simulation reveals that the short circuit current (Isc) of the field effect cell is 35% larger than that of the conventional cell, although both cells have similar open circuit voltages. This improvement in the photo-current can be attributed mainly to reduction in recombination at the ITO/p-type a-Si:H interface and in the p-type a-Si:H layer. Figure 3 shows energy band diagram of two structures along the vertical center of the cells. In the case of the conventional p-i-n cell, a negative electric field in p-type window layer caused by ITO/a-Si:H junction is clearly seen(5). Hence, most of the electrons photo-generated in the p-type layer are driven to the ITO/a-Si:H interface and do not contribute to the photo-current. On the other hand, there is no negative electric field for the field effect cell, which results in high collection efficiency for carriers. The 35% increase in Isc can be roughly accounted for by the number of photons absorbed in the 200Å thick p-type layer. This effect should be important especially for photons with high energies since most of them are absorbed near the surface. This can be confirmed by investigating wavelength dependence of quantum efficiency under bias light. As shown in Fig. 4, the quantum efficiency for light with short wavelength is dramatically improved by the use of the field effect cell. Fig. 5 shows comparison of output power between two cells. We can expect 30% improvement in the maximum output power by the use of the field effect structure. Therefore, a-Si:H solar cells with this structure are promising for future low cost solar power plants.

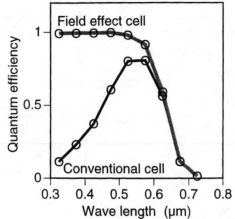

FIGURE 4 . Quantum efficiency of the solar cells.

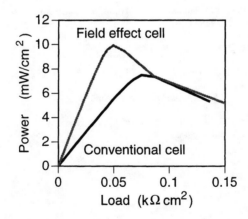

FIGURE 5 . Output power vs. load of the solar cells.

265

CONCLUSIONS

An a-Si:H solar cell which utilizes an inversion layer induced by field effect can eliminate detrimental effect of TCO and the heavily doped window layer of conventional cells. Two dimensional device simulation has revealed that the conversion efficiency of the p-i-n a-Si:H solar cells can be improved by 30% with the use of this cell structure. Hence, a-Si:H solar cells with this structure are promising for future low cost solar power plants. This improvement is mainly due to the increase in the photo-currents, which can be explained by the increase in the quantum efficiency for light with short wavelength.

ACKNOWLEDGMENT

The authors wish to thank Technology Modeling Associates, Inc. for its help in device simulation.

REFERENCES

1. Carlson,D.E., *IEEE trans. Electron Devices*, **ED24**, 449 (1977).

2. Tasaki, H., Kim, W.Y., Hallerdt, M., Konagai, M., and Takahashi, K., *J. Appl. Phys.* **63**, 550 (1988)

3. Tawada, Y., Yamaguchi, T., Nonomura, S., Hotta, S., Okamoto, H., and Hamakawa, Y., *Jpn.J.Appl.Phys.* **69**, 219 (1981).

4. Chatterjee, P., *J.Appl.Phys.*, **76**, 1301 (1994)

5. Arch, J.K., Rubinelli, F.A., Hou, J.-Y., and Fonash, S.J., *J.Appl.Phys.* **69**, 7057 (1991).

6. Hezel, R., and Schörner, R., *J.Appl.Phys.* **52**, 3076 (1981).

7. Kuwano, Y., Fukatsu, T., Imai, T., Ohnishi, M., Nishiwaki, H., and Tsuda, S., *Jpn.J.Appl.Phys.* **20**, supplement 20 (1981).

Photovoltaic Multiplicities

Hans J. Queisser

Max-Planck-Institute for Solid State Research
D- 70506 Stuttgart , Germany

Abstract. A multicell solar energy converter, produced in 1959/60 at the Shockley Transistor Corporation, is reviewed. The feasibility of this device, one of the first involving principles of Si integrated circuits, was demonstrated in anticipation of large-area Si sheets, to be pulled from Si/Pb binary melts. Secondly, the generation of multiple carrier pairs by absorption of merely one photon is discussed. Experiments on high-quality Si solar cells demonstrated this effect, which relies on inverse Auger generation. In principle, much higher maximal conversion efficiencies would be possible; novel criteria for materials optimization result. The new challenge of the *inverse band structure problem* arises. Finally, multistage optical transitions via deep centers in solar cells are briefly appraised.

INTRODUCTION

This contribution concerns three disparate topics related to solar cells, which nevertheless share the principle of multiplicity: an historical review is given for an integrated multicell, which at the time of its development anticipated availability of large-area sheets of silicon. Next, I cover the generation of multiple electron/hole-pairs by just one single photon in p-n junction solar cells, which has recently been experimentally proven to be attainable via an inverse Auger-effect. Lastly, multistage transitions via deep impurities are discussed.

MULTICELL SOLAR ENERGY CONVERTER

When I joined the laboratory of the Shockley Transistor Corporation in Mountain View, California in 1959, my first two projects were related to solar cells using silicon. William Shockley did not actually consider the production of such cells as a major business venture for his fledgling company. The availability of research grants during the time of the *Sputnik shock*, especially for the semiconductor material preferred by the Soviet competition -silicon - was of interest, since it provided us with needed opportunities to study, understand, and to harness the then ill-defined p - n junctions in Si, which was a stubborn, recalcitrant semiconductor.

CP404, *Future Generation Photovoltaic Technologies: First NREL Conference*, edited by McConnell
© 1997 The American Institute of Physics 1-56396-704-9/97/$10.00

I was put in charge of a contract (1), monitored by the US Army Signal Corps, to produce a silicon solar cell with several junctions, integrated to be in a series connection, thus delivering multiples of the output voltage of one single junction. The idea behind this rather bold attempt was to circumvent module constructions, which were -rightfully- considered as being much too complicated and too heavy for the small spacecraft of those days. Shockley proposed at that time that it might be possible to obtain very large, thin, thus lightweight, sheets of silicon, maybe not of single-crystal perfection, but at least of good polycrystalline quality. A binary melt, say of silicon dissolved in lead, might constitute a source of Si, which would float on top of the heavier Pb and might be withdrawn in very much the same fashion as a sheet of plate glass. The group-IV element Pb ought to be a harmless impurity, since it would not act as a dopant admixture. A chemical company (2) was our partner in this joint development.

Figure 1 shows a schematic of our multijunction cell, which was made out of a precariously thin wafer of Si (60 microns), to be defined by the then novel photolithography on both sides and to be diffused -for preposterous lengths of time- also from both sides. One can imagine even today how difficult a job had to be done, yet I managed to obtain about a dozen working units, thus demonstrated feasibility.

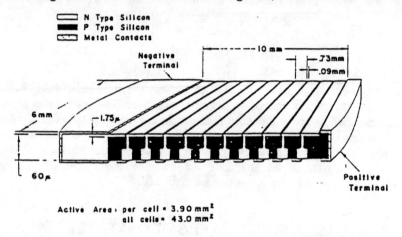

FIGURE 1. Schematic of a multijunction solar cell(1960), see Ref. 1

In retrospective, the significance of this approach seems limited for solar cell technology, in particular because the production of Si sheets proved too difficult then. However, this little cell in fact constitutes one of the first truly working examples of integration : several junctions were successfully interconnected. This fact might have had some bearing on the contemporary patent litigations for IC priority, yet the struggling Shockley Transistor Corporation had other worries to cope with. The idea of using a binary melt as a source, however, might very well be reconsidered at the present time. The store of materials knowledge, now tremendously enhanced compared to 1960, would provide a much improved starting position!

MULTIPAIR GENERATION

My second project in 1959 was of theoretical nature. William Shockley was dissatisfied with the state of solar cell theory at that time. The conventional approach was one of pragmatic engineering: find the loss mechanisms in a junction solar cell, then minimize all of them (3). This situation seemed to us as being akin to treating a steam engine prior to understanding thermodynamics and the Carnot cycle. A true thermodynamic limit of maximal efficiency had to be found for the junction solar cell! We obtained a research contract with the US Air Force (4), in order to arrive at a fundamental limit and - maybe more important for our sponsor - to provide a sound estimate about the optimal energy gap E_g for a semiconducting material, hence to identify the most promising candidate. At that time, silicon was in a severe competition with semiconductors of II-VI and III-V - types.

The principal idea of our approach had to do with the longest possible lifetime of minority carriers. This topic had previously been treated by W. van Roosbroeck with Shockley (5), who used the principle of *detailed balance* in showing that radiative recombination yielded the essential, unavoidable process for electrons recombining with holes, thus establishing an upper limit of carrier lifetime. We used the same principle (6) and published our paper in 1961 (7). This paper was initially rejected by the journal, but now -after many reprintings (8) - appears to be established and accepted. Silicon comes out as a good candidate within a rather broad maximum of cell efficiency *versus* band gap energy.

Our essential axioms for solar cell operation included the postulate that all photons above gap energy E_g create one and only one carrier pair, and that these photogenerated carriers are allowed to quickly relax, via phonon emission, to their respective band edges. I remember very well the long discussions with Shockley on the stringency of this condition. Our own work on silicon *p-n* junctions at that time placed great emphasis on reverse breakdown (9) and its vital point of carrier multiplication. We were fully aware that a sufficiently field-accelerated, *hot* carrier was able to generate additional pairs by impact ionization. However, we kept this single-pair-axiom, but cautiously included an explanatory footnote (10).

Silicon technology, especially protection for a very shallow junction, is now vastly superior to that of the beginning era in the early Sixties. Therefore, a doctoral candidate at Stuttgart, Ms. Sabine Kolodinski, was recently asked to use highly perfected silicon cells and to extend measurements of pair-generation quantum-efficiencies into the blue and ultraviolet spectral regimes.

Her measurements showed that multiple pair generation can indeed arise under favorable sample perfection (11), (thus also reconfirming earlier transport data of Tauc (12) and Vavilov (13) on bulk materials). Whenever the incident photon provides sufficient energy to at least one of the carriers of the pair, then this energetic carrier has a finite probability to evade the lossy, entropy-enhancing process of phonon generation; instead energy and crystal momentum can be conserved in a thermalization of the hot carrier by an appropriate production of a secondary electron/hole - pair, shown in Fig. 2.

This principle of improved utilization of the incident solar energy, as indicated by Fig.2, differs from the earlier, simpler arguments (9), in particular regarding an equipartition of excess energy among the two original photogenerated carriers. This approximation yields an onset of multiplication at 3 E_g. The analysis of our data, however, indicates that particularly favorable conditions arise especially under such optical absorption, where one of the two primary particles receives the *lion's share* , namely *all,* of the excess energy $E_x = h\nu - E_g$ from the absorbed photon with above-gap energy $h\nu$. This entropy-minimizing situation hinges on details of the band structure : a large combined density of states is required for this specific absorption, resulting in one cold carrier, close to its band extremum, plus one energy-hogging hot carrier, far away from band structure equilibrium.

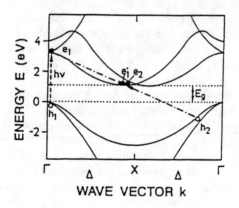

FIGURE 2. Generation of pair e_2, h_2 by hot electron e_1 , see Ref. 11

The correlation between the observed peaks in the multipair-generation with the known details of the band structure E (k) for silicon, as demonstrated in Ref. 11, leads to specific requirements for an optimal solar cell material. We thus face an entirely novel challenge in finding a material, which fulfills -or at least approaches- the detailed conditions of the band structure for multiple pair generation. This task is equivalent to finding a three-dimensional scattering potential, which causes the desired dispersion E(k). Such problems of identifying potentials from scattering data are prevalent in many branches of physics, especially for particle- and high-energy physics. The approach is called *inverse problem*, while the conventional method of calculating scattering from a given potential is termed *the direct problem*. The inverse problem is by far the more difficult one (14); in many cases it is not even proven that any solution, let alone a unique one, exists for the inverse problem. For example, one has to examine the very question to decide whether or not one deals with a so-called *ill-posed problem* !

The inverse problem for a three-dimensional Schroedinger equation, as needed in our situation, is thus a very difficult one. There are, however, glimmers of hope. Very practical methods have been developed to relate atomic configurations to band structure features (14,15); these brief, approximative methods might lend themselves more easily to inversion.

Assuming that a material with optimized band structure for multiple pair generation might be identified and actually made available, a strong enhancement of the ideal maximal conversion efficiency would be possible (16). Figure 3 displays the comparison of the ideal limits with and without multiple pair generation (16). The parameter m indicates the multiplicity of pair generation assumed, with energy conservation of course always being obeyed. Notice from these calculated curves that the optimal band gap of the hypothetical material shifts towards smaller energies. The physics behind this trend is clear: the reduced gap energy enhances the short-circuit currents because of (i) more absorption in the red part of the spectrum, and -in particular- (ii) multiple generation in the blue regime of the solar spectrum. The price to pay for this current enhancement, however, lies in the reduction of the maximal open-circuit voltage, which is a necessary consequence of lowering the gap energy. In total, the current enhancement gain clearly exceeds the voltage loss.

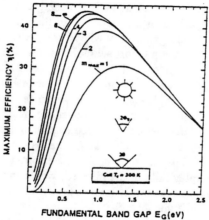

FIGURE 3. Ideal limit of cell efficiency with multiple pair generation, Ref. 16

Practical realizations of the multiple pair enhancements are not trivial. Even if a suitable material were found, it must - for example- be guaranteed that the surface perfection will suffice to suppress all types of recombination processes in favor of phonon generation. This feature is of particular severity, since those photons from which multiple pairs are expected will in principle all be absorbed very close to the surface. Surface drift fields of advantageous polarity might here be most helpful in transporting the carriers away from surfaces, into the bulk (17). The economy of such cells will also pose problems because of the required refined, hence expensive technology of manufacturing. It is, however, necessary to perform fundamental research regarding these proposals. For example, one has to provide sound and realistic estimates for the competition between the lossy phonon generation *versus* the favorable Auger generation (11). Experimental investigations with lower-gap semiconductors, such as initially performed with SiGe - alloys (18), will have to be continued - as well as with other materials. We have to find ways to increase the efficiencies of solar energy conversion, and the science community is consequently obliged to exert research efforts along all possible pathways. The considerations related to idealized thermodynamic efficiency limits should guide these efforts as a chal-lenging goal line. Recent papers have alluded to this proposal and its possible realization in actual cells (19).

MULTITRANSITION PHENOMENA

The other major waste of the incident radiative energy, beside carrier thermalization in the blue regime, arises on the other end of the spectrum: red light with photon energies below the band gap energy will be completely excluded from any photovoltaic conversion. This sharp cutoff below E_g is assumed in almost all treatments of cell efficiency; Shockley and I also followed this convention in our analysis (7). Various attempts were later made to overcome this deficiency. The obvious recipe would be to provide energy levels within the forbidden gap of the semiconductor. Optical transitions in a multistage mode might then be used, for example by lifting a valence electron into the deep level, choosing the level type so that reemission is comparatively low, such that another low-energy photon can be absorbed by the occupied deep state to expedite the electron into the conduction band. Two photons, preferably both in the infrared, would thus have created one carrier pair rather than having been totally lost (20).

This idea was discussed quite vividly in the late Sixties. The seemingly simple idea is, however, far from trivial. Deep electronic levels belong to the most difficult problems of semiconductor physics. These centers result from atoms or groups of atoms, which badly fit into the host lattice, hence are not at all describable with a simple hydrogenic model, which so neatly fits all shallow centers. The solubility of deep centers is thus in principle lower than that of shallow ones (21).On the other hand, these deep centers are usually quite effective locations for recombination and reducing lifetime, which is deleterious in solar cells. It is difficult to estimate optical absorption coefficients for these deep centers, but a delta-function binding potential is often a simple but very reasonable assumption (21).

G. Guettler and I tried to make a general estimate (22). We compared the benefit of extra optical absorption against the disadvantages of reducing the lifetime of *all*

272

the photogenerated carriers through the added recombination rates of the deep centers. Our result indicated that, under fairly general conditions, the disadvantages clearly outweighed the extra absorption. In general, it is always the most highly perfected junction and overall structure, that yields the best efficency. The junction obeying the ideal rectifier equation almost always wins over the nonideal one, which is characterized by strong space-charge layer recombination !

Much later, the idea was once again proposed and supposedly experimentally proven (23). A lively discussion ensued. To my knowledge, the claim of observing truly enhanced efficiencies - a difficult task - has not really been established as yet. Very careful absolute measurements are needed. Theoretical considerations on the beneficial role of deep impurities under particular conditions have most recently been presented, *e.g.* by Green and coworkers (24 - 26), Acevedo (27), and also by Luque (28). Some claims made from these treatises assert that there may indeed be situations in solar cell constructions and modes of operation, where the efficiency might become enhanced by the addition of such multistage processes in a cell. The predictions of these various efforts still seem controversial. Acevedo (27) arrives at a rather negative conclusion: neither multipair generation nor inclusion of deep levels will lead to essential improvements. The Australian group of M.Green *et al.* (24-26) is more optimistic; the effects of optical competition and of light trapping inside the cell body seem to favor the beneficial role of deep centers via the *impurity photovoltaic effect*. These authors have quantitatively treated indium as a deep dopant and have provided estimates for the optimal dopant concentrations. Luque (28), on the other hand, arrives at an even more positive conclusion, but based only on his assumption that only radiative (*detailed balance!*) transitions involving the centers are allowed. This assumption is somewhat difficult to appraise, because we know of the efficient capture of carriers by means of nonradiative transitions.

ACKNOWLEDGMENTS

I gratefully acknowledge the invitation by Robert McConnell to the Denver conference and the kind hospitality extended to me.

I also wish to thank the numerous colleagues with whom I had the pleasure to cooperate on studies relating to solar cells, their principles and technology; I mention especially G. Güttler, J.H. Werner, S. Kolodinski, E. Bauser, R. Bergmann, and R.Brendel. Major portions of our work at Stuttgart were financially supported by the German Federal Ministry of Research and Technology.

In remembrance of the glorious early years in *Silicon Valley*, working in Shockley's old apricot barn on 391 South San Antonio Road in Mountain View, I cordially and thankfully salute S.M. Fok, A.Goetzberger, W.W. Hooper, and R. H. Finch, who all shared the excitement of working with silicon, with solar cells being our vehicles to understand this semiconductor and its opportunities.

REFERENCES

1. Queisser, H.J., *Multicell Solar Energy Converters*, Contract No. DA 36 - 039 - SC - 85239, US Army Signal Corps, dated 6 August 1959
2. Grace Research and Development Co., Contract DA 36 - 039 - SC - 85242
3. Wolf, M. *Proc. IRE* **48**, 1246 - 1263 (1960)
4. Queisser, H.J., *A Study of Photovoltaic Solar Cell Parameters*, Sept. 1962, Contract AF 33 (616) - 7786 with Wright-Patterson Air Force Base, Dayton, Ohio
5. van Roosbroeck, W. and Shockley, W. , *Phys. Rev* .**94**, 1558-1560 (1954)
6. Queisser, H.J. and Shockley, W., *Bull. Am. Phys. Soc.* **5**, 160 (1960)
7. Shockley, W. and Queisser, H.J. *J.Appl. Phys.* **32**, 510-519 (1961)
8. Reprint, *e.g.* in: Sze, S.M. *Semiconductor Devices: Pioneering Papers*, World Scientific, Singapore (1991); also in: *Solar Cells*, Backus, C.E.,editor IEEE Press, New York (1976)
9. Shockley, W. *Solid-State Electronics* **2**, 35-67 (1961)
10. Footnote 12 on page 512 in Ref. 7
11. Kolodinski, S. *et al. Appl.Phys.Lett.***63**, 2405-2407 (1994)
12. Tauc, J., *J.Phys.Chem.Solids* **8**, 219-223 (1959); also, see Ref.10
13. Vavilov, V.S., *J.Phys.Chem.Solids* **8**, 223-226 (1959); also, see Ref.10
14. *The Inverse Problem* ,Lübbing, H., editor. Akademie-Verlag, Berlin (1995)
15. Queisser, H.J. and Werner, J.H., *Optimization of High-Efficiency Solar Cells: An Inverse Problem"* in Ref.14, 165-180 (1995)
16. Werner, J.H., Kolodinski, S., and Queisser, H.J., *Phys.Rev.Lett.* **72**, 3851-3854 (1994)
17. Geist, J., Gardner, J.L., and Wilkinson, F.J., *Phys. Rev.* **B42**, 1262 -1267 (1990)
18. Kolodinski, S., Werner, J.H., and Queisser, H.J., *Appl.Phys.***A61**, 535-539 (1995)
19. Liakos,J.K. and Landsberg, P.T., *Semic.Sci.and Technol.* **11**, 1895-1900 (1996)
20. Wolf, M., in Ref. 3, page 1359
21. Queisser, H.J., *Deep Impurities*, in: *Festkörperprobleme/Advances in Solid State Physics* **XI**, Madelung, O., editor, Pergamon/Vieweg, Braunschweig (1971), 45-64
22. Güttler, G. and Queisser, H.J., *Energy Conversion* **10**, 51-55 (1970), also: *J. Appl. Phys.* **40**, 4994-4995 (1969)
23. Li, Jm., *et al., Appl.Phys.Lett.* **60**, 2240-2243 (1990)
24. Keevers, M.J. and Green, M.A., *J. Appl. Phys.* **75**, 4022-4031 (1994)
25. Keevers, M.J. *et al. Proc.13th Eur.Photovolt.Energy Conf.* PO5A3 (1995)
26. Keevers, M.J. and Green,M.A., *Solar Energy Mat.and Solar Cells* **41/42**, 195-204 (1996)
27. Acevedo, A.M., *Rivista Mexicana de Fisica* **42**, 449-458 (1996)
28. Luque, A., private communications

THE FUTURE ROLE
OF SUPER-HIGH-EFFICIENCY
SOLAR CELLS

Straight Talk about Concentrators

Richard M. Swanson

SunPower Corporation
435 Indio Way, Sunnyvale, CA

Abstract. This paper addresses the issue of why concentrator systems have not gained significant market share. Various concentrator and flat-plate PV system approaches are compared by computing the expected cost of energy. Based on this result, some conclusions and recommendations for the concentrator industry are presented.

1. INTRODUCTION

The allure of concentrating sunlight as a means to dramatically reduce the cost of photovoltaic systems has been felt since the beginning of the DOE and EPRI terrestrial photovoltaic programs. Considerable resources, both public and private, have been expended in an effort to make concentrating systems a commercial success. From a commercial point of view, the reality has fallen quite far short of expectations. The same might be said of all forms of photovoltaic systems; however, flat-plate approaches have at least developed a foothold in remote power markets. Today, sales of concentrating systems are less than one percent of all photovoltaic system sales. This paper takes a fresh look at why this is so, and discusses the prognosis for the future of concentrating systems. Finally, recommendations for future directions in concentrating system development are presented.

The central aspect of this paper is a cost comparison among various concentrator and flat-plate approaches in order to asses their relative potential. One might ask, "why do this when so many cost calculations have been done over the years?" The reasons are several fold. The first is the use of common cost assumptions. Often, cost calculations are given for a particular technology by proponents of that technology. This makes comparisons difficult. Second, different types of systems respond differently to direct and diffuse components of sunlight. Many early cost comparisons were done prior with solar resource data that is much less accurate, particularly for the direct component, than that available today. For this paper, the data in the NREL *Solar Radiation Data Manual for Flat-Plate and Concentrating Collectors* is used. Third, various options are often compared on a cost-per-installed-watt basis. This does not fairly compare options with different annual capacity factors. For this paper, the levelized cost of electricity is computed for each option using the methodology of the DOE Five Year Plans. Fourth and finally, today we have a much better understanding of both the potential performance and the cost of various photovoltaic system components such as trackers than was the case even five years ago.

CP404, *Future Generation Photovoltaic Technologies: First NREL Conference*, edited by McConnell
© 1997 The American Institute of Physics 1-56396-704-9/97/$10.00

2. BARRIERS TO ENTRY

From the beginning, concentrators were usually envisioned for application as large power plants, producing significant quantities of non-polluting, renewable energy. In this application they compete with conventional fossil fuel plants which produce low-cost electricity, particularly when fueled by abundant, low-cost natural gas. Global concern over the need for renewable energy has waned since the discovery of new natural gas supplies. In the meantime, flat-plate modules have found a ready market for small, remote power sources. There are many existing markets where PV modules are very cost effective compared to the alternatives. These applications are typically low power, less than several hundred watts. In each case, however, the value of the PV system comes from some aspect other than savings of fossil fuel. In the case of roadside emergency phones, for example, the value comes from saving the conduit and wiring cost of distributing electricity to the phones. Concentrators are not well suited to these small applications. The total system cost including installation, storage, and ancillary equipment is usually over $15 per watt, of which the PV module costs around $5 per watt. Offering a $2 per watt concentrator module and tracker would only lower the overall cost to $12 per watt. This is not sufficient cost differential to entice most customers to accept the added complexity of concentrating systems.

For applications requiring large amounts of power, it is usually more cost effective to either make connection to utility distribution lines or, if such lines are too far away, generate the power on-site with generators powered by fossil fuel. To be competitive in these applications, PV systems must cost around $1 per watt for systems in the hundreds of megawatts to between $2 and $3 per watt for systems in the 10 kW to 100 kW range. Achieving such low cost is difficult, and this presents concentrators with an essential dilemma—the markets for which they are most suitable require low cost from the beginning because of well established alternatives.

Concentrators have additional burdens compared to flat-plate systems. Concerns over tracking system reliability are added to concerns over their obtrusive appearance and more restrictive mounting options. They are difficult to integrate into residential roof, for example. They tend to be more bulky to ship and difficult to install. Finally, they do not use the diffuse component of radiation. The effect of this is that, for a given annual energy production, concentrators produce a more erratic output. This places more burden on the storage system and/or back-up power source in non-grid-connected applications. The net impact of all these issues has been to stymie the concentrator industry and restrict most sales to demonstration projects. The purpose of the cost analysis below is to determine if there is sufficient economic advantage to warrant continued pursuit of concentrator systems and to help identify the best market applications for early commercial products.

How can one quantify the reluctance of customers to incur the added complexity of concentrating systems? For the purposes of this analysis, it is assumed that the purchase decision would be based solely on cost for systems in 100 kW to 1 MW range (to be termed medium-sized systems here). Such systems would often have Diesel generators on-site and thus maintenance personnel should be conveniently available. Alternatively, they would be located near housing or industrial parks in locations where service could be provided through local contracts in a manner similar to that for air conditioning systems. For systems in the 2 kW to 100 kW size range (termed small-sized systems here), however, customers could reasonably purchase flat-plate systems and may not have easy access to maintenance. Typical applications might be remote homes or irrigation systems. In this case, it is reasonable to assume that concentrators will have to cost less than an equivalent flat-plate system to become the purchase of choice. For this analysis it is assumed that one-axis tracking systems must cost 10% less than flat plate systems, and two-axis tracking systems must cost 20% less, because of their greater complexity and more obtrusive appearance. In the analysis below, therefore, concentrating system energy cost is increased by 1/0.9 and 1/0.8 for one-axis and two-axis tracking systems respectively. This allows direct comparison with non-tracking options. This is termed the concentrator premium.

3. COST COMPARISONS

3.1 The Market Segments

Four different market segments are analyzed and compared. These are described below.

1. Medium-Sized System, High-Resource Area. The first market is for a medium-sized systems in a high-solar-resource areas. Insolation data from Albuquerque, New Mexico is used, although the result is representative of any desert or high-insolation region. By medium-sized system, one in the 100 kW to 10 MW size is considered, and cost components applicable to such a system are used. (Large-sized systems, comparable to utility power plants are not analyzed here.)

2. Small-Sized System, High-Resource Area For this case, cost components representative of a small installation in the 2 kW to 100 kW size are assumed.

3. Medium-Sized System, Low-Resource Area. This is the same as 1) above except that the resource data from Boston, Massachusetts is used. This is representative of a low solar resource area with a high diffuse component to the insolation.

4. Small-Sized System, Low-Resource Area This is the same as 2) above except that resource data from Boston, Massachusetts is used.

3.2 The Candidate Technologies

Ten different system approaches are compared. These are discussed below. Detailed cost assumptions are included on the spread sheets in the appendix. It is not proposed that the cost data is the last word. The reader is encouraged to supply his or her favorite assumptions and, thereby, come to conclusions for which they are most comfortable.

1. Fixed Flat-Plate (FFP) This is the standard silicon module, mounted facing south at a slope equal to the latitude.

279

2. 1-axis Tracking Flat-Plate (1-axis FP) Standard silicon modules are mounted on a horizontal, north-south axis tracker. This is used in some larger installations for producing higher summer capacity factors.

3. 2-axis Tracking Flat-Plate (2 axis-FP) Standard silicon modules are mounted an a two-axis tracker which is always facing the sun during daylight hours.

4. 1-axis Tracking Parabolic Trough (Si 1-axis trough) This is a polar-axis tracking reflective dish with 50 X concentration on a photovoltaic receiver.

5. Static Concentrator A static concentrator with a concentration of 4X is assumed. It is mounted south-facing with latitude slope.

6. Thin Film The costs are for a generic thin-film module. Future module cost is assumed to be $75 per square meter with an efficiency of 12% (The year 2000 DOE goal for CIS). This gives a module cost of $0.63 per watt—which is certainly thought to be an aggressive goal for thin films. The assumed costs are thus projections on what might happen; however, there is no certainty that these goals can be met.

7. Central Receiver In this concept, a field of mirrors directs light to a high-concentration, water cooled photovoltaic panel at the top of a tower.

8. 2-axis Tracking Static Concentrator (2-axis Static) This is a static concentrator as in item 4, mounted on a two-axis tracker.

9. High-concentration Silicon Point-Focus Concentrator (Si Dish) This is a reflective dish using high-efficiency silicon concentrator cells operating at a concentration of 400X. The result also applies to a point-focus Fresnel concentrator. The analysis is not accurate enough to distinguish between these two concentrating means.

10. High Concentration GaAs Point-Focus Concentrator (GaAs Dish) This a system similar to the above but the silicon cell is replaced with a very high efficiency, multi-junction cell based on III-V (gallium arsenide related) materials. This cell does not exist as a commercial product today; however, it is possible that it will in the future and that performance and cost will meet the target in eight to ten years.

3.3 Results

The results of this calculation are shown on the following plots. The associated input assumptions are on the spreadsheets in the appendix. The taller bar represents the "present" costs and the shorter bar "future" costs, i.e., where the technology has the potential to go in ten years. In the case of systems other than flat-plate silicon, the "present" cost is an estimate of what might be achievable with a serious development effort in several years time. The technologies are sorted on the basis of future cost potential. The assumed levelized fixed charge rate is 10%. This is intended to be the current dollar fixed charge rate.

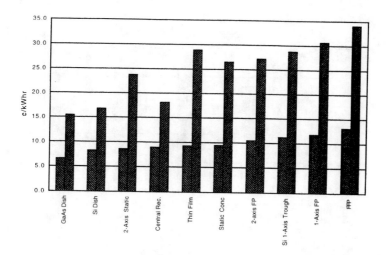

Figure 1. For the medium-sized plant in Albuquerque, the concentrator options beat all the thin-film and silicon flat-plate approaches. The quickest path to the lowest cost would appear to be to start with the silicon dish, refine the concept and reduce dish and cell costs, and finally move to GaAs cells when they are ready (in ten years or so). This path is shown by the arrows. No other approach demonstrates as low a cost, both now and in the future. Static concentrators, mounted on two-axis trackers are a close second.

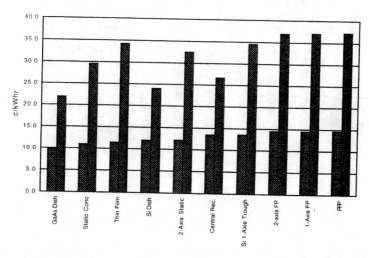

Figure 2. For small-sized plants in Albuquerque, the GaAs dish still has the lowest ultimate cost; however, the future thin-film and static concentrator have moved ahead of the silicon dish. The Si dish still represents the lowest present cost and can be viewed as a vehicle to develop the GaAs dish. Static concentrators remain the second best choice with thin films third.

281

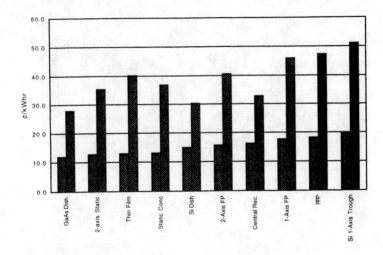

Figure 3. For medium-sized plants in Boston, the GaAs dish surprisingly maintains its lead, despite the lower direct normal solar resource. (In other words, a dish based on 35% efficient cells is something of the ultimate technology). The 2-axis static concentrator remains ahead of the thin-film approach and Si-dish stays as the lowest cost present approach.

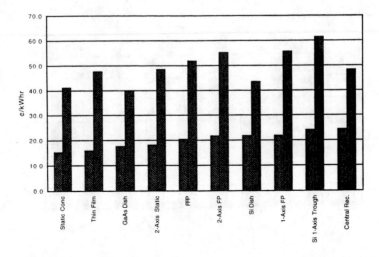

Figure 4. For small-sized plants in Boston, the non-tracking approaches finally take the lead. Nevertheless, the static concentrator remains ahead of the thin film approach. Third place is taken by the GaAs dish. This is the only case where flat-plate silicon appears to beat the Si dish, although not after the transition to GaAs. Note that, in none of the target markets does wafered silicon appear to be the lowest cost option, now or in the future.

4. Required Ingredients for Commercialization.

From the preceding analysis, it would appear that concentrators have great potential to become the low cost photovoltaic option, producing power in the 7¢ to 22¢ per kilowatt hour range, depending on system size and location. Combining the issues discussed in Sections 2 and 3, the most natural market is for medium-sized systems grid-support or PV-Diesel hybrid applications. Most of the customer concerns are best addressed with this market. What then is needed for the industry to take its place as a serious energy supplier? The following might do it.

- Infusion of significant capital, preferably from large, international corporations. The concentrator industry has been something of a garage-shop industry dependent on government contracts to date. Much more resources are needed to provide the sophisticated designs, reliability testing, automated manufacturing, service infrastructure, marketing, and customer support needed to move to the next level.

- Significant penetration of the remote Diesel markets in the form of Diesel-hybrid systems. This market appears to offer the best combination of need and suitability to concentrator systems. The industry must soon expand shipments enough to reach a critical mass before alternative technologies become low-enough in cost to squeeze out further competition.

- Government support. Governments can help in the from of support for green marketing, pollution credits, tax credits and R&D funding. Increased concern over greenhouse gasses, should it develop, will obviously make government support more likely.

- Further depletion or disruption of fossil fuel supplies. In the event that the above do not happen, or are insufficient, there always remains the eventual spark of reduced availability of fossil fuel. One may argue as to when this might occur, but lets hope we don't wait until it does.

5. Conclusions and Recommendations

Concentrators will never be the whole PV business—different products will evolve for different markets. High-efficiency concentrators will not be beat, however, for low-cost, medium-scale power in good solar resource regions. For the smallest premium power markets, the only possible competition to wafered silicon modules are thin-films and static concentrators. Static concentrators appear to have lower cost potential than thin films.

The following recommendations emerge from this analysis. First, pursue high-concentration silicon concentrating systems as a market entry vehicle and as a means to quickly evolve concentrator technology. Next, develop high-efficiency tandem cells to be introduced into high-concentration systems when cost and performance warrants. Finally Explore static concentrators in a effort to find a cost-effective, practical design.

ACKNOWLEDGEMENTS

The author is indebted to those who have helped bring concentrator systems close to commercial readiness—dedicated people from Sandia, NREL, EPRI and, of course, his colleagues at SunPower.

APPENDIX

MEDIUM PLANT-ALBUQUERQUE		GaAs Dish	Si Dish	2-Axis Static	Central Rec.	Thin Film	Static Conc	2-axis FP	Si 1-Axis Trou	1-Axis FP	FFP
Desert (Albuquerque)	KWhr/m2/day	6.566	6.566	8.624	5.025	6.336	6.336	8.624	6.08	7.41	6.336
Diffuse (Boston)	KWhr/m2/day	3.626	3.626	5.782	2.775	4.554	4.554	5.782	3.42	4.94	4.554
Albedo factor		1	1	1	1	1	1	1	1	1	1
BOS Area (low)	$/m2	70	70	70	70	70	70	70	70	70	70
BOS Area (high)	$/m2	140	140	140	140	140	140	140	140	140	140
BOS Power (low)	$/W	0.3	0.3	0.3	0.3	0.3	0.3	0.3	0.4	0.3	0.3
BOS Power (high)	$/W	0.6	0.6	0.6	0.6	0.6	0.6	0.6	0.7	0.6	0.6
Tracking (low)	$/m2	35	35	35	35	0	0	35	20	20	0
Tracking (high)	$/m2	67	67	67	67	0	0	67	40	40	0
Module (low)	$/m2	90	90	115	30	75	85	75	90	75	75
Module (high)	$/m2	160	160	230	60	150	160	150	160	150	150
Cell (low)	$/m2	30000	15000	300	20000	0	300	200	5000	200	200
Cell (high)	$/m2	100000	20000	1000	25000	30	1000	400	15000	400	400
Cell Efficiency (high)		0.3325	0.26	0.21	0.26	0.12	0.21	0.2	0.24	0.2	0.2
Cell Efficiency (low)		0.285	0.23	0.17	0.23	0.08	0.17	0.15	0.2	0.15	0.15
Operating Temp.		65	65	60	65	55	60	55	65	55	55
deta/dteta		2.20E-03	2.20E-03	3.30E-03	2.20E-03	2.00E-03	3.30E-03	3.30E-03	2.40E-03	3.30E-03	3.30E-03
Concentration		1000	400	4	400	1	4	1	50	1	1
Module Transmission		0.85	0.85	0.9	0.85	0.95	0.9	0.95	0.9	0.95	0.95
BOS eff		0.85	0.85	0.9	0.85	0.9	0.9	0.9	0.85	0.9	0.9
Conc premium		0	0	0	0	0	0	0	0	0	0
Cost-diff low	¢/KWhr	12.0	15.0	12.9	16.3	13.0	13.2	15.7	20.1	17.8	18.3
Cost-diff high	¢/KWhr	28.0	30.4	35.5	32.9	40.3	36.9	40.7	51.2	46.0	47.4
Cost-Desert low	¢/KWhr	6.6	8.3	8.6	9.0	9.4	9.5	10.5	11.3	11.8	13.2
Cost-Desert high	¢/KWhr	15.5	16.8	23.8	18.2	28.9	26.5	27.3	28.8	30.7	34.1
Cost-low	$/W	1.59	1.99	2.71	1.66	2.16	2.19	3.32	2.50	3.20	3.05
Cost-high	$/W	3.70	4.02	7.49	3.33	6.69	6.14	8.58	6.39	8.30	7.89

Figure 5. Cost assumptions for medium-sized systems

SMALL PLANT-ALBUQUERQUE		GaAs Dish	Static Conc	Thin Film	Si Dish	2-Axis Static	Central Rec.	Si 1-Axis Trou	2-axis FP	1-Axis FP	FFP
Desert (Albuquerque)	KWhr/m2/day	6.566	6.336	6.336	6.566	8.624	5.025	6.08	8.624	7.41	6.336
Diffuse (Boston)	KWhr/m2/day	3.626	4.554	4.554	3.626	5.782	2.775	3.42	5.782	4.94	4.554
Albedo factor		1	1	1	1	1	1	1	1	1	1
BOS Area (low)	$/m2	100	100	100	100	100	100	100	100	100	100
BOS Area (high)	$/m2	200	200	200	200	200	200	200	200	200	200
BOS Power (low)	$/W	0.4	0.4	0.4	0.4	0.4	0.4	0.4	0.4	0.4	0.4
BOS Power (high)	$/W	0.7	0.7	0.7	0.7	0.7	0.7	0.7	0.7	0.7	0.7
Tracking (low)	$/m2	35	0	0	35	35	35	20	35	20	0
Tracking (high)	$/m2	67	0	0	67	67	67	40	67	40	0
Module (low)	$/m2	90	85	75	90	115	30	90	75	75	75
Module (high)	$/m2	160	160	150	160	230	60	160	150	150	150
Cell (low)	$/m2	30000	300	0	15000	300	20000	5000	200	200	200
Cell (high)	$/m2	100000	1000	30	20000	1000	25000	15000	400	400	400
Cell Efficiency (high)		0.3325	0.21	0.12	0.26	0.21	0.26	0.24	0.2	0.2	0.2
Cell Efficiency (low)		0.285	0.17	0.08	0.23	0.17	0.23	0.2	0.15	0.15	0.15
Operating Temp.		65	60	55	65	60	65	65	55	55	55
deta/dteta		2.20E-03	3.30E-03	2.00E-03	2.20E-03	3.30E-03	2.20E-03	2.40E-03	3.30E-03	3.30E-03	3.30E-03
Concentration		1000	4	1	400	4	400	50	1	1	1
Module Transmission		0.85	0.9	0.95	0.85	0.9	0.85	0.9	0.95	0.95	0.95
BOS eff		0.85	0.9	0.9	0.85	0.9	0.85	0.85	0.9	0.9	0.9
Conc premium		0.2	0	0	0.2	0.2	0.2	0.1	0.2	0.1	0
Cost-diff low	¢/KWhr	17.7	15.4	16.0	21.9	18.2	24.5	24.2	21.8	21.9	20.5
Cost-diff high	¢/KWhr	39.7	41.2	47.7	43.6	48.6	48.4	61.5	55.2	55.7	51.9
Cost-Desert low	¢/KWhr	9.8	11.0	11.5	12.1	12.2	13.5	13.6	14.6	14.6	14.7
Cost-Desert high	¢/KWhr	21.9	29.6	34.3	24.1	32.6	26.7	34.6	37.0	37.2	37.3
Cost-low	$/W	2.35	2.55	2.66	2.90	3.84	2.48	3.02	4.59	3.95	3.40
Cost-high	$/W	5.26	6.85	7.93	5.77	10.25	4.91	7.68	11.66	10.05	8.63

Figure 6. Cost assumptions for small-sized systems.

284

Low Cost High Power GaSb ThermoPhotovoltaic Cells

Lewis M. Fraas, Han X. Huang, Shi-Zhong Ye, James Avery, and
Russell Ballantyne

JX Crystals Inc.
Issaquah, WA 98027

Abstract. High power density and high capacity factor are important attributes of a TPV system and GaSb cells are enabling for TPV systems. A TPV cogeneration unit at an off grid site will compliment solar arrays producing heat and electricity on cloudy days with the solar arrays generating electricity on sunny days. Herein, we project that GaSb cells generating 2 Watts each can be made in 1 MW quantities at $4 per cell. This will allow TPV circuits to be made at $2 per Watt. At this cost, the off-grid cogeneration and self-powered furnace markets will be viable.

Introduction

Over the last two decades, the photovoltaic community has focused its attention on producing low cost flat plate solar arrays. This has proved to be a very formidable task for two reasons. First, the solar intensity is very small corresponding to only 0.1 Watts per square centimeter and, second, sunshine is only available during daylight hours. A 10% efficient traditional solar cell, therefore, will only produce 0.01 Watts per sq. cm. for approximately 5 hours per day. It then takes 20,000 days or approximately 50 years for this 1 sq. cm. cell to produce I kWhr of electricity. If a solar cell were to be made for 10 cents per sq. cm., it would then take approximately 50 years for it to pay for itself, yet 10 cents per sq. cm. is a very small number compared to the cost of all other semiconductor devices. These facts have led to a focus on thin film solar cells for low cost electricity produced from photovoltaic devices.

There are, however, alternatives based on higher power density. In a first alternative, lenses are used with solar cells to concentrate the sun's energy onto the cell. For example, a concentration ratio of 100 in a desert location where one could expect 10 hours per day of sunshine would reduce the time to produce 1 kWhr to 100 days. Unfortunately, this alternative has received much less attention than the thin film option perhaps because it is considered to be a much more geographically limited option.

The truth is that all solar cells, even thin film cells, will be limited to sunny locations, and furthermore, people do live in colder climates as well as warmer climates and even in warmer climates, winters are colder with less sunshine than summers. These facts may be limiting for solar cells but they create an opportunity for thermophotovoltaic cells, a second high power density alternative. Referring to figure 1, infrared sensitive thermophotovoltaic cells can be combined with man

CP404, *Future Generation Photovoltaic Technologies: First NREL Conference*, edited by McConnell
© 1997 The American Institute of Physics 1-56396-704-9/97/$10.00

made heat sources to cogenerate electricity along with heat in colder climates and during winter months.

Perhaps the most interesting thing about the thermophotovoltaic application is that the electric power that a 1 sq. cm. cell can produce has been measured at 2 Watts. Furthermore, the generator can run day or night. This implies that a TPV cell operating 10 hours per day will produce 1 kWhr of electricity in 50 days..

From the above considerations, the economic advantages of high power density and higher capacity factor for the concentrator and thermophotovoltaic approaches are considerable. Geographically, these approaches are complimentary. Furthermore, both approaches are well founded on single crystal semiconductor device technology. So, no miraculous breakthroughs in thin film solar cell technology are required.

For the past several years, JX Crystals has been fabricating GaSb thermophotovoltaic cells. We have the equipment, facility, and know-how to grow the semiconductor crystals, convert the crystals to wafers, process the wafers into cells, and mount the cells into power producing circuits. This paper will focus on a quantitative analysis of the cost of thermophotovoltaic crystals, cells, and circuits as well as the implications of these resultant costs for various potential thermophotovoltaic market sectors.

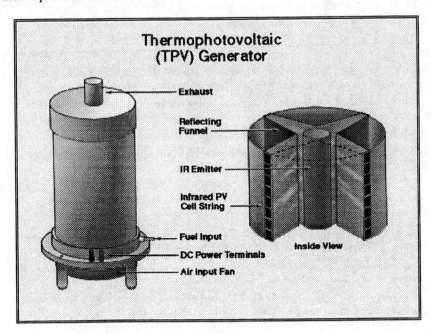

FIGURE 1: Thermophotovoltaic generator concept

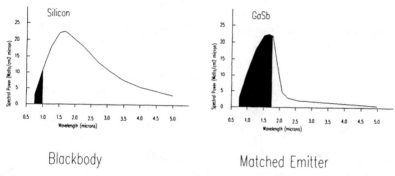

Blackbody   Matched Emitter

FIGURE 2: Silicon vs GaSb with a blackbody emitter and matched emitter.

Gallium Antimonide Thermophotovoltaic Cells

The idea of thermophotovoltaic generators originated in the 1970's. However until recently, there was a mismatch between the power spectrum emitted by man-made heat sources and the response band of the available photovoltaic cells. This problem and its solution are illustrated in figure 2. The temperature of the heated and radiating emitter in a TPV generator is limited to the 1400 C to 1700 C range by the nature of hydrocarbon combustion and by the thermal durability of appropriate ceramic materials. This means that nearly all of the energy emitted is in the infrared at wavelengths longer than 1 micron. Unfortunately, the only available photovoltaic cell until recently, the silicon cell, responds to wavelengths shorter than 1.1 microns. Referring to figure 2a, the power available to a Si cell from a 1700 C emitter, the black sliver under the power curve, is very small while the wasted infrared power, the white region under the power curve, is very large.

The solution to this problem is shown in figure 2b. In 1989, Fraas, Avery, and Girard invented the GaSb infrared cell while at Boeing. Referring to figure 2b, this GaSb cell which responds out to 1.8 microns, dramatically increases the convertible power. In 1993, JX Crystals Inc. obtained from Boeing an exclusive license to this cell technology. Subsequently in 1996, JX Crystals Inc. discovered a ceramic emitter material which emits strongly below 1.8 microns and weakly at longer wavelengths. Figure 2b shows the results of combining the GaSb cell with the new matched emitter. JX Crystals Inc. has filed a patent application on this new matched emitter and has been recently informed that the appropriate claims have now been allowed. As figure 2b shows, the combination of these two new key components now provides for higher power density and higher conversion efficiency than prior art TPV systems.

JX Crystals Inc. now has GaSb cells in low volume production Table 1 summarizes illuminated current vs voltage test data for a best cell flash tested at room temperature and a typical water cooled cell in continuous operation in front of a heated SiC emitter operating at 1380 C. In the following sections, our cost analysis will be based on cells with total area of 1.5 sq. cm.with each cell producing 2 Watts of electricity.

TABLE 1: Representative 1.5 sq. cm. GaSb cell performance data.

	FF	Voc	Isc	Imax	Vmax	Pmax
Flash tested best cell	0.84	0.52V	5.95A	5.65A	0.46V	2.59W
Cell with 1380 C glowbar	0.67	0.44V	5.87A	5.36A	0.33V	1.76W

Before presenting our detailed study of GaSb cell costs in the following section, some overview observations are appropriate. The readers initial question may be: GaSb, is the cost of Ga and Sb raw material prohibitive? The answer to this question is that the total cost of Ga and Sb in a 1.5 sq. cm. GaSb cell is 15 cents. For a flat plate solar cell, this cost would be prohibitive, but for a 2 Watt TPV cell, this cost is nearly insignificant. The real cost of a single crystal semiconductor chip is not the material cost but the processing cost. This point can be easily made by noting that carbon is cheap but diamonds are very expensive.

Given that processing cost dominate, we have made an effort to copy low cost silicon solar cell processing in fabricating GaSb cells. Specifically, we use converted silicon pullers for crystal growth and diffusions for junction formation. Nearly all of the equipment at our facility is used silicon processing equipment. Epitaxy, generally used in GaAs space solar cell fabrication, is an example of a process we have avoided. This process uses toxic gases creating a huge safety expense. Furthermore, it is very capital intensive and has low wafer throughput. The key to low cost devices is high throughput with minimal capital and labor.

Projected Wafer, Cell, and Circuit Fabrication Costs

Table 2 summarizes our projected costs at annual production volumes of 100 kW and 1 MW per year. These costs are broken down into three areas: Wafco by which we mean wafer fabrication; Cellco by which we mean cell fabrication; and Circo by which we mean power circuit fabrication. Wafco operations include crystal growth, rounding, slicing, and etching. Cellco operations include diffusion, photolithography, front and back metal deposition, filter deposition, testing, scribing, and dicing. And finally, Circo operations include solder dispense, pick and place, die attach, inner and outer lead bonding, and testing. The finished product is a 20 cell circuit that produces 40 Watts of electricity.

The data in table 2 are based on the present equipment, facility and know-how at JX Crystals. For example, referring to Wafco, we presently have two crystal pullers which together can produce one 3" diameter crystal per day where each crystal when sliced will produce 110 wafers per day. Assuming 5 days per

288

Table 2: Projected Manufacturing Costs in $ / Watt

	100 kW per year	1 MW per year
WAFCO		
Capital	0.54	0.07
Lease	0.14	0.01
Leasehold Improvement	0.07	0.02
Power	0.01	0.01
Direct Material	0.30	0.25
Indirect Material	0.07	0.05
Maintenance	0.14	0.06
Direct Labor	0.48	0.19
Benefits @ 35%	0.17	0.07
Wafco Total	**$ 1.92**	**$ 0.73**
CELLCO		
Capital	0.58	0.09
Lease	0.19	0.02
Leasehold Improvement	0.07	0.02
Power	--	--
Direct Material	0.11	0.08
Indirect Material	0.21	0.15
Maintenance	0.15	0.11
Direct Labor	0.50	0.17
Benefits @ 35%	0.17	0.06
Cellco Total	**$ 1.98**	**$ 0.70**
CIRCO		
Capital	0.07	0.10
Lease	--	--
Leasehold Improvement	--	--
Power	--	--
Direct Material	0.37	0.24
Indirect Material	--	--
Maintenance	0.03	0.02
Direct Labor	0.33	0.13
Benefits @ 35%	0.12	0.05
Circo Total	**$ 0.92**	**$ 0.54**
GRAND TOTAL	**$ 4.82**	**$ 1.97**

week and 52 weeks per year and 90% yield, this implies that we can produce approximately 25,000 wafers per year. Given that there are 20 cells per wafer at 2 Watts each, then Wafco's full capacity is 1 MW. Similarly, we chose the 100 kW throughput number because this represents the single shift present capacity for Circo as well as the probable market demand for the first year of production.

Summing the cost in table 2 gives a power circuit cost of $4.82 per Watt at a production volume of 100 kW per year dropping to $1.97 per Watt at a volume of 1 MW per year. These cost translate to a per cell cost in a circuit of $9.64 and $3.94 respectively. Note that at a production volume of 100 kW, capital and labor cost are still dominate.

Discussion of Costs

While space does not allow a complete discussion of all of the inputs in table 2, we will discuss some specific examples to provide a flavor of the manufacturing assumptions.

As a first example, we describe the front and back metallization in the Cellco process. We presently use an electron beam evaporation machine with an 18" bell jar capable of coating eighteen 3" diameter wafers at a time. A 100 kW cell annual production rate given 40 Watts per wafer is equivalent to 50 wafers per week or 10 wafers per day. Fifty four wafers can be coated per week in 3 afternoon runs for front metal and 3 morning runs for back metal. This implies a single shift equipment utilization of 60%. In order to produce 1 MW per year, we would use two machines all five days per week on three shifts.

As a second example, refer to the Ga and Sb materials cost in the 100 kW Wafco column in table 2. Note that 30 cents per Watt corresponds to 60 cents per cell which is 4 times larger than the cost of Ga and Sb in a cell. Why? It turns out that for the 4 kg that goes into a crystal, only 1 kg is incorporated in cells for our current process. This is because 1 kg is lost in rounding and 1.5 kg is lost in slicing and 0.5 kg is lost at the edge of a completely processed wafer. While this yield is poor, the total Ga and Sb costs are still insignificant at these production rates. However, this does not mean that improvements in utilization efficiency are not possible. For example, automatic diameter control during crystal growth will reduce rounding losses, a wire saw rather than an ID saw will reduce kerf losses, and larger 4" or 6" wafers will reduce the edge loss percent.

Turning now to Circo costs, we note that 100 kW per year can be accomplished in batches primarily by manual labor. However, when it comes to 1 MW per year, we plan to invest in automated equipment for the pick and place and lead bonding operations. Also note that the purchased substrate direct material cost is as large as the Ga and Sb materials cost.

For experts in semiconductor wafer fabrication, it is instructive to state costs in per wafer costs. Thus, in the 100 kW production column, we are fabricating 3" wafers at a cost of approximately $160 each, and in the 1 MW column, we are fabricating 3" wafers at $56 each. These numbers can be compared with the cost of a 4" single crystal silicon solar cell produced at approximately $5 each in much higher volumes.

Clearly, processed wafer costs decrease with increasing volume implying that still further cost reduction can occur for TPV cells beyond the projections in table 2. However, on a cost per Watt bases, note that the $5 per Watt circuit cost projection is comparable to the cost of a flat plate solar module today and note that the $2 per Watt cost projection is comparable to the target set for solar modules for the last several years, a target still not achieved. Also note that $2 per Watt for a TPV circuit will really be more cost effective than $2 per Watt peak for a flat plate solar module because of the higher capacity factor for the TPV generator. The TPV generator can generate more Watt-hrs per day.

Applications

The first commercial market for a TPV generator is as a compliment to flat plate solar panels in off-grid dwellings. Our market research indicates that there are at least 20 thousand potential customers per year purchasing solar panels in off -grid environments. These environments include mountain cabins, sailboats, and recreational vehicles. Many of these customers are accessible through solar panel distributors. Figure 3 shows a Midnight Sun Micro-Cogeneration unit designed and built by JX Crystals to address this market (patent pending). This unit produces approximately 40 Watts of electricity and 8,000 BTU per hour of heat.

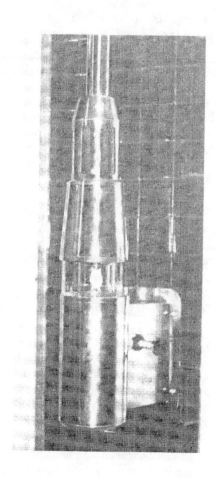

FIGURE 3: Midnight Sun Micro-Cogenerator produces 40 W of electricity and 8,000 BTU per hour of heat.

We believe that our Midnight Sun Micro-Cogeneration unit will sell well in this market because, while solar panels work well during warm summer months, when winter comes around and clouds, rain, or snow come, there is a need for heat and a need to provide for the electricity no longer provided by the solar panel. The midnight Sun unit meets these needs perfectly. Solar panels and the Midnight Sun both are designed to provide 12 Volts of DC to a battery bank. Another positive selling point will be that the customer is already familiar with and sympathetic to photovoltaic panels. Finally, there will be an additional benefit for new solar panel customers in that they can down size their solar panel and battery bank requirements since there will no longer be a need for excess capacity for several days of bad weather. This market should be accessible at the $10 per cell level.

The next largest TPV market and the first grid connected market is the self powered furnace market. A Gas Research Institute study concluded that among the 500,000 to 1,000,000 households in the Northeast and North Central US which experienced outages greater than 4 hours in duration each year, there exist a niche markets willing to pay a premium of $500 to maintain heater service during outages. This need can be met with a TPV system in which a heating furnace is redesigned to include an infrared emitter and a 200 W TPV array wired to a DC blower for forced air heating or a water pump for radiator heating. An Arthor D. Little survey indicated that 20% of new family dwellings with natural gas (70% of total) or approximately 150,000 customers per year would be willing to pay $400 extra for a self powered furnace. Note that $400 for 200 Watts corresponds to $2 per Watt and that 150,000 x 200 W is equal to 30 MW per year. This appears to be an exciting and viable market.

Finally, the biggest market for TPV is probably the home cogen market in which more TPV output is added to the home furnace for colder climates to provide the majority of electricity needed by the home. However, there will be hurdles to overcome to enter this market in that AC will probably be required and the TPV cost will need to be below $2 per Watt. Still, the following calculation indicates why this market may be interesting. At $2 per Watt and 10 hours per day, the TPV cogen unit will cost 20 cents per Watt hour per day. If electricity costs 10 cents per kWhr, the system will pay for itself in 2000 days or 5.5 years.

Conclusions

High power density and high capacity factor are important attributes of a TPV system and GaSb cells are enabling for TPV systems. It is very reasonable to project that GaSb cells generating 2 Watts each can be made in 1 MW or larger quantities at $4 per cell. This will allow TPV circuits to be made at $2 per Watt. At this cost, the off-grid cogeneration and self-powered furnace markets will be viable. It would be appropriate for the photovoltaic community to target the home cogeneration market as a huge potential market in the next century and to expand their horizon to beyond the solar PV silicon cell.

NEW MATERIALS AND DEVICE ARCHITECTURES FOR PV

A CdSe nanocrystal/MEH-PPV polymer composite photovoltaic

N. C. Greenham, Xiaogang Peng, and A. P. Alivisatos

Department of Chemistry, University of California, and

Materials Science Division, Lawrence Berkeley National Laboratory, Berkeley, California 94720

Abstract. We have prepared simple photovoltaic devices based on composite materials formed by mixing cadmium selenide or cadmium sulfide nanocrystals with the conjugated polymer poly(2-methoxy, 5-(2'-ethyl)-hexyloxy-*p*-phenylenevinylene) (MEH-PPV). When the surface of the nanocrystals is treated so as to remove the surface ligand, we find that the polymer photoluminescence is quenched, consistent with rapid charge separation at the polymer/nanocrystal interface. Transmission electron microscopy (TEM) of these quantum dot/conjugated polymer composites shows clear evidence for phase segregation with length scales in the range 10 - 200 nm, providing a large area of interface for charge separation to occur. Thin-film photovoltaic devices using the composite materials show quantum efficiencies which are significantly improved over those for pure polymer devices, consistent with improved charge separation. At high concentrations of nanocrystals, where both the nanocrystal and polymer components provide continuous pathways to the electrodes, we find quantum efficiencies of up to 12 %. The absorption, charge separation and transport properties of the composites can be controlled by changing the size, material and surface ligands of the nanocrystals.

INTRODUCTION

Composites of organic polymers and inorganic nanocrystals are particularly interesting materials in the study of electrical transport. The band gaps and offsets of typical semiconducting polymers and nanocrystals are such that charges will separate across an interface between them. This paper explores the extent of such charge separation, and the nature of charge transport in polymer/nanocrystal blends. An elementary photovoltaic based on a nanocrystal/polymer blend is described. A more complete description of this work has appeared elsewhere[1].

Electronic processes in conjugated polymers are currently the subject of intensive study, both because of fundamental interest in the nature of the electronic excitations in these "one-dimensional" semiconductors, and because they have potential applications in a range of electronic devices such as light-emitting diodes[2]. Conjugated polymers have the advantage of being easy to process to form large-area devices, and their energy gap and ionization potential can readily be tuned by chemical modification of the polymer chain. Large-area

CP404, *Future Generation Photovoltaic Technologies: First NREL Conference*, edited by McConnell

© 1997 The American Institute of Physics 1-56396-704-9/97/$10.00

thin-film photovoltaic devices based on conjugated polymers are also of interest, although devices fabricated using a single layer of polymer have been found to have low efficiencies of conversion of incident photons to electrons. Efficient collection of charge carriers requires that the neutral excited states (singlet excitons) produced by photoexcitation be separated into free charge carriers, and that these carriers are then transported through the device to the electrodes without recombining with oppositely-charged carriers. The possibility that conjugated polymer/nanocrystal composites may have the desired attributes of charge separation and transport motivates the present work.

Charge separation in conjugated polymers has been found to be enhanced at the interface with a material of higher electron affinity where it is energetically favorable for the electron to transfer onto the second material. Examples of such materials include C_{60},[3] cyano-substituted conjugated polymers, and various small organic molecules. Since the diffusion range of singlet excitons in conjugated polymers is typically in the range 5 - 15 nm, it is necessary to have a large area of interface between the two materials in order to achieve a high quantum efficiency for charge separation. Furthermore, the charge separation process must be fast compared to the radiative and non-radiative decays of the singlet exciton, which typically occur with time constants in the range 100 - 1000 ps. In composite materials where charge separation can occur, the photoluminescence is found to be strongly quenched, since the singlet exciton is no longer able to decay radiatively to the ground state. In the absence of an electric field to remove the separated charges, there must exist a non-radiative process by which recombination occurs between electron and hole on adjacent materials. Although this process is not well understood, it is likely to be much slower than the decay of the singlet exciton. The lifetime of the charge-separated species in composites of poly(2-methoxy, 5-(2'-ethyl-hexyloxy)-p-phenylenevinylene) (MEH-PPV) with C_{60} has been estimated to be of the order of milliseconds at 80 K. The problem of transport of carriers to the electrodes without recombination is a more difficult one to solve, since it requires that once the electrons and holes are separated onto different materials, each carrier type has a pathway to the appropriate electrode without needing to pass through a region of the other material. The transport must also be sufficiently fast for the carriers to be removed from the device before significant non-radiative recombination can occur at the interface between the two materials. Encouraging results have been obtained using mixtures of polymers with different electron affinities which phase separate on a length scale suitable to give effective charge separation whilst providing efficient charge transport to the electrodes.[14, 15] Recently, high photovoltaic efficiencies have been reported in C_{60}/MEH-PPV composites with high C_{60} content. In these composites, derivatization of the C_{60} molecule with a flexible alkyl group has been found to give optimum photovoltaic performance, although the detailed morphology of these composites has not yet been reported.

Nanometer-sized crystals of inorganic semiconductors are another interesting class of low-dimensional materials with useful optical and electronic properties[4,5]. When the size of the nanocrystal is smaller than that of the exciton in the bulk semiconductor, the lowest energy optical transition is significantly increased due to quantum confinement. The absorption and emission energy can thus be tuned by changing the size of the nanocrystal. High quality samples of nanocrystals of II-VI semiconductors such as CdS and CdSe can now be prepared by chemical methods. The surface of the nanocrystal is typically capped by an organic ligand which ensures solubility and passivates the surface electronically. By changing the size from 6 to 2 nm, the energy gap can be tuned from 2.6 to 3.1 eV in CdS and from 2.0 to 2.6 eV in CdSe. The ability to tune the electronic structure of the nanocrystals makes them interesting optical materials, however making contact to the nanocrystals for electrical measurements is more difficult. Nanocrystal/conjugated polymer composites offer the prospect of allowing electrical access to the nanocrystals. The electron affinity of CdS and CdSe nanocrystals is in the range 3.8 - 4.7 eV, hence they are suitable materials to act as electron acceptors when combined with conjugated polymers, where the electron affinity is in the range 2.5 - 3.0 eV. In contrast to C_{60}, the optical energy gap of these nanocrystals lies conveniently in the visible region, and it is thus possible not only to study electron transfer from polymer to nanocrystal, but also to study hole transfer from nanocrystal to polymer after excitation in the nanocrystal. The nanocrystal surface ligand can be changed without altering the intrinsic electronic properties of the nanocrystal, hence it is possible to control the charge transfer between polymer and nanocrystal, and from nanocrystal to nanocrystal. The surface ligand is also important in determining the morphology of the polymer/nanocrystal composite. Charge separation at the interface between organic molecules and nanocrystals is currently of great interest, particularly since the report by O'Regan and Grätzel of efficient photovoltaic devices based on organic dyes adsorbed on TiO_2 nanocrystalline films[6]. In these devices, the large area of TiO_2 surface allows a high optical density of dye molecules to be achieved, whilst maintaining efficient charge separation. An electrolyte solution is required to remove the holes from the organic dye after charge separation. Charge transfer between CdS nanocrystals and the hole-transporting polymer poly(N-vinylcarbazole) has previously been studied, however the CdS nanocrystals were at low concentration, and acted primarily as a sensitizer, with charge transport occurring through the polymer, rather than from nanocrystal to nanocrystal[7]. Electroluminescence in blends of CdSe nanocrystals and polymers have been studied[8,9].

TOPO Coated

Uncoated

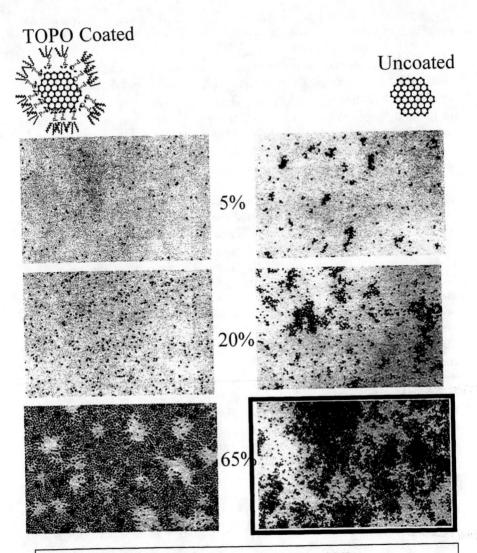

5%

20%

65%

Figure 1   Transmission electron micrpgraphs of thin films of CdSe nanocrystal/polymer blends. The nanocrystals are 5 nm diameter.

In this paper, we study the photoluminescence and photoconductivity of composite materials formed with MEH-PPV and either CdS or CdSe nanocrystals. The nanocrystals may either be in direct contact with the polymer, or, alternatively, the surface of each nanocrystal may be coated with a surfactant molecule, trioctylphosphineoxide (TOPO), which forms a barrier of 11 Å thickness between the nanocrystal core and the polymer. Figure 1. shows transmission electron micrographs of thin films (20nm thick) of various composites of 5 nm diameter CdSe nanocrystals with MEH-PPV, formed under different conditions. When the surfactant is present (left column), the nanocrystals and the polymer intermix. When the surfactant is stripped, and the film is cast from pyridine, the polar nanocrystals phase segregate from the polymer, forming interpenetrating percolating networks. This latter geometry is well suited to photovoltaic formation.

When the surfactant is stripped and the nanocrystals and polymer are in direct contact, the photoluminescence of the polymer is quenched, because electrons are transferred onto the nanocrystrals, and holes onto the polymer. If the surfactant is present, the polymer luminescence is unchanged, presumably because charge transfer across the surfactant is too slow to compete with polymer light emission. In the case of CdSe, the band gap is smaller than in CdS, and so, in addition to charge transfer, the polymer photoluminescence yield is reduced by Forster transfer as well.

Figure 2 shows the current-voltage characteristics of the polymer/nanocrystal devices in the dark and under illumination. In the presence of light the current increases. The photovoltaic response arises when photons are absorbed, electrons migrate to the nanocrystals, and hop from nanocrystal to nanocrystal until they reach the aluminum electrode. Likewise, holes are transported through the polymer layer.

It is useful for comparison to other work on photovoltaic materials to calculate the energy conversion efficiency for the devices studied here. The device described above has a quantum efficiency of 12 % at 514 nm. This value is significantly larger than values of 2 - 6 % reported for polymer/polymer mixtures,[14, 15] but not as large as the value of approximately 36 % found at this excitation intensity in blends of derivatized C_{60} and MEH-PPV.[21] In the devices studied here, the fill factor (defined as $(V I)_{max} / (V_{oc} I_{sc})$, where $(V I)_{max}$ is the area of the largest rectangle under the current-voltage curve between 0 V and V_{oc}) is 0.26, and the open circuit voltage is approximately 0.5 V, giving a power conversion efficiency of 0.6 % at 514 nm. The device begins absorbing at 650 nm, and thus covers a wider spectral range than devices where absorption occurs solely in MEH-PPV. For the solar spectrum under AM1.5 conditions, the device

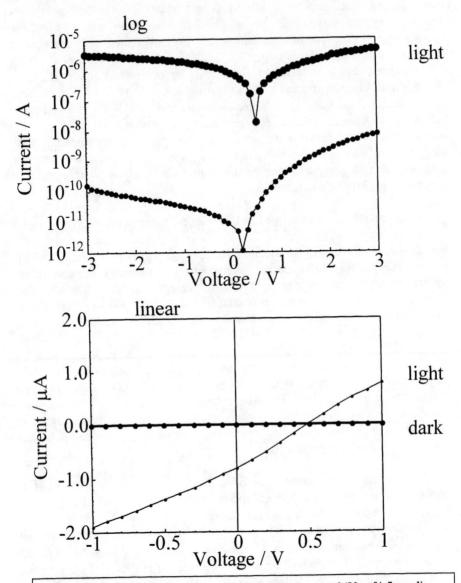

Figure 2 Current-voltage characteristics of a CdSe nanocrystal (90 wt% 5 nm dia. CdSe)/MEH-PPV device in the dark and under illumination at 514 nm. The active area of the device was 7.5 mm^2, and the maximum power density of the illumination was approximately 5 W m^{-2}.

absorbs approximately 37 % of the incident solar energy, and the energy conversion efficiency for AM1.5 conditions at 5 W m^{-2} is 0.2 %. Due to the sublinear intensity dependence of the photocurrent, under one-sun conditions (800 W m^{-2}) the solar power conversion efficiency will be approximately 0.1 %. For comparison, we estimate that the solar energy conversion efficiency for devices using blends of derivatized C_{60} and MEH-PPV with calcium electrodes[21] is no more than 0.5 % under the same conditions. We have identified above the possibility of increasing the open circuit voltage, and hence the power efficiency, by using smaller nanocrystals with lower-workfunction electrodes. Further understanding of the injection and transport of carriers of these devices may in the future allow improved fill factors to be obtained.

REFERENCES

[1]N. C. Greenham, X. G. Peng, and A. P. Alivisatos, "Charge separation and transport in conjugated-polymer/semiconductor-nanocrystal composites studied by photoluminescence quenching and photoconductivity," Physical Review B-Condensed Matter **54** (24), 17628-17637 (1996).

[2]N. Greenham and R. H. Friend, **49** ,1, Solid State Physics (Ehrenreich and Spaepen, eds.) (Academic Press, San Diego, 1995).

[3]G. Yu, J. Gao, J. C. Hummelen *et al.*, "Polymer Photovoltaic Cells - Enhanced Efficiencies Via a Network Of Internal Donor-Acceptor Heterojunctions," Science **270** (5243), 1789-1791 (1995).

[4]A. P. Alivisatos, "Perspectives On the Physical Chemistry Of Semiconductor Nanocrystals," Journal Of Physical Chemistry **100** (31), 13226-13239 (1996).

[5]M. G. Bawendi, M. L. Steigerwald, and L. E. Brus, "The Quantum Mechanics Of Larger Semiconductor Clusters (Quantum Dots)," Annual Review Of Physical Chemistry **41** (V41), 477-496 (1990).

[6]B. Oregan and M. Gratzel, "A Low-Cost, High-Efficiency Solar Cell Based On Dye-Sensitized Colloidal Tio2 Films," Nature **353** (6346), 737-740 (1991).

[7]Y. Wang and N. Herron, "Photoconductivity Of Cds Nanocluster-Doped Polymers," Chemical Physics Letters **200** (1-2), 71-75 (1992).

[8]V. L. Colvin, M. C. Schlamp, and A. P. Alivisatos, "Light-Emitting Diodes Made From Cadmium Selenide Nanocrystals and a Semiconducting Polymer," Nature **370** (6488), 354-357 (1994).

[9]B. O. Dabbousi, M. G. Bawendi, O. Onitsuka *et al.*, "Electroluminescence From Cdse Quantum-Dot Polymer Composites," Applied Physics Letters **66** (11), 1316-1318 (1995).

The 1997 Calaveras Awards
for "Leapfrog" Technologies

Paul Alivisatos

has leapt to a new level by giving

The Most Innovative Presentation

on

Binary and Ternary Nanostructures

The First Conference on Future Generation Photovoltaic Technologies
Sponsored by the National Renewable Energy Laboratory

302

Lateral Superlattice Solar Cells

A. Mascarenhas and Yong Zhang

National Renewable Energy Laboratory, Golden Colorado, 80401

J. Mirecki Millunchick, R.D. Twesten, and E.D. Jones

Sandia National Laboratories, Albuquerque New Mexico, 87185

A novel structure which comprises of a lateral superlattice as the active layer of a solar cell is proposed. If the alternating regions A and B of a lateral superlattice ABABAB... are chosen to have a Type-II band offset, it is shown that the performance of the active absorbing region of the solar cell is optimized. In essence, the Type-II lateral superlattice region can satisfy the material requirements for an ideal solar cells active absorbing region, i.e. simultaneously having a very high transition probability for photogeneration and a very long minority carrier recombination lifetime.

INTRODUCTION

In the design of solar cells one seeks to optimize performance based on choice of materials and device architecture. Invariably, the constraints for this optimization are inherently associated with availability of materials, ease of manufacture, reliability, application (terrestrial or space), and cost which in turn depends on several of the prior criteria. For terrestial flat plate applications, large area polycrystalline or amorphous thin film or polysilicon solar cells satisfy the large active area and low cost requirements, whereas for space and concentrator applications epitaxially grown alloys and tandem solar cell designs are more suitable. Whatever be the solar cell design philosophy employed, it will be based on certain design trade-offs. It is not the purpose of this paper to deal with the above mentioned issues, but to address at a much more fundamental level the constraints imposed by the trade-off between two very important solar cell material performance parameters, namely the minority carrier lifetime and the transition probability for photogeneration in conventional solar cell designs, and to examine if it is possible to overcome present day limitations.

BACKGROUND

The photo-generation of minority carriers that results from the absorption of sunlight in the active region of a solar cell must be very efficient for optimal cell performance and so the material chosen for this region should have a strong absorption coefficient. Direct bandgap semiconductors such as GaAs have strong absorption coefficients because the efficiency with which above bandgap photons are converted into photogenerated carriers is large. Indirect bangap materials such

CP404, *Future Generation Photovoltaic Technologies: First NREL Conference*, edited by McConnell
© 1997 The American Institute of Physics 1-56396-704-9/97/$10.00

as Si have weak absorption coefficients (typically three orders of magnitude lower than that of GaAs) because here the efficiency with which above bandgap photons are converted into photogenerated carriers is low. As shown in Fig. 1b, the incoming photon is only able to photo-generate carriers with the assistance of a momentum conserving phonon which lowers the probability for such processes considerably in comparison to those in Fig. 1a.

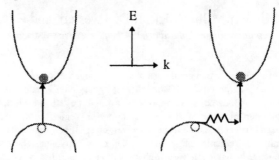

Figure 1a. Direct gap absorption   **Figure 1b.** Indirect gap absorption

In a solar cell, the photogenerated minority carriers must live long enough to be collected and so minority carrier lifetimes in the material comprising the active region should be large. As shown in Fig. 2a, in direct bandgap semiconductor alloys such as GaAs the radiative recombination of photogenerated carriers does not require the assistance of a phonon and so occurs relatively easily resulting in short recombination lifetimes (on the order of microseconds). On the other hand, this process can only occur with the assistance of a momentum conserving phonon in indirect bandgap semiconductors such as Si, making it much less probable, and thus resulting in long lifetimes (on the order of milliseconds). There thus appears to be a fundamental constraint in obtaining a material with both a high transition probability for photogeneration as well as a long recombination lifetime, in the sense that if one optimizes one of these parameters one has to trade-off on the other. This is an intrinsic limitation which is inherent in the choice of semiconductor alloys used for conventional solar cells. In the following section we examine if it is possible to overcome this limitation.

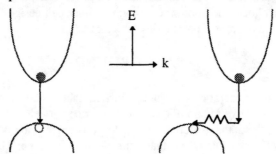

Figure 2a. Direct gap recombination   **Figure 2b.** Indirect gap recombination

LATERAL SUPERLATTICE SOLAR CELLS

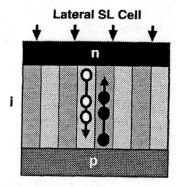

Lateral SL Cell

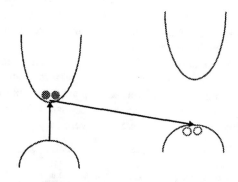

Figure 3a. Schematic of the solar cell.   **Figure 3b.** Spatial band alignment.

Consider the solar cell shown in Fig. 3a in which the active region comprises of a lateral superlattice ABABAB.. for which the band offset between the alternating regions A and B is type-II and is shown schematically in Fig. 3b. Sunlight is efficiently converted into photogenerated carriers in regions A and B since each has a direct bandgap. If the periodicity of the lateral superlattice is of the order of a few hundred Angstroms, then the photogenerated electrons quickly diffuse to the lowest conduction band regions of the superlattice whereas the holes quickly diffuse to the highest valence band regions of the superlattice.

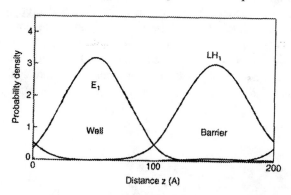

Figure 4. Spatial probability distribution of electrons and holes in a Type-II superlattice.

For the case of a lateral superlattice generated by spontaneous compositional modulation during growth of a GaP/InP short period vertical superlattice structure, a plot of the resulting spatial probability distribution(1) for photogenerated electrons and holes is shown in Fig. 4. Here a valence band offset (in the abscence of strain) of merely 10 meV was assumed to study the effect this has on the spatial localization of carriers. As is evident in Fig. 4, the spatial overlap between the wavefunctions for electrons and holes is very small and so even though the superlattice is direct in k-space it is indirect in real space. The only way for the spatially separated photogenerated electrons and holes to recombine radiatively is by lateral tunneling through their respective barriers thereby making the electron and hole wavefunction overlap integral non-zero. Since the probability for such lateral tunneling is very small in the absence of a lateral electric field, the recombination between the photogenerated electrons and holes is drastically reduce in such a superlattice structure. Such structures (see Fig. 5) thus overcome the intrinsic semiconductor material limitations discussed in the previous section and make it possible simultaneously achieve both a high transition probability for photogeneration as well as a very long photo carrier recombination lifetime.

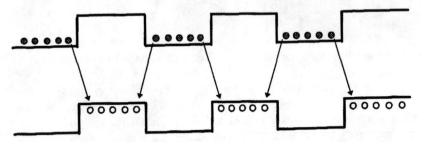

Figure 5. Spatial separation of electrons and holes in a Type II lateral superlattice.

The reason why this cannot be achieved in a conventional vertical superlattice can be seen from Fig. 6. In such a structure, even if the alternating regions of the superlattice are constructed from direct band gap semiconductors with a Type-II band alignment, the photogenerated electrons and holes are not constrained to move in spatially separated regions as they are swept towards the n and p regions of the solar cell respectively. The electron and hole paths and hence their wavefunctions have a strong spatial overlap in such a structure, and so there is no improvement as regards enhancement of photo carrier recombination lifetime.

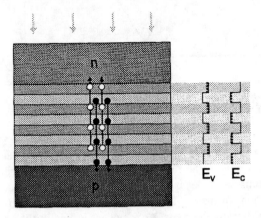

Figure 6. Schematic of a Type-II vertical superlattice solar cell

SYNTHESIS OF LATERAL SUPERLATTICE STRUCTURES

Although the synthesis of vertical superlattice structures using Molecular beam Epitaxy (MBE) and Organo-metallic Vapor Phase Epitaxy (OMVPE) techniques has been well established, the synthesis of lateral superlattice structures is relatively new and far more difficult. Techniques such as growth on V-groove patterned substrates(2) or use of metallic stripe overlayers for strain pattern transfer(3)have been employed to fabricate lateral quantum wires but they both involve elaborate and costly processing steps and are incapable of yielding small lateral periodicity's. Recently there have been several(4,5) experimental demonstrations of a relatively simple method for generating lateral superlattice structures which does not require any elaborate processing steps and is compatible with MBE and OMVPE techniques. The method relies on the phenomenon of spontaneous composition modulation in ternary semiconductor alloys ABC_2 or in vertical short period superlattice (SPS) structures $(AB)_n/(AC)_n$, where the binary constituents AB and AC are size mismatched.

Cross Sectional TEM

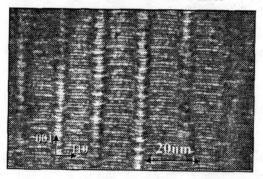

Figure 7. X-TEM image showing lateral composition modulation

As shown in Fig. 7, Mirecki et al. have demonstrated the spontaenous formation of lateral superlattices with periodicities of around 130 Å during growth of SPS structures of InAs/AlAs using MBE(5). Another group(4) has demonstrated the spontaneous formation of lateral superlattices during OMVPE growth of AlInAs epilayers on InP substrates. A similar phenomenon has been earlier demonstrated in the InP/GaP system and the InAs/GaAs system(6,7). It should be noted that in all these lateral superlattices the periodicity is not very regular but the structural quality is high as evidenced by the absence of dislocations. Although the type of band offset between the compositionally modulated regions in any of these structures has not yet been established it appears that this technique of achieving lateral superlattices could be useful for designing ideal solar cell structures in situations where the band offset between the compositionally modulated regions is Type-II. The lack of perfect regularity in the lateral periodicity exhibited in these structures is not of serious consequence for the application regarding solar cells since the only purpose of the lateral composition modulation here is to isolate the photogenerated electrons from the photogenerated holes.

CONCLUSION

It is shown that the performance of the active absorbing region of the solar cell is optimized if it is synthesized using a lateral superlattice ABABAB.... in which regions A and B comprise semiconductor alloys with direct bandgaps that are optimized to the solar spectrum and have a type-II band allignment with respect to each other. It is shown that the Type-II lateral superlattice can satisfy the material requirements for an ideal solar cells active absorbing region, i.e. simultaneously having a very high transition probability for photogeneration and a very long minority carrier recombination lifetime. The spontaneous lateral composition modulation technique offers the possibility of practically realizing such structures for solar cell applications.

Acknowledgments. This work was supported by the Office of Energy Research Material Science Division of the DOE under contract No. DE-AC36-83CH10093 and by funding from an NREL LDRD program.

REFERENCES

1. A. Mascarenhas, R.G. Alonso, G.S. Horner, and S. Froyen, *Phys. Rev. B*48, 4907-4909 (1993).
2. G. Biasiol, F. Reinhardt, A. Gustafsson, E. Martinet, and E. Kapon, *Appl. Phys. Lett.* **69**, 2710-2712 (1996).
3. K. Kash, J. M. Worlock, M. D. Sturge, P. Grabbe, J. P. Harbison, A. Scherer and P. S. D. Lin, *Appl. Phys. Lett.* **53**, 782-784 (1988).
4. Sung Won Jun, Tae-Yeon Seong, J. H. Lee and Bun Lee, *Appl. Phys. Lett.* **68**, 3443-3445 (1996).
5. J. Mirecki Millunchick, R. D. Twesten, D. M. Follstaedt, S. R. Lee, E. D. Jones, Yong Zhang, S. P. Ahrenkiel, and A. Mascarenhas, *Apl. Phys. Lett.* **70**, 1402(1997).
6. K.C. Hsieh, J.N. Baillargeon, and K.Y. Cheng, *Appl. Phys. Lett.* **57**, 2244-2246 (1990).
7. K.Y. Cheng, K.C. Hsieh, and J.N. Baillargeon, *Appl. Phys. Lett.* **60**, 2892-2894 (1992).

Hot Carrier Solar Cells

Mark C. Hanna, Zhenghao Lu and Arthur J. Nozik

National Renewable Energy Laboratory
1617 Cole Boulevard
Golden, CO 80401

Abstract.   We present our investigations of a novel concept for increasing the efficiency of solid state photovoltaic devices. We propose that hot photogenerated carriers can reduce radiative recombination in a p-i-n device structure that has an intrinsic superlattice absorption region. The reduced radiative recombination leads to higher photovoltage and conversion efficiency. We present experimental results on P-i(SL)-N devices fabricated from GaAs/AlGaAs, along with predictions of idealized hot carrier solar cell behavior obtained using a rigorous energy balance device model.

INTRODUCTION

The absorption of high-energy photons in a semiconductor produces electron hole pairs with excess kinetic energy. The excess energy of these hot photogenerated carriers in solar cells is usually transferred rapidly to the lattice by phonon emission, and is therefore unavailable for useful work. This conversion of excess energy into heat is one of the fundamental limitations on the ultimate efficiency of single bandgap solar cells. It is well known that tandem or multi-gap cells can overcome this limitation by absorbing different portions of the solar spectrum in a stack of cells with different bandgaps. A tandem cell that absorbs all photons in the solar spectrum increases the ultimate AM1.5 solar conversion efficiency from ~32% for a single gap cell to ~66% for a cell with continuously graded gaps.

Two other, more speculative, concepts for utilizing the excess energy of hot carriers in solar cells have been proposed. The first, termed multi-pair generation (1), uses the idea that one high-energy photon may be able to generate more than one electron-hole pair. To take full advantage of this "Auger generation" process requires special conditions on the band structure of the absorbing material. Multi-pair generation uses hot carriers to produce a higher photocurrent than can be obtained in normal single-gap cells. With a properly tailored charge carrier

CP404, *Future Generation Photovoltaic Technologies: First NREL Conference,* edited by McConnell
© 1997 The American Institute of Physics 1-56396-704-9/97/$10.00

multiplication process (2), the ultimate efficiency is the same (66%) as a fully selective device.

The second concept (3) to utilize the excess energy of hot carriers was proposed as a possible means to increase the efficiency of photoelectrochemical production of chemical fuels. The thermodynamics and ultimate efficiency of such an ideal hot carrier solar conversion device (HCSC) was first investigated by Ross and Nozik (4). They showed that in an ideal device in which the photogenerated carriers equilibrate among themselves at a temperature, T_H, but not with the lattice, the ultimate efficiency could reach that of an infinite gap tandem cell. The reason for an increased efficiency in a HCSC is reduced radiative recombination, which leads to a higher photovoltage from the device. The calculated ultimate efficiency for an ideal HCSC at several different hot carrier temperatures is presented in Figure 1. In the limit of very high temperatures, T_H = 3000K, the efficiency of a device that absorbs all solar photons can reach approximately 66%.

It is important to note that only modest hot carrier temperatures are required to substantially improve the ultimate efficiency in a HCSC. For example, for a HCSC with the bandgap of GaAs, a factor of two increase in the carrier temperature from 300 K to 600 K leads to an increase of 0.3 V in the photovoltage and to an increase in efficiency from 32.4% to 41.5% as shown in Figure 2.

DEVICE CONCEPT

A practical HCSC device requires an absorber material or structure with slow energy relaxation to allow the formation of a hot carrier distribution. This is very

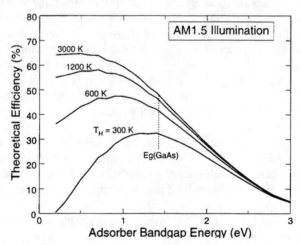

FIGURE 1. The ultimate efficiency of an ideal HCSC vs. carrier temperature and absorber bandgap.

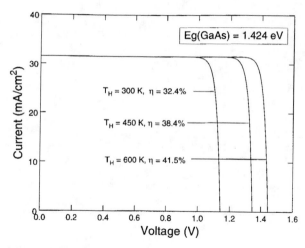

FIGURE 2. JV curves and efficiencies for an ideal HCSC with a bandgap of 1.424 eV and hot carrier temperatures of 300, 450, and 600 K.

difficult to achieve in bulk semiconductor materials due to strong carrier LO-phonon scattering which occurs on the sub-ps time scale. The excess energy of photogenerated hot carriers is typically dissipated within a few ps in bulk materials. However, in reduced dimensionality structures such as quantum wells or quantum dots, the energy relaxation of hot carriers may be much slower due to hot phonon and quantum confinement effects(3). Under high intensity excitation (photogenerated carrier densities $\Delta n > 10^{18}$ cm^{-3}), hot phonon effects come into play and can slow the effective electron energy relaxation times to several 100 ps in multiple quantum well (MQW) and superlattice (SL) structures(5-8).

In this work, we have investigated experimentally and theoretically the possibility of observing hot carrier efficiency enhancements in p-i-n devices with GaAs/AlGaAs MQW or SL intrinsic absorber regions. The p and n-type regions are made from a high bandgap material such as AlGaAs or GaInP which serve as collection regions for hot carriers produced in the intrinsic MQW or SL. A superlattice is expected to have better vertical transport properties than a MQW due to the strong coupling between wells and the formation of mini-bands. Carrier transport through the absorption region is by thermionic emission out of the wells and tunneling through the barriers. This device structure is similar to the MQW solar cell proposed by Barnham(9).

EXPERIMENTAL

We fabricated the P-I(SL)-N devices by low pressure MOCVD and studied their performance under both AM1.5 solar and high intensity monochromatic illumination. We used a Ti:S laser tuned to 750 nm as the source of the

311

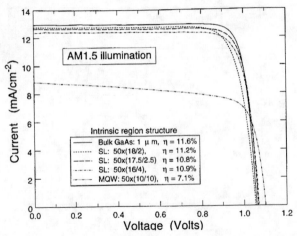

FIGURE 3. JV curves for several P-ISL-N cells with differing intrinsic structures. The notation 50x(18/2) means a SL with 50 periods of 18 nm wells and 2 nm barriers.

monochromatic light. Standard cell processing was used to make isolated, 0.25 cm² cells for AM1.5 measurements and small area (100 μm illuminated diameter) circular mesa diode structures for high light intensity characterization. Control cells with bulk GaAs intrinsic regions were also fabricated.

In Figure 3, we show the one-sun AM1.5 J-V characteristics of several devices with differing absorber structures. Each device has a 1 micron intrinsic region. The J-V characteristics were measured on 0.25 cm² devices without antireflection coatings. The three cells with SL intrinsic regions have slightly lower short circuit currents, J_{SC}, than the bulk GaAs control cell and similar efficiencies. The MQW cell with 10 nm wells and barriers has a lower J_{SC} than the other cells. As the barriers are made thicker the photocurrent decreases due to lower overall absorption, and eventually thick barriers (e.g. in the MQW device) lead to significant series resistance. The fact that the efficiencies of the SL cells are close to the bulk GaAs control cell indicates that the interfaces between the wells and barriers are of high quality.

Hot carrier effects are not observed in these one-sun J-V characteristics. A very high photogenerated carrier density is required to reach the conditions where hot phonons will slow the energy relaxation of hot photogenerated electrons. We used small mesa diode structures illuminated with monochromatic laser light to study the device performance under high light intensity. A control cell with a bulk GaAs intrinsic region (0.1 μm) and an MQW cell (5 periods of 12 nm wells and 10 nm barriers) were chosen for the high intensity studies. The J-V curves and efficiencies of these two cells were measured over a range of illumination intensities. In Figure 4, we plot the measured conversion efficiency, relative to the AM1.5 values. The relative efficiencies of both devices initially increase as the laser power is increased. The relative efficiency of the MQW cell increases faster

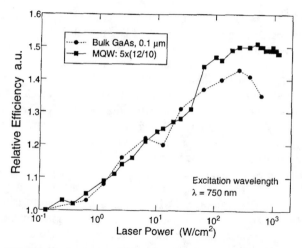

FIGURE 4. Efficiency of a control cell and a MQW cell measured as a function of excitation power. The efficiency is normalized to the AM1.5 efficiencies (0.7% for the MQW cell and 2.7% for the control cell). The efficiency of the MQW cell does not drop off at the highest intensities.

than the control cell at powers above ~60 W/cm^2 and then saturates. Above ~200 W/cm^2, the efficiency of the control cell drops due to series resistance effects. The relative efficiency of the MQW cell does not drop off at high intensities as does the control cell. The MQW cell had a higher fill factor at high intensities, which allows the efficiency to remain high. A voltage enhancement was not observed in the MQW cell as would be expected for hot carrier effects. A possible explanation for the improved fill factor in the MQW cell is given in the following section.

DEVICE MODELING

Most solar cell modeling employs a drift-diffusion (DD) model of carrier transport. In the DD model, the carriers and lattice are assumed to be in thermal equilibrium; therefore, hot carrier effects cannot be studied with the DD model. To investigate the effects of hot carriers in an idealized HCSC, we have used an extension of the DD model,(10,11) which includes an energy balance equation for electrons. This transport model allows for the possibility of hot electrons with a distribution temperature greater than the lattice temperature. The energy balance equation includes terms for electron energy loss and gain mechanisms, such as loss to phonons, loss due to recombination, change in energy from acceleration in the electric field and gain from optical generation. Also, in our device, a proper treatment of carrier transport through SL and QW regions is essential. Our model includes thermionic emission and tunneling transport over and through barriers of a QW region. We have used the energy balance model to calculate the

313

performance of an idealized P-iSL-n HCSC device and have compared the solutions to those calculated with the standard DD model. In our idealized device, we include only radiative recombination, i.e. we neglect all non-radiative recombination mechanisms, and we set the average electron energy relaxation time to be 1 ns. This artificially long energy relaxation time shows the potential beneficial influence of hot carriers on the performance of a HCSC.

In Figure 5, we show the calculated band diagram and electron temperature profile for a P-iSL-N with a 1 micron SL region. The device was illuminated with a one-sun AM1.5 spectrum and is at short circuit condition. This device structure is similar to the experimentally fabricated devices of Figure 3. With the long energy relaxation time of 1 ns, the electron distribution in the SL region is heated to over 2000 K by the excess energy of the photogenerated electrons. Furthermore, the hot electrons in the SL region easily escape from the wells by thermionic emission and are swept out of the intrinsic region to the N contact. This reduces the electron density in the SL region and results in the electron quasi-Fermi level, E_{Fn} dropping below the hole quasi-Fermi level, E_{Fp}. The hot electron depletion also lowers the radiative recombination rate in the SL region below that calculated with the DD model.

We have calculated the current versus applied voltage for this device with the energy balance and DD models and extracted the efficiency at the maximum

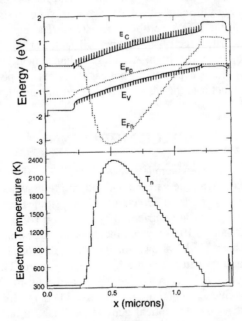

FIGURE 5. Calculated band diagram and electron temperature in an ideal P-iSL-N structure with a 50x(18/2) SL intrinsic region. The quantities were calculated under short circuit conditions, one-sun AM1.5 illumination, and with an electron energy relaxation time of 1 ns.

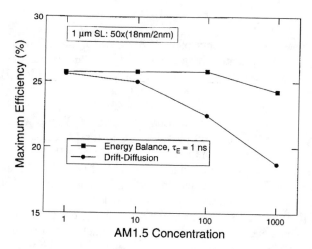

FIGURE 6. The efficiency vs. concentration calculated for the P-iSL-N structure of Figure 5. The solutions from the energy balance model are compared to those from the Drift-Diffusion model. The energy balance model predicts that hot electrons are present at high intensities at the maximum power point, leading to a slower decrease in efficiency at high intensity.

power point. We find that there is only a slight (~0.2%) increase in the maximum efficiency for the energy balance solution. Under forward bias, electrons and holes are injected into the SL region from the doped contacts. Our simulations show that these injected electrons are effective in "quenching" the hot electron distribution temperature by diluting the excess energy of the photogenerated electrons. Simulations of the device behavior under concentrated AM1.5 illumination show that higher energy input into the electron distribution can keep the electrons hot at forward bias. The hot electron density in the SL is then lower than the density calculated by the DD model, which results in lower radiative recombination in the SL. This reduction in radiative recombination with hot electrons leads to an enhancement of the calculated maximum efficiency over the DD model predictions at high concentrations as is shown in Figure 6. We see that the efficiency calculated by the DD model drops rapidly as the solar concentration is increased. This drop was found to be due to series resistance effects in the intrinsic SL regions, which degraded the fill factor of the JV curve. On the other hand, the efficiencies calculated with the energy balance model increase slightly up to 100X concentration before slowly decreasing at higher concentration. These simulations of an idealized device structure illustrate that hot electrons may help improve the performance of P-iSL-N devices under solar concentration.

SUMMARY

We have presented a concept of a HCSC based upon a P-iSL-N structure, which may have the potential of achieving high solar conversion efficiency. Thermodynamic calculations of the ultimate efficiency of an ideal HCSC have shown that reduced radiative recombination in a HCSC should lead to higher photovoltage and a higher efficiency. We have fabricated P-iSL-N cells and studied their behavior under high intensity illumination. While we have not yet observed the predicted increase in photovoltage in the SL cells, we have found that under high light intensity, the relative efficiency of a P-iSL-N cell does not decrease at high intensity as in a comparable control cell. Simulations using an energy balance device model predict similar improvements in the behavior of a HCSC under solar concentration as compared to a device without hot carriers. Further work is required to answer several outstanding questions regarding a P-iSL-N HCSC. These include a measurement of the actual carrier temperature in an operating device and a better understanding of why the energy balance model does not predict the thermodynamic efficiency in an ideal device.

ACKNOWLEDGEMENTS

This work was funded by the U.S. Department of Energy, Office of Basic Energy Sciences, Advanced Energy Projects Division.

REFERENCES

1. Werner, J.H., Kolodinski, S. and Queisser, H.J., *Phys. Rev. Lett.* **24**, 3851 (1994).
2. Spirkl, W., and Ries, H., *Phys. Rev. B* **52**, 11319 (1995).
3. Boudreaux, D.S., Williams, F., and Nozik, A.J., J. Appl. Phys. **51**, 2158 (1980).
4. Ross, R.T., and Nozik, A.J., *J. Appl. Phys.* **53**, 3813 (1982).
5. Ryan, J.F., Taylor, R.A., Tuberfield, A.J., Maciel, A., Worlock, J.M., Gossard, A.C., and Wiegmann, W., *Phys. Rev. Lett.* **53**, 1841 (1984).
6. Parsons, C.A., Dunlavy, D.J., Keyes, B.M., Ahrenkiel, R.K., and Nozik, A.J., *Solid State Comm.* **75**, 297 (1990).
7. Pelouch, W.S., Ellingson, R.J., Powers, P.E., Tang, C.L., and Nozik, A.J., *Phys. Rev. B* **45**, 1450 (1992).
8. Rosenwaks, Y., Hanna, M.C., Levi, D.H., Szmyd, D.M., Ahrenkiel, R.K., and Nozik, A.J., *Phys. Rev. B* **48**, 14675 (1994).
9. Barnham, K.W.J., and Duggan, G., *J. Appl. Phys.*, **67**, 3491 (1990).
10. Lundstrom, M., *Fundamentals of Carrier Transport*, Reading, MA.: Addison-Wesley Publishing, 1990.
11. Winston, D. W., *"Physical Simulation of Optoelectronic Semiconductor Devices"*, PhD Thesis, Univ. of Colorado, Boulder, CO (1996).

316

Photovoltaic Cells Made with Organic Composites

Gang Yu,[a] Jun Gao,[b] Cuiying Yang[b] and Alan J. Heeger[a,b]

[a]UNIAX Corporation, 6780 Cortona Drive, Santa Barbara, CA 93117
[b]Institute for Polymers and Organic Solids, University of California, Santa Barbara, CA 93106

Abstract. Blending organic semiconductors with different electron affinities results in an interpenetrating bi-continuous network of internal donor/acceptor (D/A) heterojunctions. The nano-scale D/A junctions show efficient charge separation and charge transfer. The bi-continuous network allow the separated carriers to be collected effectively at the anode and cathode contacts. These blends are soluble to common organic solvents, and are processable at room temperature. Photodiodes and photovoltaic cells were fabricated with high quantum efficiencies. The carrier collection efficiency and energy conversion efficiency of MEH-PPV:C_{60} photovoltaic cells are ~29% electrons/photon and ~3% under illumination of 20 mW/cm^2 at 430 nm, two orders of magnitude higher than that in devices with MEH-PPV alone. The photosensitivity and the quantum yields increase to 0.26 A/W and ~75 % electrons/photon at reverse bias of -2V, even higher than those in UV-enhanced Si photodiodes at the same wavelength.

1. NANOSCALE DONOR/ACCEPTOR BLENDS: IDEAL MATERIALS FOR PHOTOELECTRIC CONVERSION

Stimulated by the discovery of polymer light emitting diodes (1,2), LEDs, along with the development of processable semiconducting polymers with improved quality (3), photonic devices made with conjugated polymers as the active materials have received renewed attention (4). In addition to the early works on polyacetylene (5) and polythiophene (6), photovoltaic effects in poly(phenylene vinylene), PPV and its derivatives were studied recently (7-10).

For photovoltaic cells made with pure conjugated polymers, energy conversion efficiencies were typically 10^{-3}-10^{-2} % (5-10), too low to be used in practical applications. The recent discovery of photoinduced electron transfer in composites of conducting polymers (as donors) and buckminsterfullerene, C_{60}, and its derivatives (as acceptors) (11) provided a molecular approach to high efficiency photovoltaic conversion (12,13). Since the time scale for photoinduced charge transfer is subpicosecond, more than 10^3 times faster than the radiative or nonradiative decay of photoexcitations (11), the quantum efficiency for charge transfer and charge separation from donor to acceptor is close to unity. Thus, photoinduced charge transfer across a donor/acceptor (D/A) interface provides an effective method to overcome early time carrier recombination in organic systems and thus to enhance the optoelectronic response of these materials. For example, with the addition of only 1% C_{60}, the photoconductivity of MEH-PPV:C_{60} blend increases by an order of magnitude over that of pure MEH-PPV (12,13).

CP404, *Future Generation Photovoltaic Technologies: First NREL Conference*, edited by McConnell
© 1997 The American Institute of Physics 1-56396-704-9/97/$10.00

Although the quantum efficiency for photoinduced charge separation is near unity for a D/A pair, the conversion efficiency in a bilayer heterojunction device is limited (14): (i) Due to the molecular nature of the charge separation process, efficient charge separation occurs only at the D/A interface; thus, photoexcitations created far from the D/A junction recombine prior to diffusing to the hetero-junction; (ii) Even if charges are separated at the D/A interface, the photovoltaic conversion efficiency is limited by the carrier collection efficiency; that is, the separated charges must be collected with high efficiency. Consequently, inter-penetrating phase separated D/A networks appear to be ideal materials for photoelectric conversion (15-18). Through control of the morphology of the phase separation into an interpenetrating network, one can achieve a high interfacial area within a bulk material. Since any point in the composite is within a few nanometers of a D/A interface, such a composite is a "bulk D/A heterojunction" material. Because of the interfacial potential barrier, as demonstrated by the built-in potential in the bilayer D/A heterojunction diode (14), ultrafast photoinduced charge transfer and charge separation will occur with quantum efficiency approaching unity, leaving holes in the donor phase and electrons in the acceptor phase. This process in the MEH-PPV:C_{60} blends is illustrated in Fig. 1. If the network is bi-continuous, the collection efficiency can, in principle, be equally efficient.

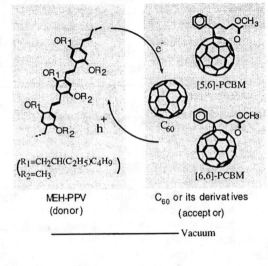

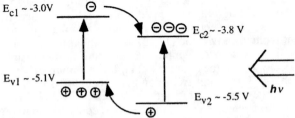

Fig. 1   Photoinduced charge transfer in MEH-PPV:C_{60}.

The morphology of MEH-PPV:C_{60} blend films has been studied systematically over a wide range of concentrations (19). Fig. 2 shows a TEM image of MEH-PPV:C_{60} blend film with 1:1 weight ratio. The image size is 0.96 mm x 0.91 mm. The dark contrast indicates C_{60} clusters with dimensions of ~10 nm. The TEM images provide direct evidence of the phase separated bicontinuous network.

Thus the charge transfer blends are novel optoelectronic materials with high interfacial area, nanoscale heterojunctions. Such a bicontinuous D/A network material is promising for use in thin film solar cells, large size photosensors as well as image sensors.

Fig. 2   A TEM image of MEH-PPV:C_{60} blend film with 1:1 weight ratio.

2. DEVICE FABRICATION AND CHARACTERIZATION

In addition to the high quantum efficiency of charge separation, the electronic structure of the bicontinuous D/A blend allows one to choose contact electrodes with work-functions which optimize the carrier collection efficiencies of holes from the donor phase and electrons from the acceptor phase. The thin film photovoltaic cell consists of a metal (Ca, Sm, Mg or Al) contact on the front surface of a blend film on a glass (or mylar) substrate, partially coated with a layer of transparent indium/tin-oxide (ITO), as shown in Fig. 3. Organic conductors such as polyanilene protonated with (±) -10-camphor sulfonic acid (PANI-CSA) can also be used as the transparent electrode (3). With PANI-CSA on PET substrates, photovoltaic devices have been fabricated in flexible form. The active area of test devices ranges from 0.1 cm^2 to 1.5 cm^2. Large size PV cells (>25 cm^2) have also been demonstrated.

Electrical measurements were performed using a Keithley 236 Source-Measure Unit. The excitation source was a tungsten-halogen lamp with a bandpass filter (centered at 430 nm, bandwidth of 100 nm). The maximum optical power at the sample was ~20 mW/cm^2. MEH-PPV:C$_{60}$ films were spin-cast from ~0.5 wt% xylene solutions which were prepared by mixing two master solutions of MEH-PPV (20) and C$_{60}$ in proper ratios. Typical film thicknesses were 1000~2000Å. The back metal electrode was thermally evaporated with thickness between 1000 and 5000Å.

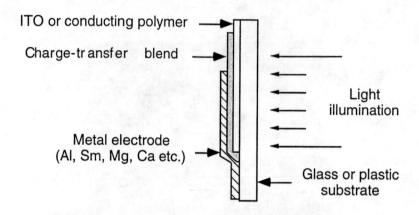

Fig. 3   Structure of organic photovoltaic cells and photodiodes.

The limited solubility of pure C$_{60}$ in organic solvents and its tendency to crystallize during film formation limit its use in high concentration blends. To overcome these problems, a series of soluble C$_{60}$ derivatives have been developed (21). The molecular structures of the two C$_{60}$ derivatives used in this study, denoted as [6,6]PCBM and [5,6]PCBM, are also provided in Fig. 1. With the soluble C$_{60}$ derivatives, we have been able to make homogeneous, stable blends containing more than 80 wt% fullerene molecules. Devices made from such blends show better photosensitivity and higher energy conversion efficiency.

Fig. 4 compares the I-V characteristics of a Ca/MEH-PPV:[6,6]PCBM/ITO device with 1:1 weight ratio and a Ca/MEH-PPV/ITO device in the dark and illuminated with 20 mW/cm^2 at 430 nm. For pure MEH-PPV devices, the exponential current turn-on and the open circuit voltage, V_{oc}, (the minimum in the I-V under photoexcitation) at 1.6 V are comparable to the work-function difference between Ca and ITO.

The forward I-V characteristics of the MEH-PPV devices (Fig. 4b) can be classified into three regions: a small shunt current ($<10^{-10}$ A/cm^2) for V < 1.2V, an exponential increase, by more than four orders of magnitude between 1.3-1.8 V, and "current saturation" in the high voltage region (the current increases, but more slowly than the exponential rate). In reverse bias, the current saturates at approximately 10^{-10} mA/cm^2 for |V|<3V. The I-V characteristics at high fields have been interpreted in terms of tunneling (22) into a thin film semiconductor depleted of carriers. Capacitance-voltage and alternating current conductivity experiments have provided evidence in support of this model (23). The

320

exponential current turn-on and the V_{oc} at high illumination level correspond to the flat band condition.

In the Ca/MEH-PPV:[6,6]PCBM/ITO device, the exponential current turn-on shifts to lower bias by ~0.8 V. This fact suggests that the lowest unoccupied molecular orbital, LUMO, of the acceptor [6,6]PCBM (corresponding to the bottom of the conduction band of a semiconductor) is ~0.8 eV lower than the Fermi energy of Ca, as illustrated at the bottom of Fig. 1. Impedance analysis have been carried out for these blend devices. A voltage independent device capacitance was observed for V<0.5 V (24). This result suggests that the blend film is depleted, similar to that observed in pure MEH-PPV devices. This fact suggests that the energy gap in the blend film remains clean and trap-free. No ground-state charge transfer occurs between the donor and the acceptor.

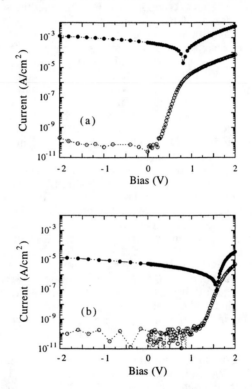

Fig. 4   (a) I-V characteristics of a Ca/MEH-PPV:[6,6]PCBM/ITO device in the dark (open circles) and under 20 mW/cm^2 illumination at 430 nm (solid circles); (b) The corresponding data from a Ca/MEH-PPV/ITO device.

Since photoinduced electrons in MEH-PPV will lower their energy by transferring to the C_{60} and photoinduced holes in C_{60} will lower their energy by transferring to the MEH-PPV, the photosensitivity (PS) is significantly enhanced in the phase separated composite (as shown in Fig. 4). The short circuit current (I_{sc}) in the MEH-PPV:[6,6]PCBM device is $I_{sc}=0.5$ mA/cm^2 under 20 mW/cm^2, corresponding to photosensitivity = 25 mA/W and $\eta_c = 7.4$ % el/ph, both approximately two orders of magnitude higher than that of MEH-PPV devices (Fig. 4b). The electroluminescence quantum efficiency of the blend device was ~3x10^{-5} % ph/el, 10^3-10^4 times *less* than in pure MEH-PPV devices, consistent with ultrafast photoinduced charge separation (14).

3. HIGH EFFICIENCY PV CELLS MADE WITH D/A BLENDS

The carrier collection efficiency (η_c) and energy conversion efficiency (η_e) of the bicontinuous D/A network material are critically dependent on the morphology and the chemical composition. By changing the solvent from xylene to 1,2-dichlorobenzene, ODCB, we have been able to cast high quality MEH-PPV:[6,6]PCBM films with methanofullerene compositions up to 1:4 weight ratio (approximately one acceptor for every polymer repeat unit). For devices made from this blend, the photosensitivity is further improved; $I_{sc} = 2$ mA/cm^2, PS = 100 mA/W and $\eta_c \approx 29$ % el/ph under 20 mW/cm^2 at 430 nm (17).

Figure 5 shows η_c and η_e as a function of illumination intensity for several blend devices in comparison with the data for MEH-PPV devices. The blend weight ratios are 3:1 for the MEH-PPV:C$_{60}$ device, and 1:1 and 1:4 for the devices using PCBM and PCBCR. For Ca/MEH-PPV:[6,6]PCBM(1:4)/ITO devices, $\eta_c = 29$% el/ph and $\eta_e = 2.9$ %, nearly independent of the incident light intensity over 5 orders of magnitude.

The photosensitivity and the carrier collection efficiency are nearly the same when Al is substituted for Ca as the cathode, although V_{oc} decreases slightly from 0.8 V to 0.7 V at 20 mW/cm^2. Typical data from Al/MEH-PPV:[6,6] PCBM/ITO devices are also included in Fig. 5. Thus, stable metals can be used as the cathode electrode to collect electrons effectively from the acceptor phase.

The significant enhancement in carrier collection efficiency achieved by using the bicontinuous D/A network material results from the large increase in the interfacial area over that in a D/A bilayer and from the relatively short distance from any point in the polymer to a charge-separating interface. Moreover, the internal D/A junctions inhibit carrier recombination and thereby improve the lifetime of the photoinduced carriers (12) so that the separated charge carriers can be efficiently collected by the built-in field from the asymmetric electrodes. Similar effects have been observed in MEH-PPV:Cyano-PPV polymer blends (16,18).

The photosensitivity in the thin film devices made with charge transfer blends (such as MEH-PPV:C$_{60}$) increases with reverse bias. External photosensitivity of 0.2-0.3 A/W and external quantum yields of 50-90% el/ph have been achieved at 430nm with -2 ~ -5V volts reverse bias (16,17). For example, PS >300 mA/W ($\eta_c \approx 90$ % el/ph) was observed in the MEH-PPV:[6,6]PCBM photodiodes biased at -2V (17). At the same wavelength, the photosensitivity of UV-enhanced silicon photodiodes is ~0.2A/Watt, independent of the biasing voltage (13). Thus, the photosensitivity of thin film photodiodes made with polymer charge transfer blends is comparable to that of photodiodes made with inorganic semiconductors.

In summary, organic D/A blends were developed with high photosensitivity. These materials are soluble to common organic solvents, and are processable at room temperature. The interpenetrating bi-continuous network of internal D/A heterojunctions at nanometer scale show efficient charge separation and charge transfer. Photovoltaic cells fabricated with these D/A blends show high η_c and η_e, ~29% el/ph and ~3% respectively under 20 mW/cm^2 at 430 nm, two orders of magnitude higher than that in devices with MEH-PPV alone. The quantum yields increase to ~75 % el/ph at reverse bias of -2V, higher than those in UV-enhanced Si photodiodes at the same wavelength. Thus, organic D/A blends provide a novel class of materials for optoelectric energy conversion. Moreover, the organic PV cells can be used in developing large size, low cost image sensors for industrial and office automation, consumer electronics, and biomedical applications.

There are parallel efforts of improving photosensitivity with inorganic nanoparticles or porous materials (see works from UC, Berkeley and NREL). However, better device performance achieved so far was from organic D/A composites. The processability of organic blends may play an important role in achieving high quality junction interfaces.

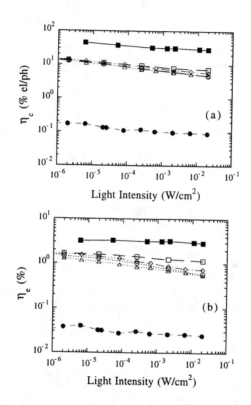

Fig. 5 η_c (panel a), and η_e (panel b) of Ca/MEH-PPV:[6,6]PCBM(1:4)/ITO (solid squares), Ca/MEH-PPV:[6,6]PCBM(1:1)/ITO (open squares), Al/MEH-PPV: [6,6]PCBM(1:1)/ITO (open diamonds), Ca/MEH-PPV:[5,6]PCBM(1:1)/ITO (open circles), Ca/MEH-PPV:C$_{60}$(3:1)/ITO (open triangles) and Ca/MEH-PPV/ITO (solid circles).

ACKNOWLEDGMENTS

We are grateful to Y. Cao, I.D. Parker, N.S. Sariciftci and C.H. Lee for valuable discussions. The work at UCSB was supported by DOE under a grant from AEPP (DOE-FG03-93 ER 12138). The work at UNIAX was supported by DOD under SBIR contracts (N62269-96-C-0009 and N00421-97-1075) monitored by the Aircraft Division of Naval Air Warfare Center.

REFERENCES

1. H. Burroughes, D.D.C. Bradley, A.R. Brown, R.N. Marks, K. Mackay, R.H. Friend, P.L. Burns and A.B. Holmes, *Nature* 347,539 (1990).
2. D. Braun and A.J. Heeger, Appl. Phys. Lett. 58,1982 (1991).
3. A.J. Heeger and P. Smith, in *"Conjugated Polymers"*, Eds. J.L. Bredas and R. Silbey (Kluwer Academic, Dordrecht, 1991) p. 141; Y. Cao, P. Smith and A.J. Heeger, Synth. Metals 48, 91 (1992); Y. Cao, G.M. Treacy, P. Smith and A.J. Heeger, Appl. Phys. Lett. 60, 1 (1992).
4. For reviews of recent progresses, see: G. Yu, Synth. Metals 80, 143 (1996); A.J. Heeger and J. Long, Optics and Photonics News, August (1996) p. 24.
5. For a review, see: J. Kanicki, in *Handbook of Conducting Polymers*, ed. T. A. Skotheim (New York, Marcel Dekker, 1986) p.543.
6. S. Glenis, G. Tourillon, F. Garnier, Thin Solid Films 111, 93 (1984).
7. S. Karg, W. Riess, V. Dyakonov, M. Schwoerer, Synth. Metals 54, 427 (1993).
8. H. Antoniadis, B.R. Hsieh, M.A. Abkowitz, S.A. Jenekhe, M. Stolka, Synth. Metals 64, 265 (1994).
9. G. Yu, C. Zhang, A.J. Heeger, Appl. Phys. Lett. 64, 1540 (1994).
10. R.N. Marks, J.J.M. Halls, D.D.D.C. Bradley, R.H. Fried, A.B. Holmes, J. Phys.: Condens. Matter 6, 1379 (1994).
11. N.S. Sariciftci, L. Smilowitz, A.J. Heeger and F. Wudl, Science 258, 1474 (1992); N.S. Sariciftci and A.J. Heeger, Intern. J. Mod. Phys. B 8, 237 (1994).
12. C.H. Lee, G. Yu, D. Moses, K. Pakbaz, C. Zhang, N.S. Sariciftci, A.J. Heeger and F. Wudl, Phys. Rev. B. 48, 15425 (1993).
13. G. Yu, K. Pakbaz and A.J. Heeger, Appl. Phys. Lett. 64, 3422 (1994).
14. N.S. Sariciftci, D. Braun, C. Zhang, V. Srdanov, A.J. Heeger, G. Stucky and F. Wudl, Appl. Phys. Lett. 62, 585 (1993).
15. N.S. Sariciftci and A.J. Heeger, U. S. Patent 5,331,183 (1994); 5,454,880 (1995).
16. G. Yu and A.J. Heeger, J. Appl. Phys. 78, 4510 (1995).
17. G. Yu, J. Gao, J.C. Hummelen, F. Wudl and A.J. Heeger, Science 270, 1789 (1995).
18. J.J.M. Halls, C.A. Walsh, N.C. Greenham, E.A. Marseglia, R.H. Fried, S.C. Moratti and A.B. Holmes, Nature 376, 498 (1995).
19. C.Y. Yang and A.J. Heeger, Synth Metals, in press.
20. F. Wudl, P.-M. Allemand, G. Srdanov, Z. Ni and D. McBranch, in *Materials for Nonlinear Optics: Chemical Perspectives*, edited by S.R. Marder, J.E. Sohn and G.D. Stucky (The American Chemical Society, Washington DC, 1991) p. 683.
21. J.C. Hummelen, B.W. Knight, F. Lepec, and F. Wudl, J. Org. Chem. 60, 532 (1995).
22. I.D. Parker, J. Appl. Phys. 75, 1656 (1994).
23. I.H. Campbell, D.L. Smith, and J.P. Ferraris, Appl. Phys. Lett. 66, 3030 (1995).
24. G. Yu, J. Gao, C. Yang and A.J. Heeger, in *Photodetectors: Materials and Devices II*, Proceedings of SPIE, Vol. 2999, 306 (1997).

Novel Material Architectures for Photovoltaics

Satyen K. Deb

Center for Basic Sciences
National Renewable Energy Laboratory, Golden, CO 80401-3393

Abstract. In this paper, an analysis is made on what can be a long-term and fruitful direction in which the new material architecture for photovoltaics may logically evolve during the next couple of decades. In spite of great advances, all of the PV material systems that are currently being developed have problems that need to be solved. Therefore, the search for new material systems will continue. It is reasonable to expect that the next generation of materials is likely to come from ternary and multinary semiconductors belonging to a class of tetrahedrally bonded diamond-like compounds. These are most promising because of their close similarity with traditional semiconductors with respect to their composition, crystal structure, nature of chemical bonds, and electronic structures.

INTRODUCTION

Since the discovery of the photovoltaic (PV) effect in silicon-based semiconductors in 1954 (1), intense research and development work, spanning over half a century, has led to many exciting developments in PV materials and device technologies. In the post-Sputnik era, the driving force for this technology was the need to find a reliable power source for space applications, where cost was not a major concern. Silicon-based PV cells successfully met that challenge. However, after the 1973 oil crisis and the adverse impact that traditional power sources have had on the global environment, the technology development for PV has entered into a new era—one in which a key requirement is cost-effectiveness. With today's rapidly growing demand for PV as a clean, alternative power source for terrestrial applications, silicon-based solar cells are the 'workhorse' of the industry and may remain so in the near future (2). Although enormous progress has been made in reducing the cost of silicon-based technology, it is not certain whether it can meet our ultimate cost goals. This has led to extensive research and development on a number of other semiconductor materials, particularly those based on thin films. The material technology has evolved from simple elementary semiconductors such as Si to more complex binary and ternary compounds. In this brief presentation, an analysis will be made on what can be a long-term, fruitful direction in which the new material architecture for PV may logically evolve during the next few decades. But, before we do that, it is important to summarize some key material parameters that must be met before a new material can be considered as a viable alternative.

CP404, Future Generation Photovoltaic Technologies: First NREL Conference, edited by McConnell
© 1997 The American Institute of Physics 1-56396-704-9/97/$10.00

MATERIAL PARAMETERS FOR PV

For a new material to be considered for PV applications, it must have some or all of the following properties: (i) strong light absorption, covering a broad range of the solar spectrum (1.0-2.0 eV); and a direct optical transition with an optimum energy gap of 1.4 eV; (ii) the feasibility to be fabricated as n- or p-type material with an efficient generation of charge carriers, low recombination loss, and good transport properties for efficient carrier collection at the front and back surfaces of the device; (iii) appropriate band-structure configurations, particularly for heterojunction devices; (iv) adjustable energy gaps and lattice parameters, especially for multijunction devices; (v) availability in abundance, at low cost and environmentally benign, with good stability and lending itself to be fabricated in thin-film form; and (vi) radiation resistance for space applications. The relationship between the band gap and cell efficiency is an important parameter. In 1960, Wysoki and Rappaport (3) published an analytical study on theoretical solar cell efficiency versus semiconductor band gap for an ideal homojunction cell as shown in Figure 1. This

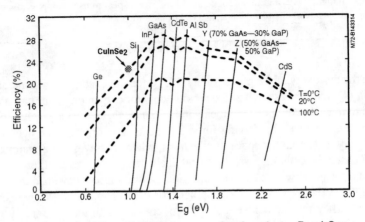

Figure 1. Silicon Cell Efficiency vs. Semiconductor Band-Gap

is still a valid preliminary guideline for choosing the right material for a single-junction device. However, for more complex multijunction devices, parameters such as the matching of energy gaps and lattice parameters need to be considered in order to maximize the conversion efficiency. This is illustrated in Figure 2 for a GaAs/GaInP$_2$ multijunction device, which has recently achieved a world record efficiency of ~30% (4). In this case, the composition and structure of the GaInP$_2$ had to be adjusted to maximize cell efficiency. Many of the materials that are currently being developed for PV applications satisfy some of these parameters, but not all. Hence, the search continues to find the most optimum material or material combinations that will have the desired properties outlined above.

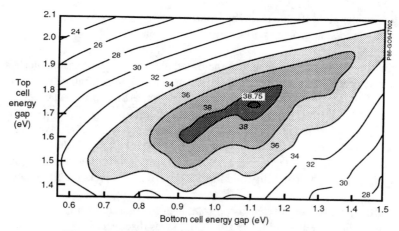

Figure 2. Theoretical Efficiency of Two-Junction Cells as a Function of Band-Gaps of Top and Bottom Cells

GENESIS OF CURRENT PV MATERIALS

Materials that have shown the greatest promise for PV applications belong to a class which evolved from a simple elementary semiconductor, like silicon, to the more complex ternary and quaternary compounds such as $CuInSe_2$, $Cu(Ga,In)Se_2$, and $GaInP_2$ as shown in Table 1. Silicon, the starting material in this evolutionary scheme

Table 1. Global views of Semiconductor Materials for PV Applications

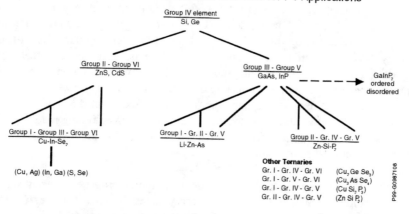

has the desirable properties of low cost, abundance, nontoxicity, stability, n- and p-type doping, a near-optimum band gap, and a vast technology base. But silicon has one outstanding drawback. The indirect nature of its energy gap makes it a relatively weak absorber of solar radiation. One way to get around this problem is to use a thin film silicon solar cell (10-20μ) with appropriate light trapping to increase the light absorption. Several approaches are currently being pursued.

The other interesting development in the broad class of silicon-based materials is the discovery of hydrogenated amorphous silicon (a Si:H), which can be doped n- or p-type. In fact, the first solar cell based on the film, a Si:H p-i-n solar cell, was fabricated in 1976 by the RCA group. Since then, enormous progress has been made on this technology and today it is the only thin-film PV material besides silicon that is produced on a large scale for power generation (~15% of all PV modules are based on this technology). This technology holds promise for very-low-cost PV power. The main issues are photostability and further improving conversion efficiency.

Shortly after the discovery of the silicon solar cell, the first thin-film heterojunction cell (based on binary semiconductors CdS/Cu_2S) was discovered. After the intense research and development of this technology for three decades, initially for space and later on for terrestrial power, it was abandoned in the early '80s when the instability of Cu_2S appeared intrinsic and, therefore, insurmountable. Several binary II-VI and III-V semiconductors, like CdTe, GaAs, and InP gradually became strong contenders for photovoltaics. GaAs is particularly interesting because it is emerging as the semiconductor material destined to replace silicon, particularly in opto-electronic applications. PV conversion efficiencies approaching 25% were achieved in GaAs homojunction devices. Efficiencies greater than 30% were achieved in several multijunction devices based on GaAs and its alloys like $GaInP_2$, $GaAlAs_2$, and $GaInAs_2$ (5). The technology based on III-V materials is considerably more expensive and therefore, appropriate for concentrator-type solar cells. Thin film GaAs solar cells on low-cost substrate is a viable approach but very little work has been done in this area. Among the II-VI material, CdTe is particularly interesting because it has a nearly optimum direct band gap (1.50 eV) for solar cells. It is also the only II-VI PV material that can be made n- and p-type. Heterojunction cells, such as CdS/CdTe, appear to be the most promising, with a reported efficiency approaching 16%. One attractive feature of CdS/CdTe cells is that several low-cost approaches can be used to fabricate them. One of the main drawbacks of these technology is the environmental sensitivity of cadmium-containing material, even though several studies show that deployment of this technology will not have an adverse impact on the environment.

The other thin-film solar cell that has made phenomenal progress in recent years is based on a more complex ternary and quaternary material such as $CuInSe_2$ (CIS) and $Cu(In,Ga)Se_2$ (CIGS). A solar cell efficiency of 17.8% has been achieved in a thin-film polycrystalline $Cu(In,Ga)Se_2/CdS/ZnO$ heterojunction device (6). Moreover, it is a very stable device that shows no degradation even after almost six years of testing under actual deployment conditions. The main issues with this technology are: (i) the reliable fabrication of large-area, high-efficiency devices with good yield, and (ii) some concern about the availability of In, especially for large-scale production (on the scale of gigawatts).

In the final analysis, all of the material systems currently being developed for PV have problems that need to be addressed. Considerable research and development will undoubtedly continue on the current materials to overcome these problems. At the same time, it is only logical to assume that a search for new material systems will continue in the future. In the following section, we propose to hypothesize on what would be a promising direction in which the new material architecture can logically evolve during the next couple of decades. This would be one of the few directions the new material technology could take, along with other scenarios that have been discussed in this workshop.

Novel Material Architecture

A cursory look at Table 1 shows that the semiconducting materials for PV applications have evolved from a simple, elementary material, silicon, to its binary analog, the II-VI and III-V compounds. The binary materials have led to ternary and quaternary compounds such as $CuInSe_2$ and $Cu(In,Ga)Se_2$, which turned out to be highly efficient and stable solar cell materials as compared to their binary analogs. In spite of material complexity, the processing of CIS and CIGS compounds for PV applications is relatively straightforward and the material is very forgiving under a wide variety of processing conditions. The immense success with CIS-based ternary material for PV opens almost limitless possibilities of finding other ternary and multinary compounds that could be even better than CIS in terms of the criteria outlined above. In fact, from band-gap consideration alone, CIS (Eg = 1.04 eV) would not have been considered as the most promising material for PV. One can, with reasonable certainty, make a prediction that the next generation of PV materials is likely to come from this class of ternary and multinary compounds. Among the many different ternary compounds, only those belonging to a class of diamond-like compounds are likely to be more promising because of their close analogy with traditional semiconductors with respect to their composition, crystal structure, the nature of chemical bonds, and therefore, electronic structure. Many of these materials crystallize in the tetragonal chalcopyrite structure. The vast majority of these compounds are valence compounds with tetrahedral structure due to sp^3 hybridization of their valence electrons, like the elemental semiconductors, silicon and germanium.

There are also many ternary compounds that show s^2p^6 hybridization, which leads to octohedral phases. These materials in the tetrahedral and octohedral phases, with an average of four electrons per atom, form a class of valence compounds that show interesting semiconducting properties appropriate for PV application. This class of materials was extensively studied in the sixties by a Russian group led by Goryunova (7), particularly with regard to their crystal chemistry, electronic structure and semiconducting properties. An excellent monograph on ternary diamond-like semiconductors by L. I. Berger and V. D. Prochachan (8), published by Consultants Bureau, New York in 1969, forms the basis for the rest of this article. Parthe (9), Borschevsky (10), Shay (11), and others, have made extensive studies on crystal chemistry and synthesis and characterization of many ternary compounds. According to Goryunova, "the tetrahedral configuration of atoms in a non-defect structure of a compound with an arbitrary number of atoms is possible if (1) the average number of valence electrons per atoms of the compound is equal to four; and (2) the valence of each of the components is equal to the number of the group in the Periodic Table to which the components belong." Goryunova's rule applies to a class of ternary diamond-like semiconductor in which many of the compounds have optical and electronic properties that may lead to excellent PV materials. The representative classes of two cation ternary compounds with four electrons per atom is shown in Table 2. They are classified under two categories, those with and without defects.

Table 2. Representative Classes of Two-Cations Ternary Compounds

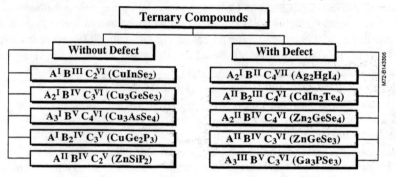

Goryunova's formula applies to defect compounds by considering these as four component systems where a vacancy is counted as an atom of zero valence. The physio-chemical basis such as phase diagrams for many of the compounds, have not been studied extensively. Therefore, one can use a qualitative approach, the well-known "Gibbs triangle," to show which ternary tetrahedral phases are likely to exist. For example, in a two-cation ternary system A-B-C (where A and B are cations), the formation of ternary compound is possible if the binary components A-B and A-C form a stable chemical compound. The early Russian work has taken this approach to predict the formation of many of these compounds. The compound formation of a typical two-cation ternary system like $A^{II}B^{IV}C_2^V$ (where A = Mg, Zn, and Cd; B = Si, Ge, and Sn; and C = N, P, As, and Sb) is shown in Table 3.

Table 3. Two Cations Ternary Compounds of the Type $A^{II}B^{IV}C_2^V$

C^V	Nitrides			Phosphides			Arsenides			Antimonides		
$\dfrac{B^{IV}}{A^{II}}$	Si	Ge	Sn	Si	Ge	Sn	Si	Ge	Sn	Si	Ge	Sn
Mg												
Zn												
Cd												

▨ Compound exists   ☐ Unknown but feasible   ▨ One of the binary component unreactive

We have only included elements that are likely to form ternaries that could be of potential interest for PV. In this series, 13 compounds are known to exist and another 17 compounds are likely to exist. Similar tables can be created for all other ternaries listed in Table 2. There are about 80 compounds from among the 5 classes of ternaries without defects that are likely to exist, a large fraction of which could be potential PV materials. Similarly, many more compounds with defects also exist and could be of potential interest for PV. If one makes quaternary alloys like CIGS from these ternaries, the list can be quite large. The optical and electrical properties of a representative group of II-VI-V$_2$ compounds are listed in Table 4. The band gaps of

Table 4. Optical and Electrical Properties of II-IV-V$_2$ Compounds

Crystal	Energy gap (eV)	Carrier conc. (cm^{-3})	Mobility (cm$_2$/V-s)	Resistivity
ZnSiP$_2$	~ 2.2	n~10^{17}–10^{18}	100	10^4–10^6
ZnSiAs$_2$	1.74	p~10^{15}	140	
ZnGeP$_2$	1.99	p~10^{13}	20	10^4
ZnGeAs$_2$	1.15	p~4×10^{18}	23	
ZnSnP$_2$	1.66	p~5×10^{16}	55	
ZnSnAs$_2$	0.73	p~10^{17} – 10^{21}	40–200	
ZnSnSb$_2$	0.30	p~10^{20}	70	
CdSiP$_2$	~ 2.4	n~10^{14} – 10^{15}	80–150	~10^6
CdSiAs$_2$	1.55	p~6×10^{15}	300–500	
CdGeP$_2$	1.72	n~10^{15}	~100	
CdGeAs$_2$	0.57	p~7×10^{15}	700–1500	
		n~10^{16} – 10^{18}	2500	
CdSnP$_2$	1.17	n~10^{15} – 10^{18}	2000	
		p-type		10^3–10^4
CdSnAs$_2$	0.26	n~10^{18}	11000	
		p~6×10^{17}	190	

these materials are in the range of 0.26 to 2.4 eV. These compounds can be doped n- or p-type with carrier densities and mobilities that could be optimized for PV devices. Obviously, the band gaps of these ternaries can be engineered to any desired value by synthesizing their quaternary alloys. Some of the low-band-gap materials with exceptionally high mobility, like $CdSnAs_2$ and $CdGeAs_2$, can be strong candidates for thermophotovoltaic devices.

Besides the ternaries, there exist a series of quaternary compounds (12) of the general composition A_2BCX_4, $A_4BC_2X_7$, $A_2BC_5X_8$, A_9BCX_5, and $ABC\ \square X_2$, as shown in Table 5. The energy gaps of a few typical representative compounds from

Table 5. Quaternary Compounds

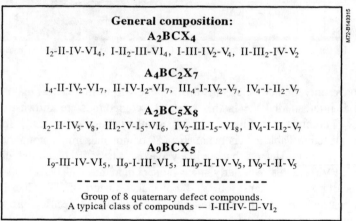

the class A_2BCX_4, which could be of potential interest for PV materials, are shown in Table 6. At this stage, the choice becomes almost limitless.

Table 6. Energy Gaps of a few Typical Quaternary Compounds

Compounds	Energy gap (eV)
$Cu_2\ Cd\ Ge\ Se_4$	1•22
$Cu_2\ Cd\ Sn\ Se_4$	1•90
$Cu_2\ Mn\ Ge\ Se_4$	1•95
$Cu_2\ Mn\ Sn\ Se_4$	1•40
$Cu_2\ Zn\ Ge\ Se_4$	1•20
$Cu_2\ Fe\ Sn\ S_4$	1•95
$Ag_2\ Mn\ Ge\ Se_4$	1•47
$Cu\ In\ Ge\ \square\ Se_4$	1•20

The apparent compositional complexity of ternary and quaternary materials makes one conclude that it will be difficult to synthesize and control the optoelectronic

332

properties of these compounds. On the contrary, the ternary compounds offer several advantages over their binary analogs: (i) relatively lower melting points make them easier to prepare, (ii) low-temperature processing reduces contamination during synthesis, (iii) they offer a wider choice of band-gap engineering by compositional changes and ordering in cation or anion sublattices, and (iv) there is a greater possibility of finding the optimum PV material with respect to cost, efficiency, stability, and environmental sensitivity. Of course, it is easier to predict which materials should be attractive for PV from their optoelectronic properties, but the development of a technology based on a given material takes years of effort. For example, it took twenty years of persistent effort by many groups to bring $CuInSe_2$ to its present state. Similarly, it took twelve years of development to bring a multijunction cell, based on an obscure material like $GaInP_2$, to a world-record efficiency of 30% (as shown in Figure 3). Undoubtedly, many interesting develop-

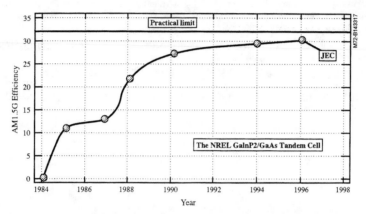

Figure 3. Efficiency of $GaInP_2$/GaAs Solar Cell as a Function of Time

ments will occur on the way to finding better PV materials in the future (13). The intent of this paper is to point out that the ternary and multinary compounds offer a multitude of options from which the ideal PV material might emerge. Those who argue that there is no compelling reason to search for new PV materials beyond what exists today should keep in mind that the technology of today becomes obsolete tomorrow.

ACKNOWLEDGMENTS

This work was supported by the U. S. Department of Energy under Contract No. DE-AC36-83CH10093.

REFERENCES

1. Chapin, D. M., Fuller, C. S., and Pearson, G. L., *J. Appl. Phys.* **25**, 676 (1954).
2. Green, M. A., *Progress in Photovoltaics*, **2**, 87 (1994).
3. Wysocki, J. J., and Rappaport, P., *J. Appl. Phys* **31**, 571 (1960).
4. Olson, J. M., Kurtz, S. R., Kibbler, A. E., P. Faine, *Appl. Phys. Lett.*, **55**, 1741 (1989).
5. Coutts, T. J. and Lundstrom, M., *J. Elec. Matls.*, **22**, 57 (1993).
6. Tuttle, J. R., Ward, J. S., Duda, A., Berens, T. A., Contreras, M. A., Ramananthan, K. R., Tennant, A. L., Keane, J., Cole, E. D., Emery, K., and Noufi, R., *Proceedings of the 1996 Spring MRS Meeting*, San Francisco, CA, 8-12 April, **426**, 193 (1996).
7. Goryunova, N. A., *The Chemistry of Diamond-Like Semiconductors*, London; Chapman & Hall, 1965.
8. Berger, L. I., and Prochukhan, V. D., *Ternary Diamonod-Like Semiconductors*, New York–London; Consultant Bureau , 1969.
9. Parthe, E., *Crystal Chemistry of Tetrahedral Structures*, New York; Gordon & Breach, 1964.
10. Borschevsky, A., "Early Work on Ternary Semiconductors," in *Proceedings of the 7th International Conference on Ternary and Multinary Compounds*, ed. S. K. Deb and A. Zunger, MRS Pub., **19**, (1987).
11. Shay, J. L. and Wernick, J. H., *Ternary Chalcopyrite Semiconductors: Growth Selectronic Properties and Applications*, Oxford; Pergamon Press, 1975.
12. Lopez-Rivera, S. A., "Quaternary Semiconducting Compounds," in *Proceedings of the International Conference on Ternary and Multinary Compounds (ICTMC-8)*, 1992, p. 266.
13. Zunger, A., Wagner, S., and Petroff, P. M., "New Materials and Structures for Photovoltaics," *J. Elect. Matls.*, **22**, 1 (1993).

334

POSTER SESSION

A Thin-Film Si Solar Cell: Deposition, Fabrication and Design

W.A. Anderson, B. Jagannathan and R. Wallace

State University of New York at Buffalo
Department of Electrical and Computer Engineering
Bonner Hall, Amherst, NY 14260

Abstract. This approach to a thin-film Si solar cell combines theoretical analysis, growth of thin-film crystalline Si (c-Si) and deposition of amorphous Si (a-Si:H) to form the heterojunction. The PC-1D model predicts a potential efficiency of > 15% for a Si thickness of 10 μm. Liquid phase growth (LPG) of the base-layer c-Si gave carrier mobility of $100 cm^2/V$-s and lifetime of 8 μs. Microwave electron cyclotron resonance (MECR) deposition of a-Si:H onto a c-Si wafer gives over 10% photovoltaic conversion efficiency. Work continues to combine MECR with LPG to give a thin-film Si solar cell.

INTRODUCTION

Progress is being made on several fronts in the area of **thin film Si photovoltaics**. Researchers in Germany have introduced the "Micromorph" cell, in which they use very high frequency glow discharge (VHF-GD) to produce microcrystal hydrogenated Si (μc-Si:H) [1]. They report a 10.7% efficiency on an a-Si:H/μc-Si:H cell. Tanaka et. al. [2] report an a-Si/poly-Si cell having a 9.2% efficiency. The poly-Si was made using solid phase crystallization of a-Si. Miyamoto et. al. [3] reported a remote PECVD method to obtain poly-Si on glass. N. Beck et. al. [4] also utilized VHF-GD and different gas dilutions to achieve μc-Si:H with drift mobilities up to 3 cm^2/V-s. Very recently, ultrathin crystalline Si on glass was formed by CVD at 1000°C [5]. Si films have also been obtained by laser crystallization of a-Si on plastic [6]. The a-Si was deposited at low temperature by PECVD. It is clear that encouraging progress is being made in the area of thin film Si for photovoltaics and for flat panel displays.

The work reported herein deals with the a-Si:H/crystalline-Si (c-Si) solar cell, where the crystalline Si may be nanocrystalline (nc-Si) or microcrystalline (μc-Si). Usually, the c-Si is p-type and the a-Si:H is layered n/i-type. This paper will

CP404, *Future Generation Photovoltaic Technologies: First NREL Conference*, edited by McConnell
© 1997 The American Institute of Physics 1-56396-704-9/97/$10.00

address design issues using PC-1D, formation of thin film c-Si, deposition of the a-Si:H and prospects for future advancement of the a-Si:H/c-Si solar cell.

PC-1D MODEL

PC-1D was used to simulate the performance of the a-Si:H/c-Si solar cell having an Si thickness of 10 μm. An efficiency of 12.1% was predicted without use of an antireflection coating. Interface recombination was introduced at the a-Si:H/c-Si interface to simulate the condition which we view as the cause for reduced efficiency. This condition of recombination velocity of 8×10^7 cm/s for electrons and 1×10^5 cm/s for holes reduced internal quantum efficiency by 20% and photovoltaic power by even more than we experimentally observed. Increasing a-Si thickness, above 500 Å, caused severe reduction in the UV portion of the quantum efficiency plot.

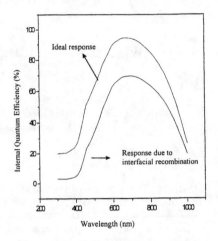

Fig. 1 Effect of interfacial recombination using PC-1D code for the a-Si:H/10μm thick c-Si

EXPERMENTAL METHODS

Liquid Phase Growth (LPG) of c-Si

This study was an attempt at forming a polycrystalline Si on a foreign substrate using a relatively low temperature process. Depositions were done in a diffusion-pumped high vacuum system. A liquid nitrogen cold finger assists in achieving a base pressure in the low 10^{-7} Torr range. A filament evaporation source deposits the solvent metal prior to sputtering of the Si by d.c. magnetron from a 2" target. Typically, a polished Mo substrate, coated with SiO_2, is first coated with a 50-500nm layer of Sn while heated to about 400°C, or at room temperature for the In/Ti prelayer. Ti is used as a

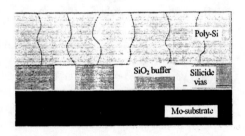

Fig. 2 Cross-section of thin-film poly-Si and substrate.

338

wetting agent to allow full wetting of the substrate surface by the liquid In layer. This is followed by the Si deposition in 2 mTorr of 5% H_2 in Ar at a rate of 1 micron per hour with the substrate at about 600°C. During deposition, a metal-Si melt is formed which is kept saturated in Si by the sputter gun. This liquid layer remains on the surface and leaves behind a poly-Si film. Chemical etching later removes the metal-Si residue and exposes the poly-Si layer. Films have been grown with thickness ranging from less than one micron to about 6 microns. The structure is shown in Figure 2. More details have been previously published [7,8].

MECR Deposition of a-Si:H

Thin film silicon growth by a microwave plasma discharge (operated in an ECR condition) in a $SiH_4/Ar/H_2/He$ type mixture addresses the need for obtaining microcrystalline Si (μc-Si) films with lower defect density at a higher rate of deposition and amorphous Si (a-Si) films with improved stability. The technique permits electrodeless deposition of silicon films at higher plasma density compared to the rf PECVD.

The schematic of the deposition system is shown in Figure 3. It consists of a Pyrex cross, mated to a quartz tube within a resonant cavity. The magnetron tube is mounted directly on the cavity, which is a rectangular Al waveguide. The input power to the magnetron can be continuously varied up to 1000 W. The two permanent magnets (885 Gauss) provide the ECR condition in the resonant cavity. The plasma is excited in a background gas (H_2, He or Ar) in the quartz tube and the process gases (2% SiH_4/He or 2% SiH_4/Ar) are fed at the front end of the quartz tube. A 2" stainless steel block with heater and thermocouple serves as the substrate holder. The chamber is pumped by a diffusion pump, and has a base pressure of $< 10^{-6}$ Torr. A cold trap is used to reduce backstreaming and reduce water vapor. Growth of μ c-Si has been achieved in both $SiH_4/Ar/H_2$ and $SiH_4/He/$ Ar/H_2 type mixtures at low substrate temperatures of 300 - 400°C at operating pressures of 1-10 mTorr. Amorphous silicon films have also been prepared with these gas combinations at temperatures of 250 °C and pressures of 40 mTorr.

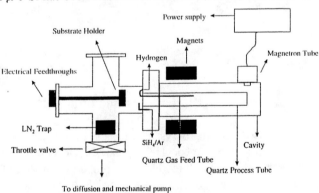

Fig. 3 MECR deposition system.

EXPERIMENTAL RESULTS

Liquid Phase Growth of c-Si

Films using the In/Ti prelayer exhibit large grain size (> 20 μm), carrier mobility up to 100 cm²/V-s, a preferred (111) orientation and carrier lifetime of 0.3 μs, which is lower than desired due to contamination by Ti. Films using the Sn prelayer have a grain size of about 1 μm, carrier mobility up to 100 cm²/V-s, no preferred orientation and carrier lefetime of 8 μs. Figure 4 shows the carrier lifetime data, taken by R.K. Ahrenkiel at NREL. Clearly, the films made with the In/Ti prelayer have 2 mechanisms, one of which we feel is caused by the Ti impurity. Figure 5 shows that the Sn-prelayer film responds well to hydrogenation with carrier mobility increases to 160 cm²/V-s.

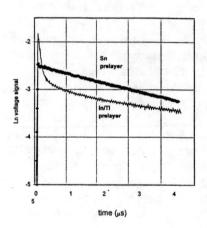

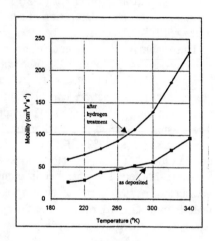

Fig. 4   Lifetime data for μc-Si films comparing Sn to In/Ti prelayers.

Fig. 5   Carrier mobility for μc-Si films using an Sn prelayer.

MECR Deposition of a-Si:H and μc-Si

The growth of microcrystalline silicon (μ c-Si) by the H_2 dilution of 2%SiH_4 (balance of either Ar or He) by an ECR-CVD process has been examined. Variations in structural and electrical properties of thin films have been evaluated with respect to growth parameters, mainly, H_2 dilution (defined here as R_H =H_2 content/(total gas content)), substrate temperature, input power and total pressure. Most of the films discussed here were deposited at 10 mTorr total pressure.

340

Structural Properties

Figure 6 shows the Raman spectra for H_2 dilution of 2% SiH_4/Ar, as it is varied from 0 to 0.55, at a substrate temperature of 300 °C, and an input power of 400 W. The dilution at a R_H of 0.55 corresponds to a condition of almost equal amounts of Ar and H_2 in the system. The films are seen to crystallize even at R_H of zero. Completely amorphous films are obtained when the input power is lowered to 200 W. As the H_2 content in the plasma increases, the amorphous contribution in the Raman spectra at 480 cm^{-1} and the shoulder at 503 cm^{-1} decrease, and for R_H levels of 0.55, only a sharp crystalline TO peak is apparent. By analysis of the Raman spectra, it was founds that ~45% of the film crystallizes when no H_2 is present, however, the grain sizes are only ~30 Å. With increasing R_H, the grain sizes and crystalline fraction increases. For R_H levels of 0.55, 70% of

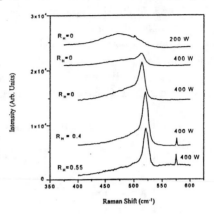

Fig. 6 Raman spectra of MECR-Si with different power and H-dilution.

the film crystallizes at grain sizes between 200-300 Å.

Figure 7 shows the hydrogen evolution in the films studied by evolved gas analysis, as the films are heated from 200-800 °C. In this method, the film is heated in high vacuum at 20°C/min, and the hydrogen release is monitored by a quadrupole mass spectrometer. For R_H of 0.55, the spectra indicates diffusion of atomic hydrogen from inside the film and its subsequent surface recombination. As the H_2 dilution reduces, the evolution shifts towards lower temperatures, and finally for R_H levels of zero, a distinct low temperature peak is evident at 350 °C. This nature seems to indicate that as the dilution reduces, the void formation in the film increases. This results in H_2 being present in these films in both molecular H_2 form (trapped in voids) as well as in atomic form (bonded to Si). For R_H levels of 0.55, the total H content in the film was ~2.2%, as compared with 16.6% for the films prepared with an R_H level of zero.

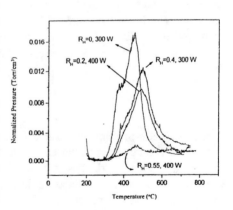

Fig. 7 Hydrogen evolution of MECR-Si.

Electrical Properties

Table 1 shows the dark/photoconductivity (σ_D/σ_{PH}) of the best films deposited by this method. The amorphous films were deposited at 40 mTorr and a substrate temperature of 250 °C, while the μ c-Si films have been deposited at 400 °C and a R_H of 0.55. Using He as the carrier gas results in films with very good σ_{PH} that also show a higher conduction activation energy compared to the films deposited with Ar, implying lower defect densities in such films. This is also seen for the μ c-Si films. In the case of the μ c-Si films, for films deposited at 400 °C, carrier transport was found to be thermally activated over many orders of magnitude, but the films deposited at 300 °C showed two transport mechanisms.

Table 1 Dark/Photoconductivity of Some ECR-CVD Films

Film Type	2% SiH₄ Carrier	σ_D (S/cm)	σ_{PH} (S/cm)	ΔE_{act} (eV)
Amorphous	Ar	8×10^{-8}	7×10^{-6}	0.53
Amorphous	He	4×10^{-8}	8×10^{-5}	0.80
Microcrystalline	Ar	3×10^{-6}	9×10^{-6}	0.30
Microcrystalline	He	6×10^{-6}	1×10^{-5}	0.49

Devices

Heterostructure solar diodes with ECR-CVD silicon/crystalline silicon (c-Si) type structures were fabricated. The ECR-CVD film was typically 300-400 Å thick. Figure 8 shows the photovoltaic characteristics of the cells. As-deposited μ c-Si/ c-Si devices show low J_{SC} and V_{OC} compared to the a-Si/c-Si type devices. However, the J_{SC} in the μ c-Si/c-Si could be increased from 12 to 24 mA/cm^2 by plasma hydrogenation of the μ c-Si layer carried out at 300 °C.

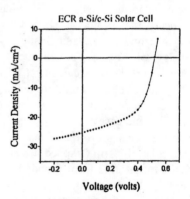

Fig. 8 Photovoltaic results for MECR-a-Si:H/c-Si wafer using 100 mW/cm^2 illumination.

DISCUSSION

The a-Si:H/c-Si thin film solar cell has a potential efficiency of 15%, based on PC-1D modeling. This work has explored the growth of thin-film c-Si using a liquid phase growth (LPG) in conjunction with a DC magnetron sputter source. Use of an Sn melt to promote LPG has led to film thickness exceeding 6 μm, carrier mobility of 100 cm^2/V-s (160 cm^2/V-s after hydrogenation), and carrier lifetime of 6 μs. On the negative side, film non-uniformity has hindered the fabrication of large-area solar cells. Solution of this problem might lie in use of a rotating substrate, introduction of a better buffer layer to improve wetting of the liquid phase or a seeding technique to control uniform nucleation.

Microwave electron cyclotron resonance (MECR) deposition of the a-Si:H has been quite successful. Films may be a-Si or nc-Si, by adjusting substrate temperature. Control of hydrogen dilution provides films of lesser hydrogen content which may improve upon light-induce degradation normally seen in a-Si:H. Solar cells fabricated using MECR-deposited a-Si:H/c-Si wafer have produced a 10% efficiency. The efficiency is limited by the a-Si/c-Si interface. We are exploring a solution to this problem utilizing hydrogenation and/or introduction of a thin nc-Si transition region between a-Si:H and c-Si. A more complex-solution would alter the structure to a *tandem* cell using (n-i-p) a-Si:H/(n-p)c-Si. This would remove the junction from the a-Si:H/c-Si interface.

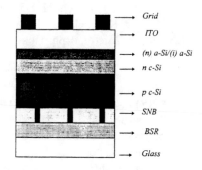

Fig. 9 Proposed design of a thin-film Si solar cell.

At some point, the MECR process will be applied to the LPG thin film to give a truly thin-film Si solar cell. Such a design is given in Figure 9 and includes a stable nucleating barrier (SNB).

CONCLUSIONS

PC-1D modeling of the a-Si:H/c-Si thin film solar cell predicts a potential efficiency of 15-18%. The liquid phase growth (LPG) process gives a thin Si film having good electrical properties and grain sizes up to 20μm or more. Improved uniformity is required which may be achievable by altering the buffer layer and/or wetting agent. MECR deposition of a-Si:H and nc-Si has produced quite promising films, the properties of which are easily controlled by process variables. Hydrogen dilution results in films having less incorporated hydrogen which may be more stable. MECR deposition of a-Si:H onto c-Si gave cells with an efficiency of 10%. Once LPG is improved, the MECR process can be used to complete the solar cell. A possible structure is shown in Figure 9.

ACKNOWLEDGMENT

The LPG work was supported jointly by National Renewable Energy Laboratory (NREL) and New York State Energy Research and Development Authority (NYSERDA). The MECR work is currently supported by NYSERDA.

REFERENCES

1. J. Meier, S. Dubail, R. Platz, P. Torres, U. Kroll, J.A. Anna Selvan, N. Pellaton Vaucher, ch. Hof. D. Fischer, H. Keppner, R. Flückiger, A. Shah, V. Shklover, K.-D. Ufert, "Towards High-Efficiency Thin-Film Silicon Solar Cells with the 'Micromorph' Concept",Preprint of the 9th PVSEC, Miyazaki, Nov. 11-15, 1996, Japan.
2. M. Tanaka, S. Tsuge, S. Kiyama, S. Tsuda and S. Nakano, "Fabrication of Polycrystalline Si Thin Film for Solar Cells", Materials Research Society (MRS) Fall Meeting, Boston, Dec. 2-6, 1996.
3. Y. Miyamoto, A. Miida and I. Shimizu, "Epitaxy-Like Growth of Polycrystalline Silicon on the Seed Crystallites Grown on Glass", MRS Fall Meeting, Boston, Dec. 2-6, 1996.
4. N. Beck, P. Torres, J. Fric, Z. Remes, A. Poruba, Ha Stuchlikova, A. Fejfar, N. Wyrsch, M. Vanecek, J. Kocka and A. Shah, "Optical and Electrical Properties of Undoped Microcrystalline Silicon Deposited by the VHF-GD with Different Dilutions of Silane in Hydrogen", MRS Fall Meeting, Boston, Dec. 2-6, 1996.
5. R. Brendel, R.B. Bergmann, P. Lölgen, M. Wolf and J.H. Werner, "Ultrathin Crystalline Silicon Solar Cells on Glass Substrates", *Appl.Phys.Lett.*, 70, 390 (1997).
6. P.M. Smith, P.G. Carey and T.W. Sigmon, "Excimer Laser Crystallization and Doping of Silicon Films on Plastic Substrates", *Appl. Phys. Lett.*, 70, 342-344 (1997).
7. R.L. Wallace and W.A. Anderson, "Thin Film Polycrystalline Si by CS Solution Growth Technique", *MRS Fall Meeting*, Boston, Nov. 28-Dec. 2, 1994.
8. R.L. Wallace and W.A. Anderson, "Solution Grown Polysilicon for Photovoltaic Devices", *25th IEEE Photovoltaic Specialists Conference*, Washington, DC, May 13-17, 1996.

Synthesis of Large-Grained Poly-Ge Templates by Selective Nucleation and Solid Phase Epitaxy for GaAs Solar Cells on Soda-Lime Glass

Harry A.Atwater, Jimmy C.M. Yang, and Claudine M. Chen

Thomas J. Watson Laboratory of Applied Physics
California Institute of Technology, Pasadena, CA 91125

Abstract. Synthesis of large-grained (20-30 μm) polycrystalline Ge films on uncoated soda-lime glass substrates is demonstrated using selective nucleation and solid phase epitaxy. The Ge crystallization kinetics and film microstructure are very similar to those previously observed for Ge on thermally-oxidized Si substrates, implying that the selective nucleation and solid phase epitaxy process is unaffected by impurities in the glass or the microstructure of the Ge/glass interface. These large-grained Ge films on glass are being investigated as templates for heteroepitaxy of GaAs/glass low-cost thin film solar cells.

INTRODUCTION

Polycrystalline thin film photovoltaic materials on low-cost substrates have received increasing attention in recent years owing to rapid progress in development of thin film materials such as CdTe, $CuInSe_2$ and related materials, as well as polycrystalline thin film solar cells with record efficiencies which are now approaching the efficiencies of the best silicon solar cells(1). Meanwhile, tandem III-V compound semiconductor thin film heterostructures epitaxially grown on high-quality single crystal substrates have achieved the distinction of the world's most efficient solar cells(2). In spite of the considerable advantages of III-V compound semiconductors (high single-crystal cell efficiencies and a relatively well-understood and mature heterojunction technology), these advantages have not been translated into a high-efficiency polycrystalline thin film solar cell, owing to the much higher sensitivity of solar cell performance to crystallographic defects in III-V compounds, as compared to II-VI and I-III-VI compound polycrystalline thin films(3). Thus achievement of efficient III-V compound polycrystalline thin film solar cells will require either (i) passivation of crystallographic defects so as to reduce photogenerated carrier recombination or (ii) dramatic reduction in the density of crystallographic defects in III-V compound semiconductor films on low-cost substrates, or ideally both.

CP404, *Future Generation Photovoltaic Technologies: First NREL Conference*, edited by McConnell
© 1997 The American Institute of Physics 1-56396-704-9/97/$10.00

Direct growth of GaAs on low-cost substrates such as glass generally yields polycrystalline films with unacceptably small grain size (a few μm) and rough surface morphology(4). Nonetheless, GaAs polycrystalline thin film solar cells grown on bulk polycrystalline Ge substrates with grain size of approximately 400 μm have demonstrated cell efficiencies of over 20%(5). Thus a means of enlarging the GaAs grain size on low-cost glass substrates by a factor of 10-100x is highly desirable. Selective nucleation and solid phase epitaxy is a technique capable of fabrication of large-grained (20-40 μm) polycrystalline Ge templates for GaAs heteroepitaxy(6). We demonstrate here a selective nucleation and solid phase epitaxy process for large-grained Ge on standard commercially-available soda-lime glass substrates at low temperatures (T < 425 °C).

SCIENTIFIC APPROACH

The approach for fabrication of large-grained Ge films on glass consists of selective nucleation and solid phase epitaxy of crystalline Ge seeds in an amorphous Ge thin film. This approach exploits the thermodynamic barrier to nucleation, as depicted in Fig. 1, whose origin lies in the size dependence of the interface and volume contributions to the crystal free energy(7). The nucleation barrier implies that growing small crystals will be temporally delayed in their evolution to macroscopic size. That is, at the onset of the transformation, there will be a finite incubation time during which no crystal nucleation occurs before the onset of steady state nucleation. The incubation time for apparent homogeneous nucleation has been observed and experimentally characterized for Si and Ge crystallization. If there are heterogeneous nucleation sites present, these can lower the thermodynamic barrier to nucleation such that the incubation time for heterogeneous nucleation is essentially insignificant as compared with homogeneous nucleation.

In practice, selective heterogeneous nucleation results from reactions between a patterned array of deposited In islands and 50-100 nm thick amorphous Ge to yield large grains (20-40 μm) in the fully crystallized film at temperatures below 425 °C. At present, the patterned array of In islands is fabricated by mechanically masked lithography. Complete film crystallization is achieved when the lateral solid phase epitaxy fronts from adjacent selective nucleation sites impinge upon each other, converting the initially amorphous Ge film to a completely crystalline film. A first anneal at 350 °C promotes selective nucleation and a second anneal at 400-425 °C enables large grain size to be achieved by lateral solid phase epitaxy, as depicted schematically in Fig 2. These temperatures are chosen so as to enable selective crystal nucleation but to suppress random nucleation. Thus the crystal size and grain boundary positions are in principle only determined by the spacing between the lithographically-defined nucleation sites.

RESULTS

Selective Nucleation

Selective nucleation is accomplished by patterning a periodic array of 5 μm diameter In islands by mechanically masked evaporation on top of an amorphous Ge film. Array periods range from 20-100 μm. Post-deposition annealing leads to heterogeneous nucleation of Ge crystals under the In pattern but not elsewhere in the amorphous Ge film. Further annealing causes the nucleated Ge crystals to seed growth of large (20-30 μm) Ge. Figure 3 is a plan-view optical micrograph illustrating the microstructure typically observed for a partially-crystallized films. By contrast, random crystallization in an amorphous Ge film typically yields average grain sizes of 0.25-1 μm.

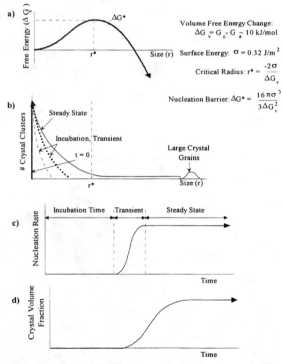

FIGURE 1. In (a) depiction of the free energy barrier to nucleation in the capillary approximation. In (b), evolution of crystal distribution with time from start of anneal to steady-state nucleation. In (c), time-variation of nucleation rate, illustrating the incubation time for nucleation. In (d), the fraction of the film transformed to the crystalline state as a function of time.

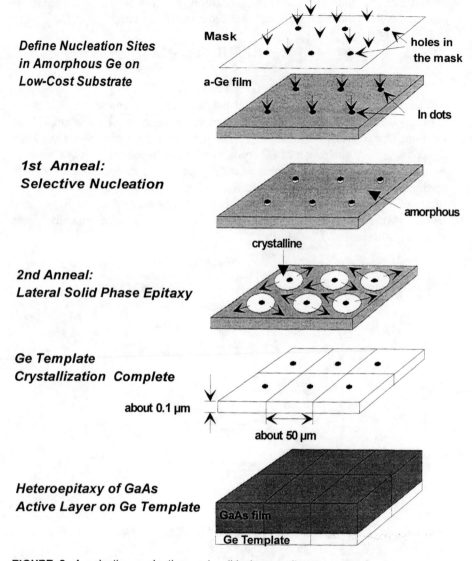

**Define Nucleation Sites
in Amorphous Ge on
Low-Cost Substrate**

Mask

holes in
the mask

a-Ge film

In dots

**1st Anneal:
Selective Nucleation**

amorphous

crystalline

**2nd Anneal:
Lateral Solid Phase Epitaxy**

**Ge Template
Crystallization Complete**

about 0.1 μm

about 50 μm

**Heteroepitaxy of GaAs
Active Layer on Ge Template**

GaAs film

Ge Template

FIGURE 2. A selective nucleation and solid phase epitaxy process for large-grained polycrystalline GaAs films on low-cost substrates is depicted schematically. Indium islands are deposited in a patterned array onto amorphous Ge films. The first anneal promotes selective nucleation and the second anneal enables large grain size to be achieved by lateral solid phase epitaxy. The large-grained Ge film then serves as an epitaxial template for growth of GaAs to form the photovoltaic device active region.

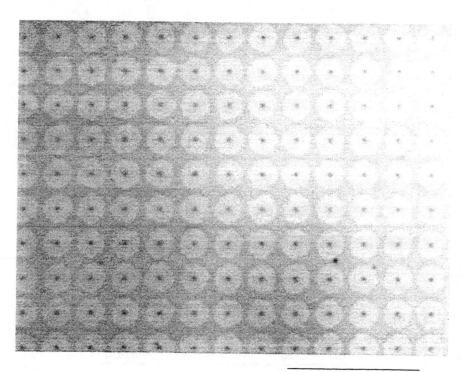

100 μm

FIGURE 3. Plan-view optical micrograph of partially-crystallized 100 nm Ge film on soda lime glass. Lightly-shaded gray circles are Ge grains; dark points in the grain centers are the selective nucleation sites. The gray background is the remaining uncrystallized amorphous Ge. The final grain size upon complete crystallization is approximately 30 μm.

A variety of materials other than In (e.g., Sn, Ni, Co, Pd, Cu and Al) have also been used for selective nucleation. At present, each selective nucleation site does not yield exactly one Ge grain; rather approximately 5-8 grains emanate from each site by lateral solid phase epitaxy. Thus the island seed regions produced by selective nucleation are polycrystalline, rather than a single grain per island. Efforts to produce a single grain in each seed region are underway.

Lateral Solid Phase Epitaxy of Ge

The kinetics of lateral solid phase epitaxy closely resemble those reported previously for Ge solid phase epitaxy in amorphous layers on single-crystal substrates(8). The apparent activation energy for lateral solid phase epitaxy in polycrystalline films is identical, to within experimental error, to the activation

349

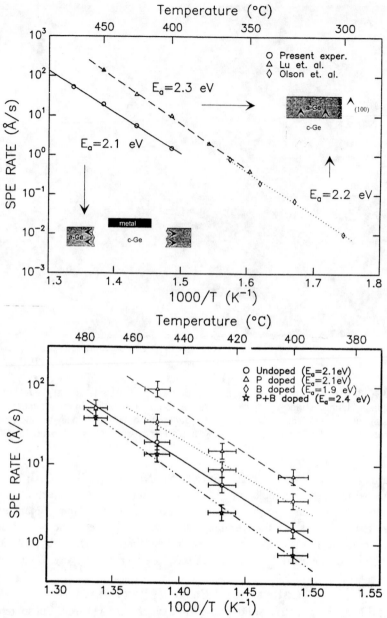

FIGURE 4. In (a), Arrhenius plot of lateral solid phase epitaxial growth rate with temperature in undoped 50 nm amorphous Ge films on SiO_2 (circles). The solid phase epitaxial growth rate of (100)-oriented single crystal Ge (triangles and diamonds, from Refs. (8,9) is also shown for comparison). In (b), Arrhenius plot of lateral solid phase epitaxial growth rate with temperature in 50 nm amorphous Ge films on SiO_2 doped with

0.6 at. % P (triangles) and 0.6 at. % B (diamonds) is compared with undoped films (circles) and films co-doped with 0.6 at. % P and 0.6 at. % B(stars).

energy for vertical solid phase epitaxy of single crystal Ge, as indicated in Fig. 3(a). However the absolute solid phase epitaxial growth rate in polycrystalline films is approximately five times lower than that for solid phase epitaxy of (100) oriented single crystal Ge. Doping of the amorphous Ge starting material with electronic dopants significantly enhances the solid phase epitaxy rate(9). Doping of a-Ge was performed by ion implantation of B and P to yield peak concentrations in the range of 0.3-0.6 at.%. For samples doped with 0.6 at. % P, an increase in the solid phase epitaxy rate by a factor of 5 relative to undoped films was found, enabling Ge grain sizes as large as 20-30 μm. For samples co-doped with both B and P at 0.6at. %, a compensation of the solid phase epitaxial growth rate enhancement is observed, i.e., equal amounts of B and P yield no enhancement –and in fact a slight decrease-- in the solid phase epitaxy rate, as illustrated by Fig 3(b).

The selective nucleation and solid phase epitaxy process leads to a restricted crystallographic texture relatively to a powder diffraction spectrum. Transmission electron microscopy and X-ray diffraction both indicate that the texture is restricted to mostly (100) and some (110) orientation, but essentially no (111) orientation. One possible explanation for this is that the In metal-induced selective nucleation process leads to a large distribution of starting crystallographic textures and orientations at the beginning of solid phase epitaxy, the fastest-growing orientation ``wins'' a competition among grain orientation as lateral solid phase epitaxy goes on. Thus this process yields large grains with a texture or textures yielding the highest possible growth rate. On the other hand, random nucleation yields a random crystallographic texture for each grain, yielding a lateral solid phase epitaxy rate which is slower than that characteristic of the fastest-growing orientation.

Ultimate Limits to the Process and Grain Size

The present results indicate that no fundamental or practical limit to enlargement of grain size has yet been encountered. Since the incubation time for nucleation is found to be quite insensitive to doping, but the lateral solid phase epitaxy rate is strongly dependent on doping, the most promising avenue for increasing the grain size is to optimize the doping concentration and thus maximize the lateral solid phase epitaxy rate. Other researchers have found that the lateral Si solid phase epitaxy rate is lower than the vertical solid phase epitaxy rate in thin Si films (< 1 mm thick), but that the lateral solid phase epitaxy rate increases with amorphous film thickness(10). The reasons for this are not well understood, but may be related to variation with thickness of the amorphous-crystalline interface microstructure, the crystallographic texture and the film stress. Little is known about these issues in Ge solid phase epitaxy.

The fundamental limitation on Ge grain size is due to randomly nucleated grains impeding the growth of selectively nucleated grains. Random nucleation could result from the intrinsic rate of homogeneous crystal Ge nucleation in amorphous Ge, or could also result from heterogeneous nucleation at sites other than the patterned selective nucleation sites. If homogeneous nucleation were limiting random nucleation, then the nucleation rate would increase and the incubation time would be expected to decrease with increasing film thickness. Therefore we investigated the dependence of the random nucleation rate on Ge film thickness. For Si thin films on SiO_2 it was observed that the number of nucleated Si crystals per unit area is constant with increasing Si film thickness, as would be expected for heterogeneous nucleation(11). For Ge films of 20, 50, and 80 nm thickness annealed in high vacuum at 475 °C for 10 min, all three film thicknesses showed about the same density of crystals, about 0.1 crystal per μm^2, as observed by transmission electron microscopy. This result suggests that the random nucleation is also heterogeneous in Ge, and probably originates from either the Ge/ SiO_2 interface or the Ge film surface, rather than within the interior of the thin film. Thus passivation of these interfaces may reduce random nucleation and thereby increase the final grain size.

At present, each predefined selective nucleation site in the Ge films produces not one, but multiple crystals, due to the structure of the multiple In islands formed at each site by mechanical masking, resulting in a polycrystalline microstructure with wedge-shaped grains emanating from the selective nucleation site(6). Several other approaches to reducing the nucleation density per site to a single grain are deemed promising. These include: i) reduction of the In island size by deposition of ultrafine nanocrystal In particles onto the amorphous Ge starting material, with each nanocrystal seeding a single Ge grain and ii) partial etchback of the deposited In islands to reduce their size and to reduce the density of Ge grains at the In island edges.

If each of these elements of the selective nucleation process is optimized, Ge grain sizes in the 100-1000 μm range may be possible. Beyond this, the ultimate limit to suppression of random nucleation and increase in grain size may substrate surface cleaning. The fact that the Ge-on-soda-lime-glass crystallization kinetics and film microstructure are very similar to those previously observed for Ge-on-thermally-oxidized-Si is encouraging since it implies that the random nucleation rate is unaffected by impurities in the glass or the microstructure of the Ge/glass interface. Future work will be aimed at exploitation of the large-grained Ge films on glass as templates for heteroepitaxy of GaAs on glass.

References

1. K. Zweibel, Progress in Photovoltaics, Vol.. 3, No. 5, Sept/Oct 1995.
2. K.A. Bertness, D.J. Friedman, S.R. Kurtz, A.E. Kibbler, C. Kramer, and J.M. Olson, I.E.E.E. Aerospeace and Electronics Systems Magazine, **9**, 12 (1994).
3. M. Yamaguchi and Y. Itoh, J.Appl. Phys., **60**, 413 (1986).
4. S. Kurtz, private communication.
5. R. Venkatasubramanian, B.C. O'Quinn, J.S> Hills, M.L. Timmons and D.P. Malta, *NREL Photovoltaic Program Annual Report* FY 1995, NREL/TP-410-21101, June 1996, pp 78-81.
6. C.M. Yang and H.A. Atwater, Appl. Phys. Lett., **68**, 3392 (1996); C.M. Yang and H.A. Atwater, Mat. Res. Soc. Symp. Proc. **403**, 113 (1996).
7. Christian, J.W., *Theory of Transformation in Metals and Alloys, Oxford, Pergamon Press, 1975, ch. 10, pp 418-475.*
8. G.Q. Lu, E. Nygren, and M.J. Aziz, J.Appl. Phys., **70,** 5323 (1991).
9. I. Suni, G. Goltz, M-A Nicolet and S.S. Lau, Thin Solid Films, **93**, 171 (1982); I. Suni, G. Goltz, M.G. Grimaldi, M-A Nicolet and S.S. Lau, , Appl. Phys. Lett., **40**, 269 (1982);
10. M. Moniwa, K. Kusukawa, E. Murakami, T. Warabisako and M. Miyao, Appl. Phys. Lett. **52**, 1788 (1988).
11. G.L. Olson and J.A. Roth, Mat. Sci. Reports, **1,** 1 (1988).

Simplification of process technology for high-efficiency solar cells

A.W. Bett, S.W. Glunz, J. Knobloch, R. Schindler, W. Wettling

Fraunhofer Institute for Solar Energy Systems, Oltmannsstrasse 5, D-79100 Freiburg, Germany
Phone (49) 761-4588 257, FAX (49) 761 4588 250, email: bett@ise.fhg.de

Abstract. The main cost factor in solar cell production is still related to the area-proportional costs. It is therefore obvious, that high-efficiency solar cells are required to minimize the overall system costs. On the other hand, high-efficiency processes for Si and GaAs are still very cost intensive. This paper addresses ways to simplify the process technology and to maintain still high-efficiency values.

INTRODUCTION

The reduction of module cost is one of the most important task for photovoltaic R+D. There are different ways to reach this goal. However, it is important to note, that in todays industrial solar module production more of 50 % of the module costs are related to wafer costs (1). The efficiencies of industrial cell production are about 12-15 %. On the other hand, several groups have reported efficiencies of more than 21 % for Si (2-4) and 25 % for GaAs solar cells (5). Those efficiency values require a sophisticated technology which is much more expensive than the technology which is common in industry. Nevertheless, for lower costs it is important to increase the efficiency and to maintain a low-cost technology. At the Fraunhofer ISE a research program has been implemented that aims at high-efficiency values using processing schemes as cost-effective as possible. This paper shows some results on the way to obtain this goal. In particular the following topics will be discussed:

i. The mesh structured emitter solar cell (MESC) is a new structure, respectively concept, for high-efficiency Si solar cells. This concept has the potential to use low cost processing steps like screen printing contacts.

ii. The random pyramid passivated emitter and rear cell (PR-PERC) structure uses a simplified technology process and obtains still efficiencies of up to 21.6 %.

CP404, *Future Generation Photovoltaic Technologies: First NREL Conference*, edited by McConnell
© 1997 The American Institute of Physics 1-56396-704-9/97/$10.00

iii. The LBSF/PERL process was adopted successfully to Si-Cz material. Efficiencies of 22 % could be realised.

iv. Multicrystalline Si solar cells (21 cm²) with unpassivated (17.2 %) and passivated (17.4 %) emitters were realised.

v. Simplified GaAs-heteroface solar cell structures without back surface field were fabricated using environmental friendly liquid-phase epitaxy (LPE). 23.3 % efficiency for one-sun illumination and 25 % (C = 100 x AM1.5d) for concentrated sunlight was measured.

vi. A simple Zn vapour-phase diffusion process was applied to GaSb in order to fabricate high-efficiency GaSb photovoltaic cells.

Si-SOLAR CELL TECHNOLOGIES

The Mesh Structured Emitter Solar Cell (MESC)

The emitter dark saturation current J_{0e} and the shadowing losses of the front contact metallization are two reasons for the still existing gap between mass production solar cells and state of the art laboratory cells. Highest efficiencies (> 20 %) are only achieved with an oxide passivated emitter exhibiting a relatively low doping in order to reduce the emitter dark saturation current and with photolithographically defined front contacts. However, the front metallization of mass production cells is fabricated using screen printing techniques, resulting in relatively broad contacts and requiring a high phosphorous doping concentration in order to reduce the contact resistance. The new cell type, the MESC (mesh-structured emitter solar cell) structure opens the possibility to extend the distance between the screen printed front contact fingers, keeping the recombination losses in the emitter at an acceptable level.

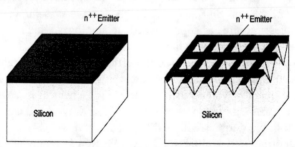

FIGURE 1. Processing scheme of the MESC-structure.

The MESC-structure is fabricated in two steps: First a relatively highly doped phosphorus emitter is diffused. In a second step inverted pyramids or other texturisation patterns known to reduce reflection losses very effectively are etched through this emitter profile, resulting in a mesh of a highly doped emitter lines (Fig. 1).

The phosphorous peak concentration in the ridge tops of the inverted pyramids is high enough to achieve a low contact resistance between screen printed contacts and emitter. On the other hand this highly doped phosphorus mesh results in a low sheet resistance, allowing to extend the distance between the front contact fingers, respectively to reduce the shadowing losses. In first experiments the recombination losses of this emitter type are found to be much lower than those of lateral uniform heavily diffused emitter.

Another motivation for the MESC-structure is the simplification of the complex processing scheme of high-efficiency solar cells with a locally diffused emitter (see Fig. 2). This emitter structure known as two-step emitter or selective emitter has led to the highest efficiencies for silicon solar cells fabricated from floating zone (FZ) silicon or Czochralski (Cz) silicon, but its processing scheme is more complex, due to the extra photolithography step for the local emitter. In the MESC processing scheme the fabrication of the local highly doped emitter can be combined directly with the fabrication of the inverted pyramids.

In order to test the MESC concept we started our experiments with a conventional mask set developed at Fraunhofer ISE for the local back surface field (LBSF/PERL) high-efficiency solar cells. We obtained maximum confirmed efficiency of 22.7 % which demonstrates the high-efficiency potential of this structure. In the next steps we omitted first the local back surface field and further the additional shallow emitter. We obtained 21.1 % and 19.2 %, respectively. The results are summarized in table 1.

It should be mentioned that these are the first results in simplifying the process technology. Further development and improvement of the MESC concept are under way at Fraunhofer ISE.

TABLE I. First results of the MESC structure (confirmed values)

	MESC without shallow emitter and local ohmic rear contacts	MESC with shallow emitter and local ohmic rear contacts	MESC with shallow emitter and local back surface fields
V_{oc} (mV)	657.8	677.0	696
J_{sc} (mA/cm^2)	39.5	39.8	40.5
FF	0.739	0.783	0.803
Efficiency	19.2	21.1	22.7

The RP-PERC process

At the Fraunhofer ISE the local back surface field (LBSF) solar cell (6) which is similar to the PERL process (7) was developed on FZ-material. The main steps of the process sequence are shown in Fig. 2 and described in more detail in (8). This high-efficiency process yielded in efficiencies of up to 23.3 % on FZ-silicon (see table II). However, as shown in Fig. 2 six photomask steps are needed during the sophisticated fabrication process. For a large scale industrial production these high-efficiency process is too far away to be cost effective. A step on the way to become more cost effective was made by developing a simplified process at the Fraunhofer ISE called RP-PERC (= random pyramids, passivated emitter and rear cell) to distinguish it from the PERC process developed at UNSW which originally has V-grooved texturization (9). The 6 main processing steps are shown in Fig. 2 in the right column to allow a direct comparison with the LBSF/-PERL type process.

In order to reduce wafer costs all experiments were performed with fairly thin wafers of 220 to 250 μm thickness which were chemically etched (not mechanically polished). This work was performed in close cooperation with the German PV companies Siemens Solar (SSG), Munich and ASE, Heilbronn. Both companies developed processing schemes that are similar, but not identical to the Fraunhofer ISE approach and that reached efficiencies between 18 and 19 % on a laboratory scale (10, 11, 12). A similar structure was developed at ISFH (13).

This RP-PERC processing scheme was first tested using FZ-Si. We used 4 inch diameter wafers of 0.5 Ω cm resistivity. The process turned out to be stable with most cells showing an efficiency above 20 %. The best cell had an efficiency of 21.6 % (AM1.5) as confirmed by the Fraunhofer ISE PV calibration lab. To our knowledge this is the highest value obtained with such a fairly simple process.

Cz-Si Material

FZ-Si material is not appropriate for a cost effective solar cell production because the price per wafer is about three times the price of a Cz-Si wafer. Therefore in 1994 we adopted our LBSF/PERL process to Cz material. It turned out that only slight variations concerning the temperature time profiles were necessary to achieve reproducible efficiency values above 20 %. This was tested for Cz-material of various resistivities between 1 and 10 Ω cm and of different vendors. The best efficiency value to date is 21.7 % for a 6.8 Ω cm material (4).

In contrast to FZ-material Cz-Si solar cells of 1 Ω cm material undergo a slight degradation of efficiency under light exposure or carrier injection (forward bias). This phenomenon which is not yet fully understood was discussed in a previous paper (4). The degradation of cell performance is attributed to a degradation of the bulk diffusion length of Cz-silicon. It was argued that it could be caused by a dissociated Fe-B pair defect (14), but more investigations are necessary to clarify this point. From a practical point of view the degradation can be avoided by using Cz-material with higher resistivity.

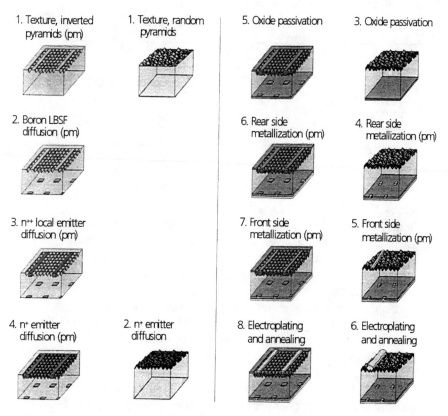

1. Texture, inverted pyramids (pm)

1. Texture, random pyramids

5. Oxide passivation

3. Oxide passivation

2. Boron LBSF diffusion (pm)

6. Rear side metallization (pm)

4. Rear side metallization (pm)

3. n++ local emitter diffusion (pm)

7. Front side metallization (pm)

5. Front side metallization (pm)

4. n+ emitter diffusion (pm)

2. n+ emitter diffusion

8. Electroplating and annealing

6. Electroplating and annealing

FIGURE 2. Comparison of the processing schemes for the LBSF/PERL solar cell (left column) and the RP-PERC cell (right column).

The simpler RP-PERC process was also applied to Cz-material. The best values achieved until now were 19.7 % but we are confident that 20 % can be reached in short time as soon as the processes are adapted to the special requirements of the material.

In Table II the best efficiency values are summarized. Solar cells of area 4 cm^2, 21 cm^2 and 42 cm^2 are listed for four types of cells: (1) LBSF/PERL cells on FZ-Si, (2) LBSF/PERL cells on Cz-Si, (3) RP-PERC cells on FZ-Si and (4) RP-PERC cells on Cz-Si. It should be emphasized again that the wafers were not specially prepared with respect to surface polishing and wafer thickness. It should also be mentioned that no special high-efficiency treatment like "alneal" (2) was used.

TABLE II. Best efficiencies for solar cells processed from FZ and Cz-Si by the LBSF/PERL and the RP-PERC processing sequences.

	FZ silicon solar cells				Cz silicon solar cells			
	LBSF/PERL		RP-PERC		LBSF/PERL			RP-PERC
Area (cm²)	4	21	4	21	4	21	42	4
V_{oc} (mV)	685	692	676.4	675.4	680.5	674.3	678.5	656.3
I_{sc} (mA/cm²)	42.0	38.6	39.6	38.7	41.8	38.7	38.9	37.7
FF	0.806	0.802	0.807	0.799	0.772	0.790	0.786	0.795
η (%)	23.3	21.4	21.6	20.9	22.0	20.6	20.8	19.7

MULTICRYSTALLINE Si

Multicrystalline silicon often is considered to be favourable to solar cell production due to the lower costs in crystallisation. The disadvantage of multi-crystalline silicon is fairly high concentration of crystalline defects like dislocations and grain boundaries which interact with residual impurities and may be electrical active. Most of the processes for mc-silicon solar cells which yield an efficiency higher than 16 % therefore rely on some form of impurity gettering and on bulk and surface passivation. Gettering and thermal oxidation typically add to the thermal load of the material and even may be of negative influence on minority carrier lifetime.

One of the problems related to mc-silicon is the inhomogenous distribution of minority carrier lifetime or diffusion lengths in the material leading to losses in open circuit voltage which has been demonstrated in an investigation by means of mini solar cell arrays (15) and which is responsible for reduced cell efficiency even in larger solar cells. Additionally, the evolution of diffusion lengths during processing may be different for various grains which shows different temperature stability during processing (16). As a consequence, for large area solar cells the temperature load should be low and the local distribution of the effective lifetime or diffusion length should be very tight. This contradicts in some way the efforts for improving minority carrier lifetime by extensive gettering and surface passivation by means of thermal oxidation.

This in turn requires gettering strategies which do not add to the thermal budget and easily are incorporated into the solar cell process. One key element in this strategy is the combined emitter diffusion and back surface field formation (17).

The back surface field is formed by aluminum alloying, precisely liquid-phase epitaxy during cooling following the phosphorus diffusion. Al gettering is segregation induced gettering due to the differences in solubiltiy in solid silicon and liquid aluminum. Thus in one single step emitter and back surface field are formed while gettering takes place.

In a study we applied different solar cell processes of varying complexity to five differently crystallized materials. The materials differed with respect to the crystallization velocity, planarity of the crystallization interface, and in terms of feedstock quality. The influence of different surface passivation schemes was investigated.

With one of the materials which used electronic grade silicon as feedstock our particularly simple process without any form of additional bulk and surface passivation and one single high temperature step (820 °C) an efficiency of 17.2 % on 21 cm² area was achieved.

With increasing process complexity the efficiency of the solar cells is increased. By adding surface passivation to the process the efficiency is increased to 17.4 %.

III-V-SOLAR CELLS

GaAs-heteroface solar cells fabricated by LPE

Nowadays, most GaAs heteroface solar cell structures are grown by metal organic vapor epitaxy (MOVPE). This technology allows to produce sophisticated structures combined with a high throughput. The main disadvantage are the highly toxic sources. In comparsion, the liquid-phase epitaxy (LPE) technology is environmental friendly but shopisticated structures are more difficult to realize. This is especially valid in combination with a high throughput demand. At Fraunhofer ISE we investigated a fairly simple isothermal LPE process which is called LPE etchback regrowth (LPE-ER) (18). The process scheme is shown in Fig. 3. We used a Gallium based melt which is in contact with a n-doped substrate for 45 minutes. During this time the pn-junction is formed by Zn-diffusion and the AlGaAs-heteroface is grown isothermally. More details of the process are given in (18). The simple structure of the solar cell is shown in Fig. 4. It is noteworthy, that the solar cell has no back surface field and in comparsion to MOPVE grown structures a thick emitter of 2 μm. In spite of this simple structure efficiencies of up to 23.3 % were achieved (see table III).

In case of a application as concentrator solar cell this cell structure has the advantage of lower ohmic losses in the relative thick emitter. An efficiency of 24.9 % at 100 suns were measured, only limited by the contact resistance (19,20).

361

In order to overcome the problem of high throughput, a special crucible was developed for the LPE-ER process (21). In the prototype of this crucible a total substrate area of 60 cm² could be successfully processed in one run.

It should be mentioned that we used the same technology to produce AlGaAs solar cells (see table III) which are suitable as top cells in mechanically stacked tandem cells. More details of this process are published elsewhere (22).

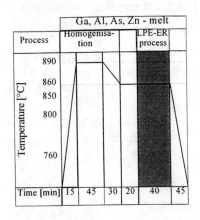

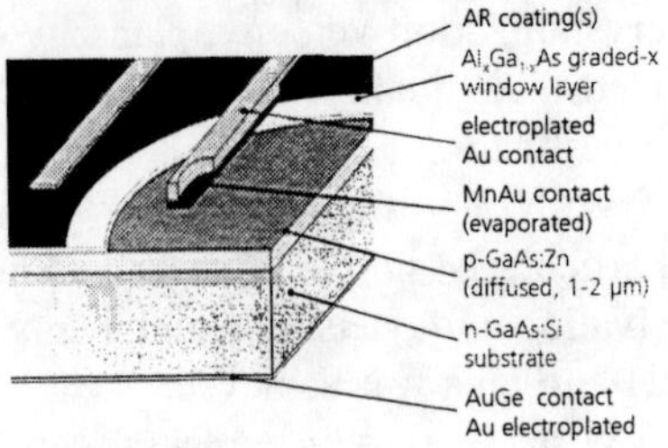

FIGURE 4. Scheme of a GaAs-heteroface solar cell.

FIGURE 3. Process sequence of the LPE-ER process.

Table III. Efficiency values for different LPE-ER grown solar cell structures

	GaAs	GaAs (C=100)	Concentratormodule C_{geo}=160	AlGaAs (C=30)
Area (cm²)	1	0.13	243 (Aperture)	0.13
V_{oc} (V)	1.027	1.132	13.300	1.362
I_{sc} (mA/cm²)	27.0	2560	350 mA	429
FF	0.841	0.852	0.832	0.879
η (%)	23.3	24.9	20.1 (793 W/m²)	17.1

GaSb photovoltaic cells

GaSb and related compounds are of large interest for photovoltaic devices, e.g.:

i. bottom (infrared) cells in high-efficiency mechanically stacked tandem solar cells with AlGaAs/GaAs top cells (23,24).

ii. photovoltaic (PV) cells in radioisotope thermovoltaic generators at relatively low temperatures (e.g. < 1250°C) (25) or in the newly developed thermophoto-voltaic electric generators with a hydrocarbon burner (26).

To meet the demands of a large-scale production, one should develop a GaSb PV cell technology which exhibits high reproducibility, simplicity and be harmless to the environment. In the case of GaSb a simple, pure diffusion process can be successfully used to fabricate PV cells. At Fraunhofer ISE the pseudo-closed box method was used to perform the diffusion of Zn into GaSb from the vapour phase (27). This method avoids the inconvenience of sealed ampoules and proved to be quite simple and reproducible.

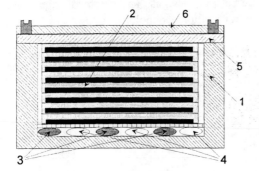

FIGURE 5. Scheme of the graphite diffusion boat for the wafers in horizontal position;
1 - body of the boat, 2 - GaSb wafers in a holder, 3 - Zn or Zn+Ga as sources of Zn vapour, 4 - Sb as a source of Sb vapour, 5 - cover, 6 - plate with screws

Zn diffusion was performed into 2.5 x 2.5 cm^2 n-GaSb:Te (n$\approx$3x10^{17} cm^{-3}) (100) substrates. The diffusion process was performed in a H$_2$ atmosphere, purified by a Pd cell, in a specially designed multi-wafer graphite boat shown in Fig. 5. The multi-wafer boat allows (i) to place 4x4 cm^2 or smaller wafers vertically or horizontally by simple changing of the substrate holder, (ii) to vary the distance between the wafers with a minimum distance of 0.5 mm, (iii) to use several separated Zn vapour sources (pure Zn or Zn and Ga melt) and Sb vapour sources. This design of the graphite boat ensures the uniformity of the Zn vapour pressure across the wafer surface, and thus the uniformity of the p-GaSb layer depth. Zinc and antimony were always used in more than sufficient quantities to provide saturation vapour pressures. A relatively low diffusion temperature (480°C) was used for the Zn diffusion.

The GaSb PV cell structure investigated is shown schematically in Fig. 6. The designated areas of the cells were 0.13, 0.92 and 3.61 cm^2.

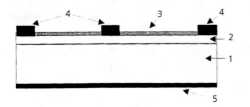

FIGURE 6. Scheme of a GaSb PV cell:
1- n-GaSb base (substrate),
2 - diffused p-GaSb emitter,
3 - single-layer AR coating
4, 5 - front (TiPdAg), back (AuGe) contacts

In the case of GaSb it is very difficult to evelute the obtained efficiencies because it is strongly dependant on the application of the PV cell. For example, in the case of bottom cell application the overall AM1.5 efficiency is not the whole truth. In the case of TPV application the efficiency is strongly dependent on the used radiative emitter. We designed our cell that the maximum quantum yield is obtained in the wavelength range between 900 and 1800 nm. Nevertheless we characterized our cells by concentrated AM1.5d illumination. Table IV summarizes the efficiencies for different grid designs. The obtained efficiencies demonstrate the high potential of this very simple technology.

TABLE IV. Efficiency values for GaSb photovoltaic cells under concentrated sunlight. The efficiency for a black-body illumination (T=1250-1750K) is as high as 20-25% if one assumes a 100% reflection of the not absorbed IR part of the spectrum (λ > 1800 nm).

Designated area (cm^2)	Concentration (suns)	Short-circuit current density (mA/cm^2)	Open-circuit voltage (mV)	Fillfactor (%)	Efficiency (%)
0.13	200	5871	456	74.0	9.9
0.92	23	790	414	69.9	9.7
0.92	1	34.4	357	69.6	8.2
3.61	1	35.5	354	65.6	8.2
3.61	5	179	398	70.5	9.4

CONCLUSIONS

In this paper the program of the Fraunhofer ISE to simplify process technology in combination with high-efficiency values was described. First experimental results of the MESC concept gives rise to the hope that 20 % efficient Si solar cells can be processed on a large scale in a cost effective way in not too far future. It was shown that solar cells with efficiencies of 20 % (AM1.5) and higher can be produced using a more simple cell process (RP-PERC process) with only two photomasking steps. Cells of 4 and 21 cm^2 area were fabricated on FZ-silicon as well as on Cz-silicon wafers. The efficiencies lie only about 1 percent (absolute) below cells processed with the much more involved LBSF/PERL processing scheme. A very simple process for multicrystalline Si solar cells resulted in efficiencies of 17.2 % on 21 cm².

In the case of III-V compound materials high efficiencies were obtained using enviromental friendly LPE-ER and vapor phase diffusion processes.

ACKNOWLEDGMENTS

The author thanks the "high-efficiency team", the "TechII group" and the "III-V group" of Fraunhofer ISE for the excellent work. This work is supported by several projects of the German Ministry for Education, Science and Technology, BMBF.

REFERENCES

1. W.Wettling, "High-Efficiency Si Solar Cell Development at Fraunhofer ISE and in Germany", in Proceedings of PVSEC-9, Miyazaki, 1996, p. 153

2. J. Zhao, A. Wang, P.P. Altermatt, S.R. Wenham, M.A. Green, "24 % Efficient Silicon Solar Cells", in Proceedings of 1st WCPEC, Hawaii, 1994, p. 1477

3. P. Verlinden, R.M. Swanson, et al., "7000 high-efficiency cells for a dream", Progress in Photovoltaics Vol. 2, 1994, p. 143

4. J. Knobloch, S.W. Glunz, D. Biro, W. Warta, E. Schäffer, W. Wettling, "Solar Cells with Efficiencies above 21 % Processed from Czochralski Grown Silicon", in Proceedings of 24th IEEE PVSC, Washington, 1996, p.405

5. J.M. Olson, S.R. Kurtz, A. Kibbler, "High-efficiency GaAs Solar Cells using GaInP$_2$ Window Layer" , Proceedings of 21st IEEE-PVSC, Kissimimee, 1990, p. 138

6. J. Knobloch, A. Aberle, B. Voß, "Cost Effective Processes for Silicon Solar Cells with High Performance", Proceedings of 9th E.C. PVSEC, Freiburg, 1989, p. 777

7. M.A. Green, S.R. Wenham, J. Zhao, J. Zolper, A.W. Blakers, "Recent Improvement in Silicon Solar Cell and Module Efficiency", Proceedings of 21st IEEE-PVSC Kissimimee, 1990, p. 207

8. J. Knobloch, A. Noel, E. Schäffer, U. Schubert, F.J. Kamerewerd, S. Klußmann, W. Wettling, "High-efficiency Solar Cells from FZ, Cz and mc Silicon Material", 23rd IEEE PVSC, Louisville, 1993, p. 271

9. A.W. Blakers, A. Wang, A.N. Milne, J. Zhao, X. Dai, and M.A. Green, "22.6 % Efficient Silicon Solar Cells", 4th PV SC, Sydney, 1989, p. 801

10. A. Münzer, Statusbericht des BMBF (1996), p. 8-1 (in German)

11. K.A. Münzer, R.R. King, R.E. Schlosser, H.J. Schmidt, J. Schmalzbauer, S. Sterk, H.L. Mayr, "Manufacturing of Back Surface Field for Industrial Application", 13th E.C. PVSEC, Nice, 1995. p. 1398

12. P. Uebele, K.-H. Tentscher, R. Kern, S. Mattes, K.-D. Rasch, W. Schmidt, F. Schomann, G. Strobl, "Pilotline Production of Cz-Si Cells Approaching 20 % Efficiency", 12th E.C. PVSEC, Amsterdam, 1994, p. 67

13. A.G. Aberle, B. Kuhlmann, R. Meyer, A. Hübner, C. Hampe, R. Hezel, "Comparison of p-n juction and inversion-layer silicon solar cells by means of experiment and simulation", Progress in Photovoltaics Vol. 4, No. 3, p. 193 (1996)

14. J.H. Reiss, R.R. King, K.W. Mitchell, "Characterization of Diffusion Length Degradation in Cz-Si Solar Cells", Appl. Phys. Lett. 68 (23) 3302, 1996

15. R. Baldner, S.W. Glunz, R. Schindler, W. Warta, W. Wettling, "V_{oc}-limits of high-efficiency mc-Si solar cells investigated by the mini solar cell (MSC) method" Technical Digest PVSEC-9 (1996) Miyazaki, p. 319

16. W. Warta, S.W. Glunz, A.B. Sproul, H. Lutenschlager, I. Reis, R. Schindler, „Properties of mc Si heat treatment by classical and RTP", in „Polycrystalline Semiconductors III", Solid State Phenomena **37-38**, 415 (1994)

17. R. Schindler, „The art of living with defects: gettering and passivation", in „Polycrystalline Semiconductors III", Solid State Phenomena **37-38**, 343 (1994)

18. A. Baldus, A. Bett, U. Blieske, O.V. Sulima, W. Wettling, "Etchback-regrowth process for AlGaAs/GaAs solar cell structures", J. of Cryst. Growth 146, 1995, p 299

19. A. Baldus, A.W. Bett, U. Blieske, T. Duong, F. Lutz, C. Schetter, W. Wettling, "GaAs one-sun and concentrator solar cells based on LPE-ER grown structures" in Proceedings 1st WCPEC, Hawaii, 1994, 1697

20. A. Blug, A. Baldus, A.W. Bett, U. Blieske, O.V. Sulima, W. Wettling, "Zn post diffusion for GaAs LPE-ER concentrator solar cells", in Proceedings 13th EC-PVSEC, Nice, 1995, p. 910

21. A. Baldus, A. Bett, U. Blieske, O.V. Sulima, W. Wettling, „AlGaAs/GaAs solar cells grown by horizontal and vertical LPE-ER process" in Proceedings 12th EC-PVSEC, Amsterdam,1994, p.1485

22. F. Dimroth, A.W. Bett, W. Wettling, "Liquid Phase Epitaxy of AlGaAs and Technology for Tandem Solar Cell Application", J. of Cryst. Growth, in press

23 L. M. Fraas, J. E. Avery, J. Martin, V. S. Sundaram, G. Girard, V.T. Dinh, T. M. Davenport, J. W. Yerkes and M. J. O'Neil, "Over 35% Efficient GaAs/GaSb Tandem Solar Cells", IEEE Transactions on Electron Devices 37, 1990, 443.

24. V.M. Andreev, L.B. Karlina, A.B. Kazantsev, V.P. Khvostikov, V.D. Rumyantsev, S.V. Sorokina and M.Z. Shvarts, "Concentrator Tandem Solar Cells Based on AlGaAs/GaAs-InP/InGaAs (or GaSb) Structures", in Proceedings 1-st World Conference on Photovoltaic Energy Conversion, Hawaii, 1994, p. 1721

25. A. Schock, C. Or and V. Kumar, "Small Radioisotop Thermophotovoltaic Generators", in: Proc. 2-nd NREL Conference on Thermophotovoltaic Generation of Electricity, Colorado Springs, CO, 1995, p. 81

26. L. Fraas, R. Ballantyne, J. Samaras and M. Seal, "A Thermophotovoltaic Electric Generator using GaSb Cells with a Hydrocarbon Burner", in Proceedings 1-st World Conference on Photovoltaic Energy Conversion, Hawaii, 1994, p. 1713

27. A.W. Bett, S. Keser, G. Stollwerck, O.V. Sulima, W.Wettling, "GaSb-based (thermo)photovoltaic cells with Zn diffused emitters", Proceedings of 25th IEEE-PVSC, Washington, 1996, p.133

Energizing the Future: New Battery Technology a Reality Today

Henry Chase, Jack Bitterly, and Al Federici

U. S. Flywheel Systems

Abstract. The U.S. Flywheel Systems' flywheel energy storage system could be the answer to a critical question: How do we replace conventional chemical batteries with a more-efficient system that lasts longer and is non-polluting? The new product, which has a virtually unlimited life expectancy, has a storage capacity four times greater per pound than conventional chemical batteries. USFS designed and built each component of the system--from the specially wound carbon fiber wheel, the magnetic bearing, the motor/generator, and the electronic control. The flywheel is designed to spin at speeds up to 100,000 rpm and deliver about 50 horsepower using a proprietary high-speed, high-power-density motor/generator that is the size of a typical coffee mug. Some of the important markets and applications for the flywheel storage system include electric vehicles, back-up power supply, peak power smoothing, satellite energy storage systems, and locomotive power.

INTRODUCTION

Even though there have been many significant advances in fields such as medicine, computers, and communications, the source for on-demand stored energy still remains the chemical battery, which was first developed more than a hundred years ago. The lead-acid battery and its sister chemical systems have been the major source for motive and stationary power requirements. The ever-increasing need for world-wide stored energy, coupled with new application requirements, are escalating the negative impact on the earth's environment of dangerous gases, toxic wastes from battery production and disposal, and global warming.

But a safe, non-polluting alternative "mechanical battery" technology now exists to store energy for the 21st century. This technology, developed by U.S. Flywheel Systems (USFS) of Newbury Park, California, combines an internally designed and integrated system with state-of-the-art capabilities in magnetic levitation, advanced electronic controller and power systems, and non-polluting composite fibers. Many of the members of the USFS team of scientists and engineers came from aerospace and defense-oriented companies in Southern California.

The USFS flywheel battery has storage capability that is more than 4 times greater per pound than lead-acid batteries. The system design, which employs magnetically

CP404, *Future Generation Photovoltaic Technologies: First NREL Conference*, edited by McConnell

© 1997 The American Institute of Physics 1-56396-704-9/97/$10.00

levitated composite wheels, provides for a virtually unlimited life estimated to approach 100,000 charge and discharge cycles. Our current modules provide for a variety of power requirements, including a proprietary high-speed and high-power-density motor generator (about the size of a coffee cup) that provides up to 50 horsepower.

The self-contained module--as shown in Figure 1 at the Future Generation conference--is about 14 inches in diameter and 16 inches high, and it provides flexibility for several power-system designs specific to the needs of the customer or application. The energy that can be stored in this module is immense and variable, depending on speed of rotation and size of the rotor. The design that USFS has developed for the stationary applications calls for the module to have up to 4 kWh of "on demand" energy.

Figure 1. U.S. Flywheel Systems' energy storage system contains a flywheel that is 14 inches in diameter. The system was connected to a set of PV modules furnished by Byron Stafford of NREL in a demonstration at the Future Generation conference.

The company demonstrated an operating system with 500 Wh of energy and power of 3 kW at the National Renewal Energy Laboratory's 1997 Future Generation Photovoltaic Technologies conference in Denver, Colorado. This demonstration included a solar array provided by NREL and USFS's complete solar energy management system, which converted the solar power to regular household power, including energy stored in the flywheel (see Figure 2). USFS was awarded "Best Poster Presentation" for the conference. Table 1 lists some of the more important attributes of the USFS stationary energy management system, which includes highly efficient electronic power management and high-speed motor-drive electronics.

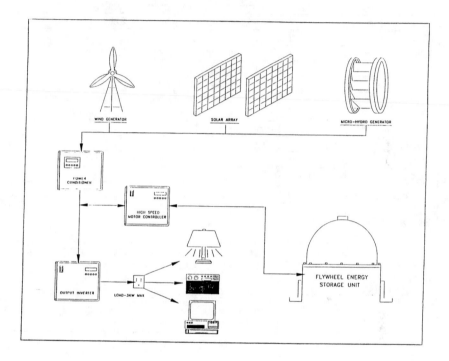

FIGURE 2. The USFS complete solar energy management system, in which solar power is converted to regular household power and energy is stored in the flywheel.

In the future, flywheel energy storage systems will provide commercial and environmental benefits in various industries. USFS has both delivered and designed operational systems for many of these applications. The Defense Advance Projects Research Agency (DARPA) has awarded grants to USFS to complete two separate programs for life-cycle testing and safety testing.

TABLE 1. Key Attributes of the USFS Stationary Energy Management System.

Power Output	3 kW continuous electrical power.
Usable Energy	500 Wh of on demand energy.
Voltage Input	60 VDC @ 10 amps, electronics can be modified to accommodate 36 VDC to 240 VDC or 110 VAC to 440 VAC single or three phase.
Voltage Output	110 VAC, electronics can be modified to accommodate 36 VDC to 240 VDC or 110 VAC to 440 VAC single or three phase. Multiple outputs available.
Charge Time	45 minutes from full stop to full capacity.
Charge Mode	System automatically charges to full capacity.
Recharge Time	30 minutes from 95% Depth of Discharge.
Recharge power	1.0 kW steady, 2.5 kW peak.
Coast power	30 W required to maintain full charge.
Discharge Mode	System automatically senses load voltages and responds to changes beyond preset limits.
Flywheel Module Size	14" diameter by 18" long.
Flywheel Module Weight	140 lbs.
Electronic module size	3 – 15" x 13" x 8" weatherproof boxes.
Electronics module weight	20 lbs.
Design life	20 years.
Maintenance Cycle	5 years.
Operating Temperature Range	-50° F to +180° F
Integral System Safety Features	System sensors to protect the module from excess temperature, imbalance, voltage, current and speed. "Crow-bar" shorts are automatically isolated to prevent damage to system and personnel.
Efficiency	Power electronics exceeds 93%; Round trip system exceeds 80%.

SATELLITE POWER SYSTEMS

Flywheel energy systems have distinct advantages over conventional battery systems in that they can deliver the required high power and energy density for a much longer period of time than chemical battery systems. NASA has identified certain space missions that could only be accomplished with flywheel energy storage. In addition to energy storage and power management, flywheel batteries can provide attitude control required for satellite operation without the use of extra propulsion systems requiring fuel. USFS is delivering operational systems to aerospace companies in preparation for actual use in commercial and defense space applications.

BACK-UP POWER

Without the protection provided by standby power sources, crucial data, life-supporting equipment, communications capabilities, and government functions could be lost during times of power failure. Flywheel battery systems can provide the dependable pollution-free source of power required by these applications by replacing toxic and dangerous lead-acid batteries. In addition, flywheel energy storage systems can be used with photovoltaic power generation systems to provide reliable, maintenance-free, non-toxic renewable energy to remote villages or homes anywhere in the world.

PEAK POWER SMOOTHING

Electric utility companies must maintain sufficient generator capacity to handle power demands during peak use. An energy storage system of many USFS flywheels would allow a utility to quickly store and release required peak energy at a much lower cost than constructing and maintaining extra generating facilities.

LOCOMOTIVES

The development of electric locomotives offers an excellent application for flywheel battery technology. A hybrid locomotive consisting of a typical diesel and flywheel power source would offer several advantages, including reduced fuel consumption of 30%, burst speed for inclines, reduced emissions, and energy recovery through regenerative braking.

AUTOMOTIVE

Both electric and hybrid-electric vehicles will benefit greatly from our flywheel technology. Either application will reduce the emissions to below standards set in California for "zero" emission vehicles. Prior to the end of this century, the company plans to demonstrate a vehicle integrating our flywheel energy technology.

The overall efficiency, power and long life of these "kinetically charged" flywheel systems already make this technology commercially viable and a competitive alternative that the chemical storage industry will have to reckon with.

This technology is capable of initiating social and economic changes starting in this decade. The competitive edge this new technology will give U.S. industry is very exciting. But even more, people in developing countries will enjoy the benefit of renewable energy without the usual infrastructure. Coupling energy from wind turbines and solar panels with flywheel energy storage will deliver reliable and non-polluting energy far into the future.

The 1997 Calaveras Awards
for "Leapfrog" Technologies

Henry Chase

has leapt to a new level by giving

The Best Poster Presentation

on

Demonstration of a Flywheel Energy Storage Unit
for Photovoltaics

The First Conference on Future Generation Photovoltaic Technologies
Sponsored by the National Renewable Energy Laboratory

Refractive Spectrum Splitting Optics for Use with Photovoltaic Cells: a Research Plan and Qualitative Demonstration

Alexander Kaplan Converse

Physica, Madison, Wisconsin 53701-1349 USA

Abstract. A new optical design has been proposed to refractively disperse sunlight and concentrate different portions of the spectrum onto an array of photovoltaic cells with suitable band gaps. Since the proposal of this design, a research plan has been developed for prototype construction, quantitative testing, and further design work, and a model has been constructed that qualitatively demonstrates the concept.

INTRODUCTION

A Refractive Spectrum Splitting design has been proposed (1) as a solution to the problem in photovoltaic energy conversion of matching the semiconductor band gap to the solar spectrum. This concept uses refractive dispersion to direct narrow portions of the solar spectrum to photovoltaic cells with appropriate band gaps thus improving the conversion efficiency. Compared to the most commonly used spectrum splitting technique, stacking cells, the present method may be advantageous in that each cell can be individually optimized without regard for concerns such as substrate transparency and lattice matching. Compared to spectrum splitting methods using dichroic filters or diffraction gratings, this innovation may be advantageous in that it can be inexpensively manufactured using techniques similar to those found in Fresnel lens production.

Real photovoltaic cells lose energy through a number of mechanisms, including electron-hole recombination, internal resistance, reflection, and incomplete absorption. However, even before considering these losses, only half of the energy in the solar spectrum is available to an ideal single material semiconductor cell because of the band gap matching problem. Compared to a photovoltaic conversion system using only 1 material, ideal systems using 2, 3, and 4 cells can produce 37%, 54%, and 64% more energy, respectively, when exposed to the direct normal AM 1.5 spectrum.

PRELIMINARY DESIGN

In the preliminary design presented in (1), sunlight first passes through a linear array of dispersing prisms and then through an array of concentrating prisms, whose apex angles are chosen to focus

CP404, *Future Generation Photovoltaic Technologies: First NREL Conference*, edited by McConnell
© 1997 The American Institute of Physics 1-56396-704-9/97/$10.00

photons of a particular energy onto a reference line on an array of photovoltaic cells. Higher energy photons tend to be refracted beyond this line, where they fall on high band gap photovoltaic cells, while lower energy photons tend to be refracted to the near side, where they fall on low band gap cells.

A conversion device using this preliminary optical design and four ideal PV materials was simulated using a Monte Carlo ray tracing routine. The PV efficiency was found to exceed that of a single-cell device by 37%. This falls short of the 64% for an ideal four-cell device mentioned above due to incomplete separation of the spectral components of the sunlight. The efficiency of the preliminary Refractive Spectrum Splitting design is further reduced due to losses in the second optical element, so its overall efficiency advantage in comparison to an ideal single cell Fresnel concentrator was found to be 23%.

RESEARCH PLAN

A research plan has been developed to address three questions: Are the optical calculations described above correct? What would be the optimized efficiency of a commercial prototype using real PV materials? What would be the mass production cost of such a device?

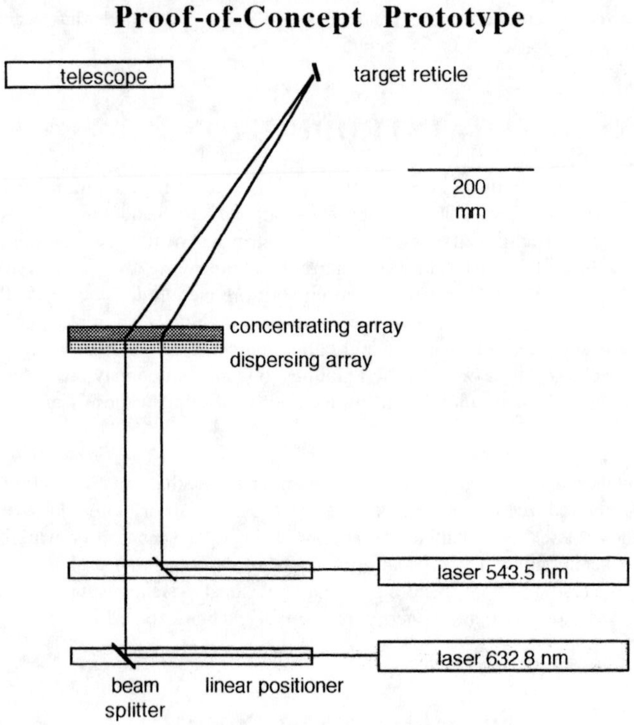

Proof-of-Concept Prototype

Figure 1. Experimental arrangement.

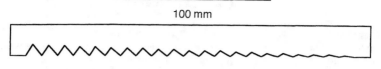

100 mm

Figure 2. Proof-of-concept prototype concentrating array.

A proof-of-concept prototype of the optical portion of the system would be constructed following the design presented in (1), and it would be used to test the ray tracing calculations to ± 0.5 mm. Once the calculations are verified, the next step would be to proceed with additional design work and plans for a commercial prototype.

The optical component of the simulation software would be tested using the experimental arrangement shown in Fig. 1. A detailed view of the prototype concentrating array is shown in Fig. 2. For ease of construction, the prototype concentrating and dispersing arrays would have 10 mm wide facets instead of the 1 mm wide facets specified in (1). A PV conversion system built using 10 mm wide facets would require more optical material and have lower efficiency, however such a prototype is adequate for testing the optical calculations.

The ray tracing would be checked to ± 0.5 mm absolute and ± 0.1 mm relative (green vs. red). The expected distribution of solar photons at 543.5 nm (2.282 eV) in the PV plane is $\mu = -7.0$ mm, $\sigma = 3.7$ mm, and at 632.8 nm (1.960 eV) it is $\mu = -3.2$ mm, $\sigma = 3.7$ mm. In addition to the paths of individual rays, these distribution predictions would be checked. Furthermore, transmission predictions would be checked to ±1.0% using a laser power meter.

Design Work

Having completed the proof-of-concept prototype test, additional design work would be undertaken. The following design tasks are intended to more realistically model the system and to improve the separation of the solar spectrum, thereby increasing the conversion efficiency:

• Extend the current model to include 3-D ray tracing.
• Include index of refraction and absorption data for the 0.5 - 3.25 eV energy range.
• Consider additional optical materials, e.g. polycarbonate which has higher dispersion but may have unacceptable wear characteristics.
• Include a PV model that accounts for the intensity dependence of the conversion efficiency and the quantum efficiency of PV cells made from materials such as germanium, silicon, gallium arsenide, cadmium telluride, and zinc telluride.
• Consider the effect of different geometries to improve efficiency, e.g. vary the algorithms that determine prism apex angles, place the facets on a curved rather than a flat surface, and set the PV cells in different planes.
• Model a point focus arrangement consisting of the line focus Refractive Spectrum Splitting system with a crossed line focus concentrator.
• Determine the range of efficiencies expected using measured solar spectra and modeled spectra accounting for different atmospheric conditions.
• Design a commercial prototype using three real photovoltaic cell materials and predict its conversion efficiency to ± 0.03.

Cost Analysis

Material cost estimates would be required for the three main components of a Refractive Spectrum Splitting module: the optics, the photovoltaic cells, and the mounting system. Assembly cost estimates would also be obtained. The information acquired would be used to predict the manufacturing cost of Refractive Spectrum Splitting modules to ± $0.5/Wp.

QUALITATIVE DEMONSTRATION

A model has been constructed using commercially available acrylic optical components that qualitatively demonstrates the Refractive Spectrum Splitting concept. The model, shown in Fig. 3 consists of three equilateral prisms and a line focus Fresnel lens. The entrance aperture is 75 mm x 66 mm, and a 75 mm x 10 mm dispersed line focus is produced 150 mm past the Fresnel lens. A second Fresnel lens may be used to produce a 10 mm x 10 mm dispersed point focus.

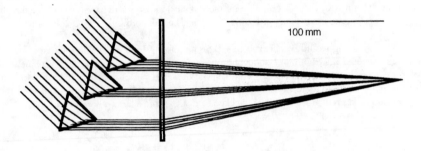

100 mm

Figure 3. Qualitative demonstration model.

SUMMARY

A new optical design has been proposed to refractively disperse sunlight and concentrate different portions of the spectrum onto an array of photovoltaic cells with suitable band gaps. Since the proposal of this design, a research plan has been developed for prototype construction, quantitative testing, and further design work, and a model has been constructed that qualitatively demonstrates the concept. A preliminary production cost analysis would be helpful in judging the value of pursuing the research plan.

REFERENCES

1. Converse, A. K., "Refractive spectrum splitting optics for use with photovoltaic cells," in *Proceedings of the Twenty-fifth IEEE Photovoltaic Specialists Conference*, 1996, pp. 1299-1302.

Low Cost, Single Crystal-like Substrates For Practical, High Efficiency Solar Cells

A. Goyal[1], D. P. Norton[2], M. Paranthaman[3], E. D. Specht[1], Q. He[2],
F. A. List[1], D. M. Kroeger[1], D. K. Christen[2] and J. D. Budai[2]

[1]Metals & Ceramics Division, [2]Solid State Division, [3]Chemical and Analytical Sciences Division,
Oak Ridge National Lab, Oak Ridge, TN 37831-6116.

Abstract. It is well established that high efficiency (20%) solar cells can be routinely fabricated using single crystal photovoltaic (PV) materials with low defect densities. Polycrystalline materials with small grain sizes and no crystallographic texture typically result in reduced efficiencies. This has been ascribed primarily to the presence of grain boundaries and their effect on recombination processes. Furthermore, lack of crystallographic texture can result in a large variation in dopant concentrations which critically control the electronic properties of the material. Hence in order to reproducibly fabricate high efficiency solar cells a method which results in near single crystal material is desirable. Bulk single crystal growth of PV materials is cumbersome, expensive and difficult to scale up. We present here a possible route to achieve this if epitaxial growth of photovoltaic materials on rolling-assisted-biaxially-textured-substrates (RABiTS) can be achieved. The RABiTS process uses well-established, industrially scaleable, thermomechanical processing to produce a biaxially textured or single-crystal-like metal substrate with large grains (50-100 μm). This is followed by epitaxial growth of suitable buffer layers to yield chemically and structurally compatible surfaces for epitaxial growth of device materials. Using the RABiTS process it should be possible to economically fabricate single-crystal-like substrates of desired sizes. Epitaxial growth of photovoltaic devices on such substrates presents a possible route to obtaining low-cost, high performance solar cells.

INTRODUCTION

Significant progress is being made in the photovoltaics industry towards large volume production. Currently, photovoltaics is a $ 1 Billion dollar per year business worldwide with more than 20 percent annual growth (1). Further increase in market share of the photovoltaic industry will require significant reductions in cost compared to most other forms of electricity generation typical in the utility industry. Higher efficiency cells made using more automated manufacturing processes are expected to lead the way to cost reduction. As prices decrease, usage of photovoltaic cells will increase, specially because of the environmetal benefits offered by this technology compared to other forms of energy generation. Further technological innovations have the potential to cause the economic and commercial breakthrough necessary to lower prices to $ 0.10 per kilowatt-hour by the year 2000 and possibly by $ 0.04 per kilowatt-hour by 2020. In a recent overview

CP404, *Future Generation Photovoltaic Technologies: First NREL Conference*, edited by McConnell
© 1997 The American Institute of Physics 1-56396-704-9/97/$10.00

article, Zweibel has pointed out that thin-film photovoltaics (PV) have a significant advantage over the traditional wafer-based crystalline Si cells (2). The primary advantage of thin films is cheaper materials and manufacturing costs and higher manufacturing yields compared to single-crystal technologies. Thin films use 1/20 to 1/100 of the material needed for crystalline Si PV and appear to be amenable to more automated, less expensive production (2). Thin-film PV modules have a very low projected potential cost (under $ 50/m^2) and reasonable module efficiencies (13-15% or more), implying potential module costs well under $ 0.5/Wp. (2). Currently, three film technologies are receiving significant interest from the industry for large scale PV: amorphous Si, CuInSe$_2$ and CdTe. Module efficiencies are closely related to cell efficiencies, with minor losses (~10%) due to some loss of active area and some electrical resistance losses (2). However, even today's best laboratory-level modules are about 8-10% efficient (2). When product-level technology and process development has adopted most technical capabilities observed in individual laboratory experiments, best laboratory modules are expected to be ~ 90% of efficiency of best cells. Furthermore, off-the-shelf commercial modules are expected to be about 90% as efficient as the best prototype modules (2). Based on the above considerations, a PV thin film technology that can produce cells with efficiencies in the range of 20% in a controllable and reproduceable manner is required. An efficiency of 17.7% using a thin-film, polycrystalline CuInGaSe$_2$ (CIGS) has recently been demonstrated (1). In order to further increase the efficiency and to be able to reproduceably fabricate high effciency cells, further research is required to understand the microstructural features which limit the performance.

Current research at obtaining high efficiency cells is aimed at controlling the microstructural features of the film closely. While a complete understanding of the microstructral features which limit the performance are still unclear, it is reasonably well established that recombination at grain boundaries, intragrain defects and impurities is critical. In an effort to minimize the effect of grain boundaries, films with large grains are an objective. It has been estimated by Kurtz and McConnell that to achieve over 20% efficiency in GaAs solar cells, grain sizes close to 50-100 μm are required (3). This assumes grains with crystalline perfection less than 5 x 10^6/cm^2 (3). While efficiencies close to this (18-20%) have been recently attained by Venkatasubramanian et al. (4) in epitaxial GaAs films on cast, optical grade Ge substrates with grain sizes close to 1mm, the cost of raw optical grade Ge is high and it is not clear whether it can be reduced to an acceptable range for even the 1-sun , terrestrial solar cells (3). Furthermore, since all thin-films discussed above are polycrystalline, they do not have a well-defined crystallographic orientation (both out-of-plane and in-plane). Crystallographic orientation can have two important effects. The first is the effect of orientation of the growth surface on incorporation of dopants, intrinsic defects, and other impurities (3). Studies on a wide variety of dopants have shown that variations of 1 to 2 orders of magnitude can occur based on crystallographic orientation (5,6). An extreme effect of anisotropic doping is Si doping in GaAs films (3). Si doping in GaAs films, causes n-type conduction on (111)B-type GaAs, but p-type on (111)A-type GaAs (7). The second effect of crystallographic orientation is a variation in growth rate (3). Both experiments as well as simulations have shown that under certain conditions growth rates can vary by 1 to 2 orders of magnitude as a function of crystallographic orientation (8). Uncontrolled crystallographic orientation in PV materials with large grain sizes may therefore result in reproducibility problems and hence lower yields during high volume production.

Most of the microstructural features currently thought to be limiting polycrystalline, thin-film, solar cell performance can be avoided by growing epitaxial films on lattice-matched, single crystal substrates. However, the high costs of single crystal substrates prohibits their use for realistic applications. We present here a low cost substrate for potential growth of epitaxial PV films, which very closely resembles a single crystal crystallographically. The Rolling-assisted-biaxially-textured-substrates (RABiTS) technique has been recently proposed by the authors (19-21). The technique uses well established, industrially scaleable, thermomechanical processes to impart a strong biaxial texture to a base metal. This is followed by vapor deposition of epitaxial buffer layers (metal and/or ceramic) to yield chemically and structurally compatible surfaces. Epitaxial superconductor films grown on such substrates have critical current densities approaching 10^6 A/cm^2 at 77K in zero-field and have a field dependence similar to epitaxial films on single crystal ceramic substrates. Deposited conductors made using this technique offer a potential route for the fabrication of next generation high temperature superconducting wire, capable of carrying very high currents in high magnetic fields and at elevated temperatures. The basic substrate made by this technique can also be used for epitaxial deposition of elemental, II-VI, III-V, and I-III-VI semiconductors like Si, Ge, GaAs, ZnSe, ZnTe, CdTe and CIGS. We summarize here key microstructural features of substrates made using the RABiTS technique and how they could be used for epitaxial growth of semiconductors. As is illustrated in this paper, substrates made using this technique have large grains (> 100 µm) and primarily low angle grain boundaries.

The effect of grain boundaries can be circumvented in polycrystalline photovoltaic thin films if the grain sizes are large enough (grain size at which effects on properties are minimal depend among other things on the doping level). However in thin-films, grain growth is typically restricted to only twice the thickness of the film. Hence, grain boundaries in polycrystalline films have a dominant effect on efficiencies. A large number of studies have reported the effects of grain boundaries on photovoltaic properties. In order to establish the potential of biaxially oriented substrates, an overview of grain boundaries in semiconductors and their effects on photovoltaic properties is first presented.

GRAIN BOUNDARIES IN SEMICONDUCTORS AND THEIR EFFECTS ON PHOTOVOLTAIC PROPERTIES

Grain boundary (GB) effects in semiconductors can be classified as intrinsic or extrinsic in nature. "Intrinsic" GB effects correspond directly to disruptions in lattice bonding with a lattice relaxation to accommodate the misfit at the GB. Lattice relaxations result in the generation of dislocations arrays at the GB plane and in the covalent diamond lattice are accompanied by dilated, contracted and in some cases dangling bonds. These bonding effects can result in electronic states in the

379

semiconductor bandgap. Hence intrinsic effects which occur due to GBs are bond distortion and dangling bonds. Moreover, the one dimensional periodicity along the dislocation core can lead to banding of these states. "Extrinsic" GB effects occur due to the interaction of the boundary with other lattice defects such as point, line and other planar defects. Furthermore, GB segregation and diffusion can result in variations of impurities or dopant atom concentrations at the GB relative to the bulk. This can have a direct effect on the chemical potential (Fermi level) associated with the boundary region. This in turn can result in localized electronic states in the bandgap and also alter the transport properties in the vicinity of the GB due to non-uniform dopant or impurity concentrations. Electronic states in the bandgap created due to such intrinsic and extrinsic defects can act as charge carrier traps and recombination centers and result in band bending, reduced carrier lifetime, and leakage currents associated with the GB.

Although many studies probing the effect of grain boundaries in various semiconductors have been made, no unified theory applicable to all types of materials exists (9-18). Some of the measured effects of grain boundaries are summarized here. Much of the early GB studies focussed on Ge. It has been established that the GB behaves as though it contained acceptor-like electronic states in the bandgap giving rise to band bending in such a fashion so as to create a barrier to minority carrier transport across GBs in n-type Ge. These states were also found to create a hole-like conductivity in the GB plane that is independent of temperature. The structure of the GB as predicted by the crystallography has been shown to dictate the bonding structure and hence influence the properties. In contrast to Ge, GB potential barriers have been found to occur in both n-type and p-type Si. This may be due to pinning of the Fermi level by the GB states at different positions or the ability of these states to act both as electron and hole traps. Detailed transport studies have shown how GB states give rise to the back-to-back diode characteristic for thermionic current flow across a grain boundary. However, not all grain boundaries were found to be electrically active. For example, it has been observed that high-angle coherent twin boundaries are electrically inactive. Twin boundaries have a high degree of coincidence and hence a lower energy than other high-angle boundaries. This points to the importance of GB energy, its effect on GB structure and resulting electronic properties. Enhanced impurity diffusion along GBs has also observed in Si and Ge. Low-angle GBs in both Si and Ge have been shown to be composed of discrete dislocations (11,12). This description is valid for [110] tilt boundaries with misorientation angles of less than 5° (11). For twist GBs this model fails for misorientation angles in the range of 3° to 8° (12). A change in the GB structure that occurs with the transition from low-angle to large misorientations is indicated by the electronic properties of Ge GBs. For example, the anisotropic nature of transport in the grain boundary plane, indicative of the presence of discrete dislocations disappeared as the misorientation angle increased (13). In another study, tilt and twist GBs, both having a misorientation angle of 6°, displayed the same properties although their dislocation structures should be different (14). This is consistent with the transition from low angle regime observed to occur at approximately 5° (11,12). In some high angle boundaries, microfacets are known to form leading to GB plane orientations that contain a high planar density of coincident lattice sites (CSL). This arrangement is achieved by a relative translation of one grain with respect to another and can be physically interpreted as the reconstruction of the bonding structure at the interface resulting in the formation of five, six, and seven member rings at the GB (15). No translation is associated with the coherent twin boundary with an exact coincident structure (15).

GB potential barriers have also been shown to exist in both n-type and p-type GaAs, the barrier however is significantly lower in n-type material. An experimental study by Salermo et al. (9) studied the characteristics of bicrystal prepared using the lateral overgrowth technique. They studied n-type GaAs bicrystals containing [110] GBs with misorientations of 0°, 2.5°, 5°, 10°, 24° and 30°. Current-voltage, capacitance-voltage and deep level transient spectroscopy (DLTS) were used for electrical characterization of the GBs. It was found that the rectification associated with tilt GBs increased as a function of misorientation angle. For GBs greater than 10°, the potential barrier height was found to remain constant. Moreover, the slight rectification associated with the 2.5° and 5° GBs indicates that the donor density of these layers is high enough to essentially the GB trap states.

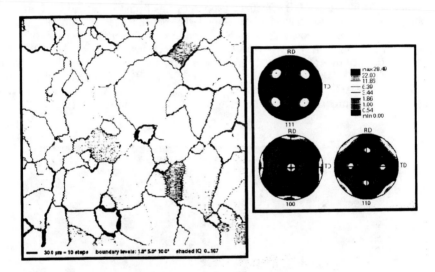

FIGURE 1. (a) Orientation image micrograph from a macroscopic region of a RABiT substrate (see text for details). BKD patterns were obtained in a hexagonal grid with a step size of 3 μm. Variations in intensity are a reflection of pattery quality or intensity of the Kikuchi bands. Three types of GBs are indicated in the figure, thin GBs denote boundaries with misorientations greater than 1° and less than 5°, thicker lines denote GBs with misorientations greater than 5° and less than 10° and the thickest boundaries denote boundaries with misorientations greater than 10°; (b) (111), (100), (110) pole figures of the orientations used to construct Figure 1 (a). Only a single orientation, {100}<001> cube texture is present.

Furthermore, the breakdown or saturation voltage was found to be a strong function of misorientation angle. These data indicate that the density of GB states increases with misorientation angle until saturation occurs in the range of 24° to 30°, suggesting that GB states in GaAs are directly related to GB structure.

GB effects on device characteristics can be described by considering the two cases of current flow and parallel and perpendicular to the boundary. For most photovoltaic materials, bulk as well as thin films, the grain structure is columnar with the grain boundary plane approximately normal to the junction, thus the case of current flow parallel to the GBs is of interest. Typically the polycrystalline device is modelled as both bulk and GB p-n junctions in parallel, with the bulk and GB regions as having different bandgaps and lifetimes. For current flow parallel to the boundary, GBs act as high conduction paths and lower the open circuit voltage and fill factor of the device. For current flow perpendicular to the GB, band bending at the boundary results in a potential barrier and therefore majority carrier transport across the GB encounters high resistance. This can result in increase in the series resistance of the device and hence reduce the short circuit current and the fill factor. Modelling of GB effects on properties has been performed by Card and Yang (16), who systematically developed the dependence of minority carrier lifetime, τ, on doping concentration, N_d, grain size and interface density of states (N_{is}) in Si. Judging from past work on heterojunctions, they postulated that the interface density of states can vary non-linearly by factors of 10^2 to 10^3 between low (~10^{10} cm^{2}eV^{-1}) and high angle (>10^{13} cm^{2}eV^{-1}) GBs. Since minority carrier lifetime is inversely proportional to N_{is}, a transition from low to high angle GB results in 3 to 4 orders of magnitude change in τ. It was shown that for poly-Si, τ

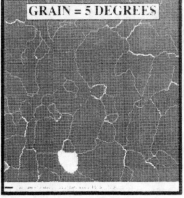

FIGURE 2. Orientation image micrographs shown in Figure 1a, shaded with the criterion that a given color represents a percolative region within (a) 2° and (b) 5°. Clearly most of the substrate is well connected by boundaries less than 5°.

varies by 4 orders of magnitude, from 10^{-6} to 10^{-10} as the grain size varied from 1000 to 0.1 μm (for a constant N_d). For a constant grain size, τ varied by 2 orders of magnitude with N_{is}. Similar calculations have been extended to other materials including GaAs (17) and CuInSe$_2$ (18).

The above discussion establishes the importance of controlling GB types in photovoltaic thin fims. Biaxially textured, columnar films with large effective grain sizes would be ideal. An effective grain size is one where grain boundaries within a certain misorientation can be tolerated. From the discussion above, it appears that for PV materials, this may be in the vicinity of 4-5°. Hence, substrates with GBs in this regime would be ideal. A second approach would be to have substrates with very large grains, so that grain boundary effects are minimal. Kurtz and McConnel have shown that for GaAs, this grain size is in the vicinity of ~100 μm. The RABiTS technique combines both these approaches and achieves very large individual grains (>100 μm), with primarily low angle grain boundaries less than 5°.

ROLLING-ASSISTED-BIAXIALLY-TEXTURED-SUBSTRATES (RABiTS)

RABiTS is a technique to produce macroscopically biaxially textured substrates (19-21). The method employs thermomechanical processing of base metals such as Cu or Ni to obtain a very sharp, well developed {100}<100>, cube texture. This is followed by deposition of appropriate chemical and structural buffer layers on the textured base metal. Substrates with biaxially textured, chemically and structurally compatible surfaces for epitaxial growth of superconducting or other electronic devices are referred to as rolling-assisted-biaxially-textured-substrates (RABiTS). Substrates made this way can be thought of as very large, defected, single crystals.

Biaxially textured Ni substrates were formed by consecutive rolling of a polycrystalline, randomly oriented high purity (99.99%) bar to total deformations greater than 90% (19). By controlling the surface condition of the work rolls, it was possible to obtain substrates with surfaces as smooth as those obtained by

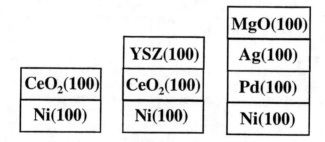

FIGURE 3. Schematic representations of cross-sections of three RABiT multilayer structures, CeO$_2$ / Ni; YSZ / CeO$_2$ / Ni; MgO / Ag / Pd / Ni.

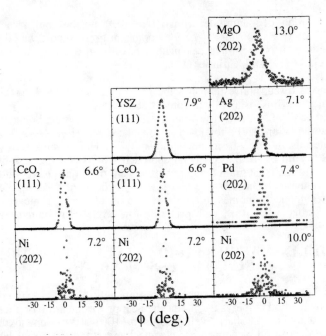

FIGURE 4. X-ray φ-scans showing the transference of texture through the various multilayers of a RABiT substrate, starting from the rolled and recrystallized metal to the top oxide layer: a) CeO$_2$ <110> ‖ Ni <100>; b) YSZ <110> ‖ CeO$_2$ <110> ‖ Ni <100>; and c) MgO <100> ‖ Ag <100> ‖ Pd <100> ‖ Ni <100>.

mechanical and chemical polishing, with rms roughnesses of ~ 10nm (21). The surface condition of a substrate can greatly affect epitaxy and integrity of buffer layers. Obtaining substrates with surfaces adequate for film growth without the need for a cumbersome polishing step is important for scale up to long lengths or large areas. Subsequent annealing of the substrates in a wide temperature range results in the formation of a sharp {100}<100> cube texture. Figure 1a shows an orientation image micrograph of a macroscopic region of a substrate recrystallized at 1000°C for 4 hrs in a vacuum of ~ 10^{-6} Torr. The micrograph was obtained using electron backscatter kikuchi diffraction (BKD). Gray level shading on the micrograph is a reflection of the pattern quality or intensity of the kikuchi bands observed at each point. Grain boundaries give rise to multiple diffraction patterns and hence have a poor pattern. Similarly, poor patterns are observed from any other crystallographic defect or strained region. BKD patterns were obtained on a hexagonal grid with a spacing of 3 μm. Total number of patterns obtained in the 0.5 mm x 0.5 mm region were close to 30,000. Indexing of the pattern at each location gave a unique measure of the orientation at that point. A hypothetical hexagonal lattice with a grain size of 3 μm was superimposed at each point from where a pattern was obtained. Grain boundary misorientations were then calculated for all the resulting boundaries using standard techniques. The micrograph was then regenerated with certain grain boundary criteria. In Figure 1a, three sets of GBs are indicated. The thin boundaries are boundaries with misorientation angles greater than 1° and less than 5°. Thicker boundaries have misorientation angles greater than 5° and less than 10°, and the thickest boundaries have misorientations greater than 10°. Regions where no boundary is indicated, but the contrast is dark correspond to boundaries with misorientations less than 1°. Figure 1b shows (111), (200) and (110) pole figures constructed using the data from Figure 1a. The presence of a sharp, well-developed, single component cube texture is evident. Typical samples have X-ray ω- and φ-scans with full-width-half-maximum (FWHM) of 6° and 7° respectively. The texture is found to be stable up to the melting point of Ni. Figures 2a and b show coloring of the same region as shown in Figure 1a, according to a criterion that a single color represents a contiguous or percolative region of orientation less than 2° and 5° in Figure 2a and b respectively. It can be seen that most of the substrate is percolatively connected within 5°. After recrystallization, the average grain size is approximately equal to the thickness of the substrate, which in this case was 125 μm. Thus the substrate can be thought to be comprised of a columnar structure of grains, with the columns aligned with the (100) plane parallel to the surface of the columns and the [100] direction aligned along the rolling direction. Typically, it is expected that for metal sheets with columnar grains, grain growth saturates at approximately twice the thickness of the substrate. In this case, the highly reduced mobility of low angle boundaries appears to saturate the grain size to approximately that of the substrate thickness.

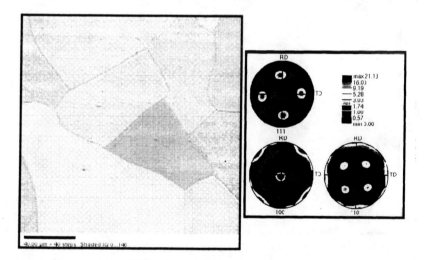

FIGURE 5. (a) Orientation image micrograph from a macroscopic region of a RABiT substrate of configuration CeO$_2$ <110> ‖ Ni <100>. BKD patterns were obtained in a hexagonal grid with a step size of 1 μm. Variations in intensity are a reflection of pattern quality or intensity of the Kikuchi bands; (b) (111), (100), (110) pole figures of the orientations used to construct Figure 5 (a).

In order to grow high quality epitaxial device films on the biaxially textured Ni substrate, a chemical and structural buffer layer is required. Typically, the desired buffer layers for superconductor film growth are oxides. Hence the task of fabricating a suitable substrate for epitaxial deposition of the superconductor involved epitaxial deposition of oxide buffer layers on Ni. This is difficult because of the ease of surface oxide formation on Ni under the typical oxidizing conditions required for oxide film growth. Although the surface oxide on (100) Ni can be epitaxial, it typically forms a (111) textured NiO layer with three equivalent epitaxial relationships, resulting in many high angle boundaries. We have found two methods so far that have proven successful in producing a single orientation, cube-on-cube epitaxial oxide buffer layer films on rolled and recrystallized Ni. The first involves epitaxial deposition of noble metal layers on Ni followed by deposition of oxides (19) and the second involves deposition of oxides directly on Ni under reducing conditions (20). A recent overview article summarizes the work on fabricating epitaxial superconductors using the RABiTS technique (21).

Oxide buffer layers can also be used for growth of epitaxial semiconductors. Figure 3 shows several possible multilayer structures that could be of use in the growth of epitaxial semiconductors like Si or GaAs. These are as follows: CeO_2/Ni, YSZ/CeO_2/Ni, MgO/Ag/Pd/Ni. Figure 4 shows X-ray phi scans of the corresponding multilayer structures. Such structures have been grown using laser

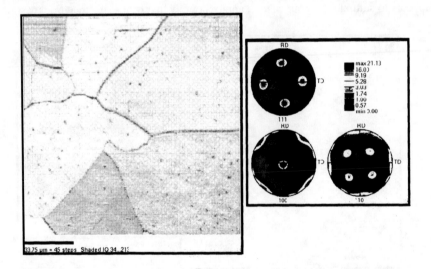

FIGURE 6. (a) Orientation image micrograph from a macroscopic region of a RABiT substrate of configuration YSZ <110> ‖ CeO_2 <110>‖ Ni <100>. BKD patterns were obtained in a hexagonal grid with a step size of 0.75 μm. Variations in intensity are a reflection of pattery quality or intensity of the Kikuchi bands; (b) (111), (100), (110) pole figures of the orientations used to construct Figure 6 (a).

386

FIGURE 7. Orientation image micrographs shown in Figure 6a, shaded with the criterion that a given color represents a percolative region within (a) 2° and (b) 5°. Clearly most of the substrate is well connected by boundaries less than 5°.

ablation (19-21), electron beam evaporation (22) and sputtering (23). Structures grown by laser ablation so far have resulted in evenly spaced cracks in the CeO_2 layer (21). However, by controlling the deposition conditions and thickness of the CeO_2 layer, crack free buffer layers have been obtained. Figure 5a shows an orientation image micrograph of a CeO_2 layer approximately 175 nm thick on Ni, grown using electron beam evaporation. Grain boundaries in the Ni substrate below are transferred through the oxide layers and are discerned in the orientation image since the pattern intensity from the boundary layer is poor. Figure 5b shows the corresponding (111), (100) and (110) pole figures from the data in Figure 5a. Single orientation cube-on-cube epitaxy, with a rotation of 45° with respect to the Ni substrate is obtained. Figure 6a shows an orientation image from the YSZ layer in a configuration shown in Figure 4. Corresponding pole figures from this region are shown in Figure 6b. A single epitaxial orientation with a cube-on-cube with respect to the CeO_2 layer is observed. If the region shown in Figure 6a is colored according to a criterion that a given color represents a percolative region of 2° and 5°, then the corresponding images are shown in Figure 7a and b respectively. The entire region is of a single color in Figure 7b, indicating the near single crystal nature of the oxide surface. Chemical analysis of the buffer layers using electron microscopy indicates that for a deposition temperature of 780°C, there is little diffusion of Ni into the CeO_2 layer.

EPITAXIAL GROWTH OF SEMICONDUCTORS ON OXIDES

Table 1 shows a compliation of epitaxial studies of oxides on Si (24-40). It is likely that similiar epitaxial relationships can be expected for growth of Si on

Table 1. Epitaxially grown oxides on Si.

Material	Out-of-Plane Orientation	In-Plane Orientation	Method	Lattice Mismatch	Ref.
YSZ	(001)‖(001)	[100]‖[100]	EB, PLD	-5.8%	(25)
					(26)
	(111)‖(111)	[110]‖[110]	PLD	-5.8%	(27)
PrO$_2$	(111)‖(111)	[110]‖[110]	PLD	0.7%	(28)
CeO$_2$	(111)‖(111)	[110]‖[110]	EB	0.4%	(29)
	(110)‖(001)	[100]‖<100>	EB	0.4%	(30)
Y$_2$O$_3$	(111)‖(111)	[110]‖[110]	EB	-2.4%	(31)
	(110)‖(001)	[100]‖<100>	EB	-2.4%	(32)
MgAl$_2$O$_4$	(001)‖(001)	[100]‖[100]	CVD	0.7%	(33)
g-Al$_2$O$_3$	(001)‖(001)	[100]‖[100]	MOMBE	-2.4%	(34)
	(111)‖(111)	[112]‖[112]	MOMBE	-2.4%	(34)
MgO	(001)‖(001)	[100]‖[100]	PLD	3.4%	(35)
SrO	(001)‖(001)	[100]‖[100]	MBE	-5.7%	(36)
	(111)‖(111)	[112]‖[112]	MBE	-5.7%	(36)
(Sr$_x$Ba$_{1-x}$O)	(111)‖(111)	[112]‖[112]	MBE	-5.7-1/6%	(37)
BaO	(001)‖(001)	[100]‖[100]	MBE	1.6%	(38)
SrTiO$_3$	(001)‖(001)	[110]‖[100]	FEB	1.7%	(39)
LiNbO$_3$	(001)‖(111)	-	Sputtering	-	(40)

Note: EB = electron beam deposition; PLD = pulsed laser deposition; CVD = chemical vapor deposition; (F)EB = (focussed) electron beam deposition; (MO)MBE = (metaloragnic) molecular beam epitaxy.

oxides. The table shows the crystallographies, method of fabrication and lattice mismatch. The lattice mismatch refers to the near coincident site lattice mismatch for each epitaxial relationship. As reflected in the table, the three surfaces shown in Figure 2 may be well suited for growth of Si. Other surfaces like BaO, SrTiO$_3$ etc could also be grown on RABiTS.

A key problem encountered in the growth of oxides on semiconductors has been that of oxidation of the semiconductor surface. This is a problem similar to that encountered during growth of oxides on Ni. In the case of Ni, formation of NiO is prevented by either depositing intermediate noble metal layers, or performing the deposition under a reducing 4%H_2-96%Ar background. This may also be possible during the growth semiconductors on some oxide surfaces, like YSZ and CeO_2. On the other hand, the problem of Si oxidation during growth of $BaTiO_3$ was overcome by depositing a thin layer of BaO, which is stable against reduction by Si (38). Similar results have also been obtained if BaO is substituted with MgO (35) or SrO (39). Depending on the type of oxide, different approaches involving multilayer structures are possible. For example it is possible to grow $BaTiO_3$/Ge/Si (100) but not $BaTiO_3$/Si (48). This is because of the differences in free energy of formation of SiO_2 (-167 kcal/mol, 1000°C) and GeO_2 (-90 kcal/mol, 1000°C) compared to that of decomposition of $BaTiO_3$.

Table 2 lists a compilation of epitaxial oxides on GaAs. Clearly, the most studied oxide is MgO which has a very small lattice mismatch with GaAs. Cube-on-cube orientation of MgO on GaAs has also been observed as indicated in the table. Cube-on-cube has also been observed for YSZ and In_2O_3.

Table 2. Compliation of Epitaxial Oxides on GaAs.

Material	Out-of-Plane Orientation	In-Plane Orientation	Method	Lattice Mismatch	Ref.
YSZ	(001)‖(001)	[100]‖[100]	PLD	-9.5%	[41]
In_2O_3	(001)‖(001)	[100]‖[100]	PLD	0.6%	[42]
MgO	(001)‖(001)	[100]‖[100]	PLD	-0.65%	[41]
MgO	(110)‖(001)	[110]‖[110]	EB	-0.65%	[43]
MgO	(111)‖(001)	[110]‖[110]	PLD	-0.65%	[42]
MgO	(111)‖(111)	[110]‖[110]	PLD	-0.65%	[45]
PLZT	(110)‖(001)	-	Sputtering	2.06-27.8%	[46]
$LiNbO_3$	(001)‖(111)	[110]‖[211]	PLD	-0.86%	[47]

EFFECT OF INTRAGRAIN DEFECTS ON PHOTOVOLTAIC PROPERTIES IN EPITAXIAL, MULTILAYER DEVICES

Epitaxially grown multilayer semiconductor thin films can have a large concentration of defects due to lattice mismatches and thermal expansion mismatches between layers and also due to growth processes. Surface defects on the substrate or underlayer film can also extend into subsequently deposited layers, the most common of these being dislocations. Dislocations, like grain boundaries, provide charged deformation regions for carrier scattering. Another intragrain defect is the stacking fault, a planar defect. In a study of epitaxial growth of GaAs on Si single crystals, dislocation densities were controllably manipulated by Vernon et al. (49). They find that a dislocation density less than 5×10^6/cm^2 is required to achieve a 20% efficiency. As the dislocation density increased to 10^7/cm^2, the open-circuit voltage, the short circuit current and the fill-factor decreased from their ideal values by 14%, 5% and 9% respectively (49). Kurtz and McConnel have summarized these results and based on data for efficiency versus dislocation densities and lifetime versus dislocation densities, have shown that a minority carrier lifetime > 10ns is required to achieve 20% efficiency in GaAs (3). Hence in order to fabricate high efficiency devices on RABiTS, innovative multilayer sequences need to be grown to realize an intargranular defect density less than $10^6/cm^2$. This is perhaps the most challenging task for this approach of device fabrication. Kurtz and McConnel point out that besides GB recombination, perimeter recombination is also important and dominates the loss mechanism in single crystal films (3).

Since thin-films have a very high surface-to-volume ratios, when the thickness of the film becomes comparable to the mean free path of the carriers, the scattering of electrons and holes from the film surface can have a significant effects on transport properties (10). If the scattering process is elastic, i.e. specular reflection from the surface where only the velocity component perpendicular to the substrate is reversed, the energy remains constant and there is no loss. However, if after scattering, the carriers emerge from the surface with velocites independent of their incident ones, a change in momentum that occurs is reflected in the conductivity. This is known as diffuse or inelastic scattering. Since real surfaces are associated with some atomic disorder, inelastic scattering dominates. Hence surface passivation is normally employed. In polycrystalline films, GB passivation may also be required depending on the nature of the GBs.

PASSIVATION OF SURFACE, INTERGRAIN AND INTRAGRAIN DEFECTS

If one assumes that the primary effect of the defect and surface is intrinsic in nature, then passivation involves mitigation of dangling and distorted or dilated bonds. With respect to dangling bonds, heat-treating in hydrogen can be effective in hydrogenating these bonds in both Si (50-52) and GaAs (53). Similar results have been observed with P implanted in poly-Si (54). With respect to distorted or dilated bonds, the solution may be to replace the atoms at the GB with atoms forming stronger covalent bonds, e.g. C in Si and P in GaAs (10). This could be viewed as cladding the GB with a higher bandgap material or, alternatively, strengthening the distorted bond. Elements forming strong covalent bonds are, for example, H, B, C, P, and S. Such passivation can be performed either by post heat-treatment or by doping and subsequent segregation of impurities at grain boundaries. For free surfaces, S, Se and N treatments have been effective in passivation (3). Higher bandgap layers also passivate the surface (3). Surface passivation can be expected to be easier than grain boundary passivation, since the normal lattice density and bonding configurations are preserved at the GB, in contrast to severe structural perturbations present at free surfaces.

Not many studies reporting the passivation of intragrain defects, low-angle GBs and CSL GBs have been performed. During commercial growth of single crystal Si, it appears that passivation of intragrain defects and very low-angle GBs is routine and simple to achieve (55). It is not however readily apparent why passivation of such low energy surfaces would be simple to achieve, since low energy boundaries would have less void space associated with them. In the presence of high energy interfaces, segregation of impurities and dopants to low energy boundaries may be restricted. However, this may not be the case when only low energy interfaces are present. Clearly, more detailed studies examining whether passivation of low-angle GBs and dislocations can be performed easily are required. Should this be the case, then epitaxial semiconductor films on RABiT substrates may be an ideal route to a high-efficiency, solar cell.

SUMMARY

In order to sustain and further the application of photovoltaic materials, devices with high efficiencies are required. Since cost is a major issue, thin film devices have considerable advantages over bulk materials. In case of thin films, choice of the substrate material dictates the performance of the device. For epitaxial films on single crystal substrates, very high efficiencies are obtained. However their cost precludes their use for large-scale applications. Amorphous and polycrystalline substrates with no controlled crystallographic texture and grain boundary character distribution, result in films with many grain boundaries. Since the grain size is difficult to increase to levels above where the effect of grain boundaries is negligible, the grain boundaries in these polycrystalline films reduce the effciencies to low levels. A review of studies relating the effect of grain boundaries on photovoltaic materials, suggests that control of grain boundaries may be one way to obtain high-effciency cells. Furthermore, anisotropy of growth rates and diffusion coefficients of impurity and dopant elements along differnt crystallographic directions, suggest that biaxial texture in these films and hence in the substrate may be desirable. An ideal substrate would be one that closely resembles a large single crystal. RABiT substrates formed by epitaxial growth of oxides on biaxially textured metals offer a potential route to a low cost, high efficiency cell. Such substrates have grains larger than 100 μm and contain primarily low-angle grain boundaries, less than 5°. A review of grain boundaries studies in semiconductor materials suggests that this may be a cross-over point from low to high for photovoltaic properties. The substrates also have a very sharp crystallographic texture in all directions. A summary of studies relating to epitaxial growth of oxides on Si and GaAs was also included. Several commonly studied oxides which have been epitaxially deposited on these semiconductors, have also been deposited on RABiTS. Innovative sequencing of multilayers will however be required to ensure a low intragrain defect density in the photovoltaic film by matching lattice constants, thermal expansions etc. of the film with the substrate.

ACKNOWLEDGEMENTS

Research sponsored by U.S. Department of Energy, Office of Efficiency and Renewable Energy, Office of Utility Technology - Superconductivity Program and the Office of Energy Research, Basic Energy Sciences, managed by Lockheed-Martin Energy Research Corporation for the U.S. Department of Energy under contract DE-AC05-96OR22464.

REFERENCES

1. NREL News release, NR-01296, May 10, (1996), http://www.nrel.gov/hot-stuff/press/thinfilm.html.
2. Zweibel, K., "Thin Films: Past, Present, Future," Progress in PV, The future of Thin Film Solar Cells, **V. 3**, # 5, 279-294, (1995); Zweibel, K. and Kline, D., "Flat-Plate, Thin Film PV Systems", preprint, (1997).
3. Kurtz, S. and McConnell, R., "Requirements for a 20%-Efficient Polycrystalline GaAs Solar Cell", preprint, this conference (1997).
4. Venkatasubramanian, R., O'Quinn, B. C., Hills, J. S., Sharps, P. R., Timmons, M. L., Hutchby J. A., Field, H., Ahrenkiel, R. and Keyes, B., "18.2% efficient GaAs Solar Cell on Optical-grade Polycrystalline Ge Substrates", in Proceedings of the 25th IEEE Photovoltaic Specialists Conference, 1996, 31-36.
5. Bhat, R., Caneau, C., Zalt, C. E., Koza, M. A., Borner, W. A., Hwang, D. M., Schartz, S. A., Menocal S. G. and Favire, F. G., "Orientation dependence of S, Zn, Si, Te, and Sn doping in OMCVD growth of InP and GaAs - application to DH lasers and lateral p-n junction arrays grown on non-planar substrates," J. Cryst. Growth, **107**, 772-778   (1991).
6. Kondo, M., Anayama, C., Okada, N., Sekiguchi, H., Domen, K. and Tanabashi, T., "Crstallographic orientation dependence of impurity incorporation into III-IV compound semiconductors grown by metalorganic vapor phase epitaxy," J. Appl. Phys., **76**, 914-927 (1994).
7. Pavesi, I., Piazza, F., Hernioi, M and Harrison, I., Orientation dependence of the Si doping of GaAs grown by molecular beam epitaxy," Semicond. Sci. and Tech., **8**, 167-171 (1993).
8. Jones, S. H., Salinas, L. S., Jones, J. R. and Mayer, K, "Crystallographic orientation dependence of the growth rate for GaAs low pressure organometallic vapor phase epitaxy," J. of Electron. Mater., **24**, 5-14 (1995).
9. Salermo, J. P., Fan, C. C., McClelland, R. W., Vohl, P., Mavroides, J. G., Bozler, C. O., "Electronic Properties of Grain Boundaries in GaAs: A Study of Oriented Bicrystals Prepared by Epitaxial Lateral Overgrowth," MIT Technical Report **669**, May 10, 1984.
10. "Polycrystalline and Amorphous Thin Fims and Devices", edited by Kazmerski, L. L., Academic Press, New York, 1980.
11. Bourret, A. and Desseaux, J., Phil. Mag., **A39**, 405 (1979); Bourret, A. and Desseaux, J., **A39**, 413 (1979).
12. Carter, C. B., Foll, H., Ast, D. G., Sass, S. L., Phil. Mag., **A43**, 441 (1981); Carter, C. B., Rose, J., Ast, D. G., Inst. Phys. Conf. Ser., **60**, 153 (1981).
13. Matukura, Y. and Tanaka, S., J. Phys. Soc. Japan, **16**, 833 (1961).
14. Mueller, R. K., J. Appl. Phys., **32**, 640 (1961).
15. "Grain Boundaries in Semiconductors," edited by Leamy, H. J., Pike, G. E. and Seager, C. H., North-Holland, New York, 1982.
16. Card, H. C. and Yang E. S., "Electronic Processes at Grain Boundaries in Polycrystalline Semiconductors Under Optical Illumination," IEEE Trans. Electron Devices, **ED-24**, 397 (1977).
17. Singh, R., Bhar, T. N., Shewchun, J. and Loferski, J. J., "Effect of Grain Boundaries on the Performance of Polycrystalline Tunnel MIS Solar Cells," J. Vac. Sci. & Tech., **16**, 236 (1979).

18. Kazmerski, L. L., "The Effects of Grain Boundaries and Interface recombination on the Performance of Thin-Film Solar Cells," Solid-State Electron, 21, 1545 (1978); Kazmerski, L. L., Sheldon, P. and Ireland, P. J., Thin-Solid Films, **58**, 95 (1979).

19. Goyal, A, Norton, D. P., Budai, J. D., Paranthaman, M., Specht, E. D., Kroeger, D. M., Christen, D. K., He, Q., Saffian, B., List, F. A., Lee, D. F., Martin, P. M., Klabunde, C. E., Hatfied, E. and Sikka, V. K., "High critical Current Density Tapes by Epitaxial Deposition of $YBa_2Cu_3O_x$ Thick Films on Biaxially Textured Metals," Appl. Phys. Lett., Sept., **69**, 1795 (1996).

20. Norton, D. P., Goyal, A., Budai, J. D., Christen, D. K., Kroeger, D. M., Specht, E. D., He. Q., Saffian, B., Paranthaman, M., Klabunde, C. E., Lee, D. F., Sales, B. C., and List, F. A., "Epitaxial $YBa_2Cu_3O_x$ on Biaxially Textured Nickel (001): An approach to Superconducting Tapes with High Critical Current Density," Science, November, **274**, 755 (1996).

21. Goyal, A, Norton, D. P., Kroeger, D. M., Christen, D. K., Paranthaman, M., Specht, E. D. Budai, J. D., He, Q., Saffian, B., List, F. A., Lee, D. F., Hatfied, E., Klabunde, C. E. and Martin, P. M., "Conductors with Controlled Grain Boundaries: An Approach to the Next Generation, High temperature Superconducting Wire," To be published in the J. of Materials Research, Special 10th Anniversary Issue, November, (1997).

22. Paranthaman, M., Goyal, A, List, F. A., Specht, E. D., Lee, D. F., Martin, P. M., He, Q., Christen, D. K., Norton, D. P. and Budai, J. D., "Growth of Biaxially Textured Buffer Layers on Rolled-Ni Substrates by Electron Beam Evaporation," Physica C, **275**, 266 (1997).

23. He. Q., Christen, D. K., Budai, J. D., Specht, E. D., Lee, D. F., Goyal, A, Norton, D. P., Paranthaman, M., List, F. A. and Kroeger, D. M., "Deposition of Biaxially-oriented Metal and Oxide Buffer-layer Films on Textured Ni Tapes: New Substrates for High-current, High-temperature Superconductors," Physica C, **275**, 155 (1997).

24. Fork, D. K. in Pulsed Laser Deposition of Thin Films, edited by Chrisey, D. B., and Hubler, G. H., John-Wiley and Sons, Inc., (1994).

25. Fukumoto, H., Imura, T. and Osaka, Y., Jpn. J. Appl. Phys. **27**(8), L1404-L1405, (1988).

26. Fork, D. K., D. B. Fenner, G. A. N. Connell, et al. (1990), Appl. Phys. Lett. **57**(11), 1137-1139.

27. Fork, D. K., D. B. Fenner, R. W. Barton, et al. (1990), Appl. Phys. Lett. **57**(11),1161-1163.

28. Fork, D. K., D. B. Fenner, and T. H. Geballe (1990), J. Appl. Phys. **68**(8), 4316-4318.

29. Inoue, T., Y. Yamamoto, S. Koyama, et al. (1990), Appl. Phys. Lett. **56**(14), 1332-1333.

30. Inoue, T., T. Ohsuna, L. Luo, et al. (1991), Appl. Phys. Lett. **59**(27), 3604-3606.

31. Fukumoto, H., T. Imura, and Y. Osaka (1989), Appl. Phys. Lett. **55**(4), 360-362.

32. Harada, K., H. Nakanishi, H. Itozaki, et al. (1991), Jpn. J. Appl. Phys. **30**(5), 934-938.

33. Mikami, M., Y. Hokari, K. Egami, et al. (1983), Ext. Abstract, 15th Conference on Solid State Development and Materials, Tokyo, pp. 31-34.

34. Sawada, K., M. Ishida, T. Nakamura, et al. (1988), Appl. Phys. Lett. **52**(20), 1672-1674.

35. Fork, D. K., F. Ponce, J. C. Tramontana, et al. (1991), Appl. Phys. Lett. **58**(20), 2294-2296.

36. Kado, Y., and Y.Arita (1987), J. Appl. Phys. **61**(6), 2398-2400.

37. Kado, Y., and Y. Arita (1988), Ext. Abstract, 20th Conference on Solid State Development and Materials, Tokyo, pp. 181-184.

38. McKee, R. A., F. J. Walker, and J. R. Conner, et al. (1991), Appl. Phys. Lett. **59**(7), 782-784.

39. Mori, H., and H. Ishiwara (1991), Jpn. J. Appl. Phys. **30**(8A), L1415-L1417.

40. Rost, T. A., T. A. Rabson, B. A. Stone, et al. (1991), IEEE Trans. Ultrason., Ferroelectr. Freq. Control **38**(6),640-643.

41. Nashimoto, K., D. K. Fork, and T. H. Geballe (1992), Appl. Phys. Lett. **60**(10), 1199-1201.

42. Tarsa, E. J., J. H. English and J. S. Speck (1993), Appl. Phys. Lett. **62**(19), 2332-2334.

43. Huang, L. S., L.R. Zheng, and T. N. Blanton (1992), Appl. Phys. Lett **60**(25), 3129-3131.

44. Tara, E. J., M. de Graef, D. R. Clarke, et al. (1993), J. Appl. Phys. **73**(7), 3276-3282.

45. Fork, D. K., K. Nashimoto, and T. H. Geballe (1992), Appl. Phys. Lett. **60**(13). 1621-1623.

46. Ishida, M., S. Tsuji, K. Kimura, et al. (1978), J. Cryst. Growth **45**, 393-398.

47. Fork, D. K., and G. B. Anderson (1993), Appl. Phys. Lett. **63**(8), 1029-1031.

48. Jacobs, J. G., Rho, Y. G., Pinizzotto, R. F., Proceedings of the MRS Symposium on Pulsed Laser Ablation, Dec. 1992, Boston, MA., **285**, 379 (1993).

49. Vernon, S. M. and Tobin, S. P., "Experimental Study of Solar Cell Performance Versus Dislocation Density," Proceedings of tha 21st IEEE Photovoltaic Specialists Conference, 211 (1990).

50. Robinson, P. H. and D'Aiello, R. V., Appl. Phys., **39**, 63 (1981).

51. Redfield, D., Appl. Phys. Lett., **38**, 174 (1981).

52. Seager, C. S. and Ginley, D. S., J. Appl. Phys., **52**, 1050 (1981).

53. Pearton, S. J. and Tavendale, A. J., J. Appl. Phys., **54**, 1154 (1983).

54. Andrews, J. M., Electron. Mater. Conf., 20th, santa Barbara, California, June 28-30, (1978).

55. Solarex, private communication at conference on Future Generation PV Technologies.

Formulation of a Self-Consistent Model for Quantum Well *pin* Solar Cells

S. Ramey and R. Khoie
Department of Electrical and Computer Engineering
University of Nevada, Las Vegas
Las Vegas, NV 89154
ark@ee.unlv.edu, sramey@ee.unlv.edu

ABSTRACT

A self-consistent numerical simulation model for a *pin* single-cell solar cell is formulated. The solar cell device consists of a $p - AlGaAs$ region, an intrinsic $i - AlGaAs/GaAs$ region with several quantum wells, and a $n - AlGaAs$ region. Our simulator solves a field-dependent Schrödinger equation self-consistently with Poisson and Drift-Diffusion equations. The emphasis is given to the study of the capture of electrons by the quantum wells, the escape of electrons from the quantum wells, and the absorption and recombination within the quantum wells. We believe this would be the first such comprehensive model ever reported.

The field-dependent Schrödinger equation is solved using the transfer matrix method. The eigenfunctions and eigenenergies obtained are used to calculate the escape rate of electrons from the quantum wells, and the non-radiative recombination rates of electrons at the boundaries of the quantum wells. These rates together with the capture rates of electrons by the quantum wells are then used in a self-consistent numerical Poisson-Drift-Diffusion solver. The resulting field profiles are then used in the field-dependent Schrödinger solver, and the iteration process is repeated until convergence is reached.

In a $p-AlGaAs\ i-AlGaAs/GaAs\ n-AlGaAs$ cell with aluminum mole fraction of 0.3, with one 100 $A°$-wide 284 meV-deep quantum well, the eigenergies with zero field are 36meV, 136meV, and 267meV, for the first, second and third subbands, respectively. With an electric field of 50 kV/cm, the eigenenergies are shifted to 58meV, 160meV, and 282meV, respectively. With these eigenenergies, the thermionic escape time of electrons from the $GaAs$ Γ-valley, varies from 220 pS to 90 pS for electric fields ranging from 10 to 50 kV/cm. These preliminary results are in good agreement with those reported by other researchers.

I. INTRODUCTION

The conversion efficiency of a single cell *pin* solar cell can be enhanced by

CP404, *Future Generation Photovoltaic Technologies: First NREL Conference*, edited by McConnell
© 1997 The American Institute of Physics 1-56396-704-9/97/$10.00

incorporating one or more quantum wells in the intrinsic region of the device.(1) The incorporation of the quantum wells has two counteracting effects: the short-circuit current is increased because of the additional absorption of the low-energy photons in the lower bandgap quantum well; and the open-circuit voltage is decreased because of the increase in the recombination of the photoexcited carriers trapped in the quantum well. Experimental results have shown, nevertheless, that the additional photocurrent resulting from the extension of the absorption spectrum to lower energies can outweigh the accompanying drop in the open-circuit voltage, and devices with quantum wells incorporated in the i-region are indeed more efficient than the corresponding base-line bulk cell (2)-(3). The improvement in conversion efficiency is dependent on, among other parameters, the width of the quantum wells. Ragay el. al., (4) reported that both the open-circuit voltage and short circuit current increase with increasing well width.

Along with these experimental studies, a number of theoretical investigations have been performed. Corkish and Green (5) studied the effects of recombination of carriers in the quantum well and concluded that although the increased recombination reduces the open-circuit voltage, but limited enhancement in the conversion efficiency can be obtained with incorporation of the quantum well, albeit not as much as previously reported by Barnham and Duggan (1). Spectral response modeling of Paxman, et. al., (6) with emphasis on the absorption in the quantum wells showed good agreement with thier experimental results. Araujo, et. al. (7) used detailed balance theory and predicted that the conversion efficiency of the quantum well cell would not exceed that of the single cell base-line device. The results of photoresponse calculations by Renaud el., al., (8) revealed that introducing the quantum wells in the intrinsic region can lead to improved photocurrent without much degradation of the open-circuit voltage. Most recently, Anderson (9) presented an ideal model for the quantum well solar cell device, incorporating the recombination and generation in the quantum wells. Anderson concluded that under AM0 illumination the maximum conversion efficiency of the pin solar cell device increases from the base-line value of 27.5% to 29.8% with the well depth of 200 meV, but decreases to 27.3% when the quantum well is 400 meV deep.

The need for a comprehensive model is rather obvious, now that there seems to be an unsolved debate as to the ultimate advantage of incorporating quantum wells in the intrinsic region of a pin solar cell. In this paper we present formulation of one such model. In this model we have included the effects of several phenomena known to be of siginificant importance in the absorption of irradiation, and the transport of the photoexcited carriers. These phenomena include: 1) capture of electrons by the wells, 2) escape of electrons from the wells, 3) absorption of light in the wells, and 4) recombination of carriers in the wells. Our model is developed for a pin single-cell solar cell consisting of a $p - AlGaAs$ region, an intrinsic $i - AlGaAs/GaAs$ region with several quantum wells, and a $n - AlGaAs$ region. Our simulator solves a field-dependent Schrödinger equation self-consistently with Poisson and Drift-Diffusion equations. The field-dependent Schrödinger equation is solved using the transfer matrix method.(10) The eigenfunctions and eigenenergies obtained are used to calculate the escape rate (11)-(12) of electrons from the quantum wells, and the non-radiative recombination rates of electrons at the boundaries of the quantum wells. These rates together with the capture rates (13)-(14) of electrons by the quantum wells are then used in a self-consistent numerical Poisson-Drift-Diffusion solver. The resulting field profiles are then used in the field-dependent Schrödinger solver, and the iteration process is repeated until convergence is reached.

II. SELF-CONSISTENT MODEL

A. The *pin* Solar Cell

The solar cell device consists of a $p - AlGaAs$ region, an intrinsic $i - AlGaAs/GaAs$ region with several quantum wells, and a $n - AlGaAs$ region. The energy band diagram of the device with four quantum wells in the intrinsic region is shown in Fig. (1). In our simulations we vary the following design parameters: the thickness of the regions, the doping levels of the n and p regions, the mole fraction of Al in $AlGaAs/GaAs$ quantum wells (the depth of the wells), the thickness of the quantum wells, and the number of quantum wells. The input parameters to our simulation programs include irradiation spectra, the transport parameters, band structure, and optical and absorption characteristics of the materials. The search for $i - v$ characteristics of the device begins with assigning an applied voltage and solving the system of differential equations for the terminal current densities. The simulation model also predicts the internal distributions of carriers densities, current densities, and electrostatic potentials. Several intermediate results will also be produced which includes the eigenenergies and eigenfunctions of electrons in the wells, capture and escape rates, and recombination lifetimes in the quantum wells.

B. Bulk Transport

The steady-state transport of electrons and holes in the *pin* structure are described by two current continuity equations written for the bulk regions as:

$$\frac{\partial n_b}{\partial t} = G - U + R_e^n - R_c^n + \frac{1}{q}\frac{dJ_n}{dx} = 0 \tag{1}$$

$$\frac{\partial p_b}{\partial t} = G - U + R_e^p - R_c^p - \frac{1}{q}\frac{dJ_p}{dx} = 0 \tag{2}$$

In these equations n_b and p_b are bulk electron, and hole densities, respectively, whereas J_n and J_p are electron, and hole current densities, respectively. The terms R_e^n and R_e^p are the rates of escape of electrons and holes from the bulk, and R_c^n and R_c^p are the capture rates of electrons and holes by the bulk. The escape and capture rates of carriers for the bulk system are indeed (assumed to be) identical to capture and escape rates of carriers for the quantum well system. In other words, escape from the bulk is capture by the quantum wells, and capture by the bulk is escape from the quantum wells. That is why escape rates have positive contributions (as in generation) whereas capture rates have negative (as in recombination). These rates are related to the escape and capture times as: $R_e^n = \frac{n_w}{\tau_e^n}$, $R_e^p = \frac{p_w}{\tau_e^p}$, $R_c^n = \frac{n_w}{\tau_c^n}$, and $R_c^p = \frac{p_w}{\tau_c^p}$. where τ_e^n and τ_e^p are the electron and hole escape times, respectively, and τ_c^n and τ_c^p are electron and hole capture times, respectively. Details of the calculations of these escape and capture times are described in the following Section. The recombination in the bulk is assumed to be radiative recombination given by: $U = B(n_b p_b - n_{b0} p_{b0})$, where B is the radiative recombination constant. G is bulk generation rate and is given by: .

$$G = \int_0^{\lambda_c} \alpha(\lambda) . N_{ph} . \left[exp(-\int_0^x \alpha(\lambda) dx) \right] d\lambda \tag{3}$$

where λ_c is set to correspond to the bandgap of the material.

The current density equations are written as:

$$J_n = q n_b \mu_n E + q D_n \frac{dn_b}{dx} \tag{4}$$

$$J_p = q p_b \mu_p E - q D_p \frac{dp_b}{dx} \tag{5}$$

Assuming low to moderate doping levels, and under 1 sun illumination we use the Einstein relationship for carrier mobilities and diffusion constants as given by: $D_{n,p} = \frac{kT}{q} \mu_{n,p}$. The above equations are solved together with Poisson equation:

$$\frac{dV^2}{dx^2} = \frac{q}{\epsilon} [N_A - N_D + n_b + n_w - p_b - p_w]. \tag{6}$$

The values of n_w and p_w are set to zero for the bulk systems. N_D and N_A are donors and acceptors doping levels.

C. Quantum Well Transport

Calculation of the absorption spectra in the quantum wells as well as the escape rate of carriers from the quantum well require the eigenfunctions and eigenenergies of the carriers in the quantum wells, which are obtained from a field-dependent Schrödinger equation given by:

$$\left[\frac{\hbar^2}{2} \frac{d}{dx} \frac{1}{m^*(x)} \frac{d}{dx} + V(x) \right] \psi(x) = E_i \psi(x) \tag{7}$$

where $\psi(x)$ is the envelope function, E_i are the eigenenergies, and $V(x)$ is the potential profile. Non-constant effective mass $m^*(x)$ are assumed. The Schrödinger equation is solved using Transfer Matrix method (10) in which wave functions of the type:

$$\psi(x) = A_j e^{p_j(x - x_j)} + B_j e^{-p_j(x - x_j)} \tag{8}$$

are searched by finding constants A and B through matching the wave functions and their derivatives at the boundaries.

The current continuity equations for the quantum well system are written as rate equations in which net balance of four rates give rise to the current. These rate equations are given by:

$$\frac{dn_w}{dt} = \frac{n_b}{\tau_c^n} - \frac{n_w}{\tau_e^n} + G_w - U_w + \frac{1}{q}\frac{dJ_n}{dx} = 0 \tag{9}$$

$$\frac{dp_w}{dt} = \frac{p_b}{\tau_c^p} - \frac{p_w}{\tau_e^p} + G_w - U_w - \frac{1}{q}\frac{dJ_p}{dx} = 0 \tag{10}$$

In the above equations τ_c^n and τ_e^n are electron capture and escape rates, whereas τ_c^p and τ_e^p are hole capture and escape rates. The recombination term U_w is a modified Shockley-Read-Hall recombination rate given by :

$$U = \frac{\sigma_n \sigma_p v_{th} N_t [pn - p_0 n_0]}{\sigma_n[n + n_t] + \sigma_p[p + p_t]} \tag{11}$$

where the capture cross sections, σ_n and σ_p are modified based on $\psi_i(x)$ for the first three subbands in the quantum wells.(15)-(16) The generation term G_w is calculated from Equation (3) with the bulk absorption coefficient replaced with that of the quantum well. The absorption coefficients of the quantum wells are calculated by a model presented by Stevens, et. al.(17):

$$\alpha(\hbar\omega,\varepsilon) = M_{cv}^2(\varepsilon).q_{exciton}.L(\hbar\omega, E_{cv}^{1,1} - E_b) \tag{12}$$

$$+ \int_{E_{cv}^{1,1}}^{\infty} M_{cv}^2(\varepsilon).Nq_{con}K(E', E_{cv}^{1,1})L(\hbar\omega, E')dE'$$

where E_b is exciton binding energy, $q_{exciton}$ and q_{con} are exciton and continuum oscillator strengths, respectively and $E_{cv}^{1,1}$ is the $n = 1$ transition energy. The term $M_{cv}^2(\varepsilon)$ is the electron-hole wavefunction overlap integral and is calculated from the wavefunctions obtained from the Schrödinger solver. For more details of the quantum well absorption model, see Stevens, et. al.(17)

The escape rates of carriers are calculated using the model reported by Moss, et. al. (11):

$$\frac{1}{\tau_e} = \frac{1}{Nd}\sqrt{\frac{kT}{2\pi m_\Gamma^*}}\left\{1 + \alpha_L exp\left[(\Delta E_b^L)u(-\Delta E_b^L)/kT\right]\right\}exp\left(-\Delta E_b/kT\right) \tag{13}$$

where $u(x)$ is the Heaviside step function, L and Γ refer to specific valley, and d is the well width. For more details on Eq. (13) see Moss, et. al. (11) Capture rates are extrapolated from theoretical data reported by Blom, et. al. (14).

III. PRELIMINARY RESULTS

The system of equations described in Section II are solved using a finite difference numerical scheme. For a $p-AlGaAs$ $i-AlGaAs/GaAs$ $n-AlGaAs$ cell with aluminum mole fraction of 0.3, with one 100 $A°$-wide 284 meV-deep quantum well, the eigenenergies with zero field are 36meV, 136meV, and 267meV, for the first, second and third subbands, respectively. The corresponding wavefunctions are shown in Fig. 2. With an electric field of 50 kV/cm, the eigenenergies are shifted to 58meV, 160meV, and 282meV, respectively, and the amplitudes of the eigenfunctions are amplified as illustrated in Fig. 3. Using these eigenenergies, the thermionic escape time of electrons from the $GaAs$ Γ-valley, as a function of electric field, is calculated and plotted in Fig. 4. The capture rates of electrons, extraplolated from the data reported by Blom, et. al.(14), are reproduced in Fig. (5) as a function of energy difference $(E_c - E_i)$. The varaiations of $(E_c - E_i)$ with well width are shown in Fig (6).

We are reporting the development of a self-consistent Poisson-Schrödinger Drift-Diffusion solver for a pin quantum well solar cell. The effects of escape and capture of electrons and holes in the quantum wells are being studied. The degradation of open-circuit voltage due to increased recombination is being investigated by studying the recombination of carriers in quantum wells. Simulations are being performed to generate the $i - v$ characteristics of the cells under illumination.

References

1. Barnham K., and Duggan J., "A new approach to high-efficiency multi-band-gap solar cells", *Journal of Applied Physics,*. Vol. 67, pp. 3490-3493, 1990.

2. Barnham K., Braun B., Nelson J., and Paxman M., "Short-circuit current and enrgy efficiency enhancement in a low-dimensional structure photovoltaic device," *Applied Physics Letters* Vol. 59, pp. 135-137, 1991.

3. Ragay F., Wolter J., Marti A., Araujo G., "Experimental analysis of the efficiency of MQW solar cells," *Proceedings of the 12th European Community Photovoltaic Solar Energy Conference*, pp. 1429-1433, 1994.

4. Ragay F., Wolter J., Marti A., Araujo G., "Experimental analysis of GaAs-InGaAs MQW solar cells" *Proceedings of the 1994 IEEE First World Conference on Photovoltaic Energy Conversion, Hawaii"*, IEEE, NY, NY, pp. 1754-1758, 1995.

5. Corkish R., and Green M., "Recombination of carriers in quantum well solar cells", *Proceedings of the 23rd Photovoltaics Specialist Conference*, pp. 675-680, 1993.

6. Paxman M., Nelson J., Braun B., Connolly J., and Barnham K., "Modeling the spectral response of the quantum well solar cell", *Journal of Applied Physics,*. Vol. 70, pp. 614-621, 1993.

7. Araujo G., Marti A., Ragay F., and Wolter J., "Effeciency of multiple quantum well solar cells" *Proceddings of the 12th European Community Photovoltaic Solar Energy Conference*, pp. 1435-1439, 1994.

8. Renaud P., Vilela M., Freundlich A., Bensaoula A., and Medelci N., "Modeling p-i(MQW)-n solar cells: A contribution for a near optimum design", *Proceedings of the 1994 IEEE First World Conference on Photovoltaic Energy Conversion, Hawaii"*, IEEE, NY, NY, pp. 1787-1790, 1995.

9. Anderson N., "Ideal Theory of Quantum Well Solar Cells," *Journal of Applied Physics*, Vol. 78, pp.1850-1861, 1995.

10. Jonsson B., and Eng S., "Solving the Schrodinger equation in arbitrary quantum-well potential profiles using the transfer matrix method,", *IEEE J. Quantum Electronics*, Vol. 26, pp.2025-2035, 1990.

11. Moss D., Ido T., and Sano H., "Calculation of photo-generated carrier escape rates from GaAs/AlGaAs quantum wells,", *IEEE Journal of Quantum Electronics*, Vol. 30, pp. 1015-1026, 1994.

12. Nelson J., Paxman M., Barnham K., Roberts J., and Button C., "Steady-State carrier escape from single quantum Wells," *IEEE Journal of Quantum Electronics*, Vol.29, pp.1460-1465, 1993.

13. Rosencher E., Vinter B., Luc F., Thibaudeau L., Bois P., and Nagle J., "Emission and capture of electrons in multiquantum-well structures," *IEEE Trans. Quantum Electronics*, Vol. 30, pp. 2875-2888, 1994.

14. Blom P., Smit C., Haverkort J., and Wolter J., "Carrier capture in a semiconductor quantum well", *Physical Review B*, Vol. 47, pp. 2072-2081, 1993.

15. Gurioli M., Vinattieri A., Colocci M., Deparis C., Massies J., Neu G., Bosacch A., and Franchi S., "Temperature dependence of the radiative and non-radiative recombination time in GaAs/AlGaAs quantum-well structures," *Physical Review B.*, Vol. 44, pp. 3115-3124, 1991.

16. Krahl M., Bimberg D., and Baur R., "Enhancement of non-radiative interface recombination in GaAs couple quantum wells," *Journal of Applied Physics*, Vol. 67, pp. 434-438. 1990.

17. Stevens P., Whitehead M., Parry G., and Woodbridge K., "Computer modeling of the electric field dependent absorption spectrum of multiple quantum well material," *IEEE Journal Quantum Electronics*, Vol.24, pp.2007-2015, 1988.

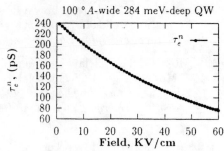

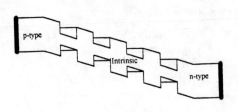

Figure 1: Energy band diagram of the solar cell with four quantum wells.

Figure 4: The thermionic escape time of electrons from the $GaAs$ Γ-valley to $AlGaAs$ Γ-valley as a function of electric field.

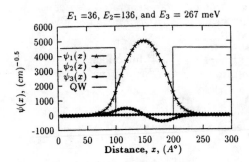

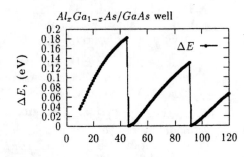

Figure 2: Eigenfunctions of a 100 $A°$-wide 284 meV-deep quantum well with no field.

Figure 5: The capture time of electrons by the quantum well as a function of $E_c - E_i$.

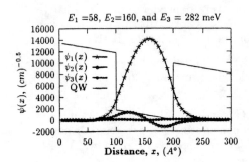

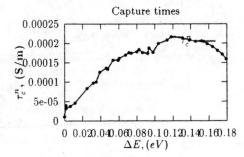

Figure 3: Eigenfunctions of a 100 $A°$-wide 284 meV-deep quantum well with 50 KV/cm field.

Figure 6: The energy difference $E_c - E_i$ as a function of well width.

Next Generation CIGS for Solar Cells

A. Rockett[a], R. Birkmire[b], D. Morel[c], S. Fonash[d], J-Y. Hou[d], M.
Marudachalam[b], J. D'Amico[c], P. Panse[c], S. Zafar[c],
and D.J. Schroeder[a]

a: Department of Materials Science, University of Illinois, 1304 W. Green St. Urbana, IL 61801
b: Institute for Energy Conversion, University of Delaware,
c: University of South Florida,
d: University of Pennsylvania

Abstract. An overview is presented of the PV-ACIST project. The goal of this
project is to demonstrate proof-of-concept of next generation $CuIn_{1-x}Ga_x(S_ySe_{1-y})_2$
(CIGS)-based solar cells. The first phase of the project began with a theoretical
analysis of the potential of tandem-junction devices. This analysis showed that
efficiencies could exceed 25% based constituent junctions comparable to the best
achieved in single CIGS junctions. Subsequent efforts have focused on development
and characterization of the materials and processes necessary to achieve the wide-gap
device portion of the tandem structure and on modeling of device results based on both
wide and narrow gap materials. Results to date have included demonstration of wide
gap devices, characterization of the diffusivity of Ga in CIGS under various process
conditions, characterization and modeling of the conduction properties of ideal CIGS,
and model results explaining the "red kink" effect in certain high-performance CIGS
solar cells.

INTRODUCTION

Photovoltaic devices with energy conversion efficiencies exceeding 10%
are routinely available and are in production. Individual devices with
significantly higher efficiencies have been achieved. To achieve a true
breakthrough in photovoltaic power production, a thin film device technology
with greater than 20% efficiency will be required. Because fixed costs of cell
manufacturing such as the costs of substrate materials provide a lower limit to
the price of individual modules, dramatic reductions in the cost per peak Watt
will require dramatic improvements in device efficiency.

$CuInSe_2$ (CIS) and alloys with similar materials promise to provide high
performance (18%), reliable photovoltaic devices. To make this promise a
reality in large area devices, it will be necessary to develop and improve
understanding of the materials. Current single junction devices use $CuIn_{1-x}Ga_x(S_ySe_{1-y})_2$ (CIGS) alloys with low Ga and S contents. To achieve their full
potential, multijunction devices with higher Ga and/or S content materials will

CP404, *Future Generation Photovoltaic Technologies: First NREL Conference*, edited by McConnell
© 1997 The American Institute of Physics 1-56396-704-9/97/$10.00

be required. The projects described here are intended to develop the new materials and processing technologies necessary to fabricate this next generation of devices and to improve the fundamental understanding of the materials through characterization of the constituent semiconductors and through detailed modeling of the results.

The research results presented below were developed under coordinated EPRI-sponsored projects with input from industrial advisors. The projects and the advising group are collectively known as PV-ACIST. The goal of PV-ACIST is to demonstrate proof-of-concept for second-generation CIGS-based photovoltaics. The research groups focus on three topics: design of production methods for second-generation solar cells, materials development and characterization, and device modeling. Efforts at The University of South Florida (USF) and the Institute for Energy Conversion at the University of Delaware (UDel) concern designing advanced methods of fabrication of wide-gap $CuGaSe_2$ solar cells. In addition, these groups are conducting materials development, and characterization of both the materials and resulting devices. The work at the University of Illinois (UIUC) produces model CIGS materials as epitaxial layers from which input data essential for modeling are being obtained. For the third leg of the PV-ACIST effort, the Pennsylvania State University (PSU) group is applying fundamental device simulation methods to provide understanding of CIGS device performances. The industrial advisors to PV-ACIST include all of the major US corporate stake-holders in the $CuInSe_2$ solar cell community. Siemens Solar Industries (SSI) and International Solar Electric Technologies (ISET) are represented by their senior $CuInSe_2$ development directors. The National Renewable Energy Laboratory (NREL), represented by senior researcher scientists also advises the project.

Although EPRI was responsible for launching the PV-ACIST project and has managed it, both EPRI and NREL acknowledge significant synergies between PV-ACIST and the NREL CIS efforts, presently. A successful ultrahigh efficiency tandem-junction technology requires both high-efficiency wide-gap and narrow-gap solar cells (~17% each). Research in the PV-ACIST groups thus includes two focuses: one (greatly leveraging NREL-supported work) on improved understanding and manufacture of high-efficiency narrow-gap devices; and one on the wide-gap portion of the tandem structure. Integration issues are also being addressed as part of the programs supported by both organizations. Efforts at the University of Illinois, The University of Delaware, and the University of South Florida directly link these programs through coordinated jointly funded research programs.

RESULTS

Tandem Junction Device Performance Potential

The PV-ACIST project began with estimates from the PSU group of the potential for ultrahigh efficiency photovoltaic devices using the AMPS computer code. The analysis made basic assumptions about the availability of acceptable contact schemes and compatible processes leading to the formation of a two, three, or four contact tandem structure photovoltaic device. The individual devices were presumed to have performances comparable to the best (>17%) efficiency solar cells achieved to date based on x<0.25, y=0 CIGS devices. The results of this study indicated that device performances exceeding 25% should be achievable if satisfactory materials and processes can be identified. The wide-gap device component of the tandem structure would have an optimum energy gap of ~1.65 eV.

Materials and Processing for Wide-Gap Photovoltaic Devices

Because high-performance narrow gap (< ~1.2 eV) devices have already been demonstrated, the focus of the project shifted to consideration of the wide-gap portion of the tandem. This effort was divided primarily between the USF and UDel groups. The groups have focused on synthesis and optimization of pure $CuGaSe_2$ (CGS) devices and on device performance as a function of Ga content beginning from existing moderate-Ga-content materials and processes, respectively.

The USF program is based on selenization of Cu/Ga precursor layers in elemental Se under a variety of process conditions and has focused on optimization of these. The process has resulted in single phase $CuGaSe_2$ absorber layers on Mo-coated sodalime glass substrates. These have then been fabricated into solar cells using standard techniques in which CdS is deposited by the chemical bath method followed by formation of a transparent conductive oxide. Preliminary work investigating S-containing materials has also been conducted at USF but is currently being given lower priority.

Active devices have been achieved with short circuit currents J_{sc} between 13 and 15 mA/cm^2 under AM1.5 simulated solar radiation, open circuit voltages V_{oc} between 500 and 700 meV, and fill factors between 0.5 and 0.6. The internal quantum efficiency of the devices obtained from spectral response measurements have been modeled using standard models.[1] Device performance is assumed to be driven by Shockley-Read-Hall recombination in the space charge region. Satisfactory fits to device data were obtained using previously reported values for key CGS parameters. The analysis suggested a built-in voltage of the heterojunction of 1.3 eV, an electron diffusion length of 0.5 μm, and an interface recombination time of 10^{-13} sec. Examples of current voltage curves and model results may be found in Fig. 1.

High gap devices have so far failed to achieve the levels of performance obtained in low gap devices. To accelerate the understanding of what limits high gap structures, the UDel group is modifying their existing low-gap process to produce higher gap materials. Analysis and modeling of the resulting devices (see below) provides a systematic method for exploring the performance loss and means of correcting it. At UDel, CIGS structures with 0<x<1 and y=0 have been fabricated by selenization and by evaporation. Results described here are for films produced by deposition of stacked In/Ga/Cu metal layers selenized in H_2Se at 450°C for 90 min. This results in a bilayer $CuInSe_2/CuGaSe_2$ structure with the $CuGaSe_2$ segregated to the back of the structure. A subsequent high-temperature heat treatment at 600°C for 60 min was found to mix the $CuInSe_2$ and $CuGaSe_2$ layers formed by selenization resulting in a uniform film composition. The mixed films were single phase as determined by X-ray diffraction. The kinetics of the Ga diffusion during this process was determined and a Ga diffusivity of 4×10^{-11} cm^2/sec and an In diffusivity of 1.5×10^{-11} cm^2/sec were obtained.[2] These values are in good agreement with results obtained at UIUC and described below.

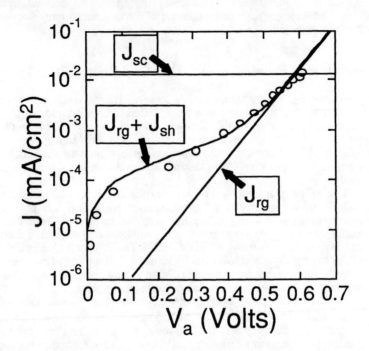

FIGURE 1. Current/voltage results for a CuGaSe$_2$/CdS solar cell produced at USF along with model results. Curves indicate the currents at short circuit, J_{sc}; generation/recombination current, J_{rg}; and shunt current J_{sh}.

Spectral response measurements on devices with varying Ga contents are shown in Fig. 2. The response function when the absorption coefficient is high is unchanged from one device to the other. Only the absorption edge is shifted. This indicates equivalent carrier collection in the two structures. The energy gap shift is 0.35 eV and is consistent with the amount of shift that would be expected for the Ga content obtained.[3] Further optimization of the processing sequence and analysis of the devices is in progress.

In addition to studies of polycrystalline material described above, the group at UIUC is growing and analyzing CIGS single crystal epitaxial layers as a function of composition on GaAs substrates.[4] Layers with $0 < x \leq 1$ were formed. In all cases except x=1, the Ga in the film resulted from diffusion out of the substrate. By fitting the diffusion profiles, the Ga diffusivity was measured as a function of composition and was found to increase rapidly and then to saturate with modest deviations from stoichiometry for Cu/In ratios above and below 1.0, see Fig. 3. The diffusivity at 700°C was found to be between 10^{-11} and 6×10^{-11} cm^2/s for Cu/In ratios away from 1.0 in good agreement with the UDel data. For Cu/In ratios very near 1.0 the diffusivity dropped dramatically to ~5x10^{-13} cm^{-3} indicating a much lower density of vacancies through which diffusion presumably occurred. The resulting films were examined by XRD. Because of the change in Ga content near the back surface of the films, a range of compositions was found. In most cases the film showed a predominant composition and a well-defined diffraction peak associated with the majority composition of the films. Compositions with higher Ga contents were also found extending to compositions approximately lattice matched to the GaAs.

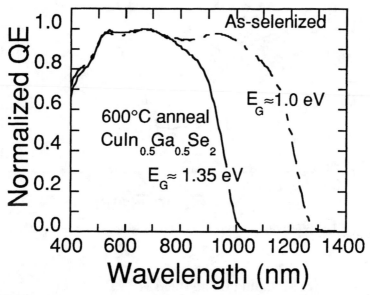

FIGURE 2. Spectral response functions for two solar cells produced at UDel with x=0.0 and 0.5.

Hall-effect measurements on the epitaxial layers showed the presence of two acceptor states and a donor state in the films; the deeper (~170 meV above the valence band edge) state having a concentration of 4x10^{17} cm^{-3} and the shallower (~45 meV above the valence band) state at 1x10^{15} cm^{-3} in CuInSe$_2$.[5] Both acceptor concentrations increased approximately exponentially with x to 5x10^{21} cm^{-3} and 2x10^{18} cm^{-3} for the deep and shallow states, respectively. Simultaneous decreases in state depth suggest that the basic states were remaining at a fixed energy with respect to the valence band as the band Ga content changed but that the widths of the resulting impurity bands were increasing. The fixed depth of the states indicates approximately hydrogenic behavior. Hole mobilities were 245±33 cm^2/V-sec and were not an obvious function of any variable monitored. The highest mobilities for holes were 310 cm^2/V-sec at 300 K and 1540 cm^2/V-sec at 50 K, both in the pure CuGaSe$_2$

sample. Finally, samples have been treated with Na to determine the effect of this contaminant on the electrical properties of $CuInSe_2$. The results indicate that Na reduces donor compensation in CIGS alloys but does not act as a dopant.

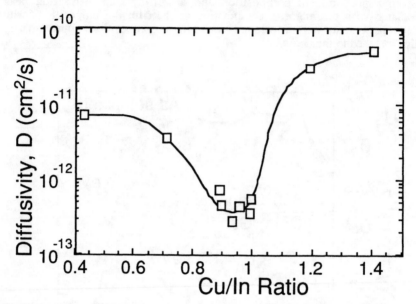

FIGURE 3. Ga diffusivity as a function of composition in single crystal epitaxial layers.

Device Modeling Using the AMPS Program

The PSU group has done extensive modeling of results obtained to date on devices produced and characterized at UDel, NREL, and Solarex using the AMPS computer code developed at PSU. Data simulated included current/voltage measurements as a function of light intensity and wavelength, and overall spectral response. Self-consistent fits for a number of devices with and without band-gap grading have been obtained. The model assumed that each device consisted of a CIGS absorber layer of thickness ~3 μm on ohmic back contacts. A thin n-type $CuIn_3Se_5$ layer was simulated on top of the bulk of the p-CIGS and which formed the heterojunction with the CdS window layer. The CdS was covered with ZnO which was taken to consist of two layers, one with high and one with low doping concentrations. These layers should adequately represent the actual device structures. Experimental data for solar spectrum, absorption coefficients, and materials parameters were used when available.

The model results provided excellent fits to the device data. The model indicates that the undoped CdS has very high resistivity in the dark but exhibits substantial photoconductivity under white light illumination. However, illumination under red light produces no photoconductivity. This leads to the "red-kink effect (see Fig. 4). Model results show that the negative voltage section of the curve is similar to the white light result but with current loss. However, the positive voltage section is modified by the low conductivity CdS layer resulting in the kink in the curve.

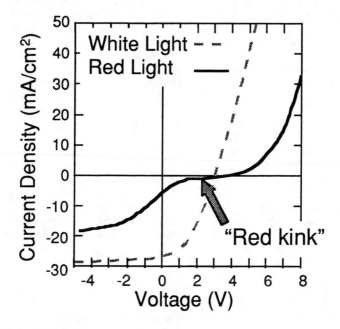

FIGURE 4. Simulated photovoltaic response for a solar cell illuminated with white light or red light only.

Additional results included demonstration that the electron mobility in the CIGS must be of the order of 300 cm^2/V-sec, that an induced homojunction occurs in the CIGS rather than the junction being dominated by the heterojunction, that there is a high defect density layer near the heterojunction and that the back contact has a negative effect on carrier collection. Reduction or elimination of the high recombination velocity associated with the high defect-density layer near the front of the CIGS and at the back contact should significantly improve device performances.

CONCLUSIONS

Model results indicate that tandem device structures can be highly efficient. Both theoretical and experimental results indicate that the CIGS material in current devices is of high quality in spite of being significantly off stoichiometry. Characterization of hole mobilities indicates that pure $CuGaSe_2$

films exhibit hole transport properties at least as good as in pure $CuInSe_2$. This supports the argument that ultimately devices based on pure $CuGaSe_2$ should be capable of performances comparable to those of the low-gap devices. This, in turn indicates that very high efficiency tandem-structure devices should be achievable.

The characterization of the pure CGS devices indicates that minority carrier transport is adequate in the materials to produce acceptable devices but that the interface quality of structures to date has limited the V_{oc} and fill factor of the devices. This is confirmed by AMPS model results which indicate significant degradation in device performance due to unoptimized junctions at both the front and back interfaces of the CIGS. It is expected that further improvements in device performance in both the wide and narrow-gap devices will be achieved based on further optimization of the processing conditions.

ACKNOWLEDGMENTS

The authors wish to thank the Electric Power Research Institute, the National Renewable Energy Laboratory, and the Department of Energy for their continuing support of this project. In addition, we wish to thank the advisors to the project: Vijay Kapur of International Solar Electric Technologies, Joseph Morabito of Lucent Technologies, Terry Peterson of the Electric Power Research Institute, Dale Tarrant of Siemens Solar Industries, and John Tuttle of the National Renewable Energy Laboratory for their advice and thoughtful analysis of the results.

REFERENCES

1) See, for example, S.M. Sze (John Wiley and Sons, New York, 1981).

2) M. Marudachalam, H. Hichri, R. W. Birkmire, J. M. Schultz, A. B. Swartzlander and M. M. Al-Jassim, Proceedings of the 25th IEEE Photovoltaic Specialists Conference, Washington, D.C. (IEEE, NewYork 1996), p.805.

3) M. Marudachalam, H. Hichri, R.W. Birkmire, W. N. Shafarman and J. M. Schultz, Appl. Phys. Lett. **67**, 1995, pp. 3978-3980.

4) D.J. Schroeder, G.D. Berry, and A. Rockett, Appl. Phys. Lett. **69**, 1, 1996.

5) D.J. Schroeder, J.L. Hernandez, G.D. Berry, and A. Rockett, J. Appl. Phys., submitted.

Pathways to High-Efficiency GaAs Solar Cells on Low-Cost Substrates

Rama Venkatasubramanian[+], Edward Siivola[+], Brooks O'Quinn[+],
Brian Keyes[*] and Richard Ahrenkiel[*]

[+]Research Triangle Institute, RTP, NC 27709 and [*]National Renewable Energy Laboratory,
Golden, CO 80401.

Abstract. Development of ~20% efficient GaAs solar cells on sub-mm grain-size, optical-grade poly-Ge substrates is reported. The techniques developed in the growth of GaAs layers on the various crystalline orientations of a poly-Ge substrate and for the amelioration of the detrimental effects of grain-boundaries in GaAs solar cells are relevant to the development of high-efficiency GaAs solar cells on other low-cost substrates. The development of Ge thin-films on glass for the nucleation and growth of GaAs active layers is described. The minority-carrier lifetimes measured in thin-film GaAs-AlGaAs double-hetero (DH) structures grown on Ge-on-glass templates are rather low in the range of 60 psec. However, molybdenum appears to be a better alternative to Ge-on-glass for the growth of GaAs thin-films with improved minority-carrier lifetimes. In our preliminary optimization studies, we have observed longer minority-carrier lifetimes (~170 psec) in GaAs DH structures on molybdenum. Initial results also suggest that Se doping for the GaAs layers lead to longer lifetimes, consistent with our current understanding of high-performance p[+]-n GaAs solar cells with Se-doping for the n-base on poly-Ge substrates.

INTRODUCTION

High-efficiency single-junction GaAs (25.7% under AM1.5G) [1] and tandem GaInP$_2$/GaAs (29.5% under AM1.5G) [2] solar cells have been demonstrated on single-crystal GaAs substrates. However, the cost of single-crystal GaAs substrates are prohibitive for the use of these solar cells in flat-plate terrestrial systems. Several low-cost substrate alternatives have been considered recently- one approach is to use single-crystal Si substrates; however, this has remained a challenge due to the lattice and thermal mismatch between GaAs and Si substrates. The reported best 1-sun efficiency for a 0.25-cm^2-area GaAs solar cell on single-crystal Si substrate is 17.14% [3]. In contrast, single-crystal Ge substrates have demonstrated to be an effective lower cost, light-weight, alternative to single-crystal GaAs substrates, especially for space applications. In addition, the use of Ge substrates lead to enhanced yield in manufacturing when compared to using GaAs substrates. However, the cost associated with single-crystal Ge substrates are also high for flat-plate terrestrial applications. Therefore, it appears necessary to investigate the development of high-efficiency GaAs cells on low-cost, large-area substrates for flat-plate applications.

Large-area poly-Ge substrates with an average grain size of sub-mm to several mm are commercially produced by a cast process for optical-window applications and appear attractive for meeting the need of *low-cost substrates for GaAs solar cells in the near-term.* The availability of large-area (up to 24") cast

CP404, *Future Generation Photovoltaic Technologies: First NREL Conference*, edited by McConnell
© 1997 The American Institute of Physics 1-56396-704-9/97/$10.00

poly-Ge substrates along with large-scale GaAs deposition by metallorganic chemical vapor deposition (MOCVD) could reduce cell-processing costs for approaching the cost-goal for flat-plate applications. Towards this end, we describe our recent progress in the development of ~20% efficient GaAs solar cells on sub-mm grain-size poly-Ge substrates. These GaAs solar cell efficiency results have motivated us to consider the development of high-efficiency *GaAs solar cells on glass or moly for the long-term* by applying the understanding developed in the amelioration of the detrimental effects of grain-boundaries in GaAs cells on sub-mm grain-size poly-Ge substrates. We have considered the development of Ge thin-films on glass as a template for the nucleation and growth of GaAs active layers. We have also explored the use of thin moly foils as an alternative to Ge-on-glass, for the growth of GaAs films. Our preliminary studies suggest that longer minority-carrier lifetimes in GaAs DH structures are achievable on moly than on Ge-on-glass substrates.

KEY ISSUES WITH POLY SUBSTRATES AND TEMPLATES

There are two key issues in the development of solar cells on polycrystalline substrates and templates. A polycrystalline template could be a regular array of polycrystalline grains on a non-crystalline substrate such as glass, shown schematically in Figure 1a. One is a *growth issue*-concerns the deposition of uniform layered materials across the various grains. Second is a *device issue*-related to the minimization of the deleterious effects of grain boundaries in photo-carrier recombination and dark-current generation in a solar cell.

We can envisage two cases in the growth of materials on polycrystalline substrates-an ideal, two-dimensional, uniform, layered growth across the various orientations so that we essentially preserve the grain-structure of the starting substrate and do not create additional grain boundaries in the active regions of the solar cell as indicated in Figure 1b. A non-ideal, three-dimensional, non-uniform growth across the various orientations leads to the formation of additional grain boundaries (Figure 1c). The key to minimization of the grain-boundary effects in deposited polycrystalline materials appears to being able to develop uniform, two-dimensional, layered growth across the various crystalline orientations.

From a device point of view, the grain boundaries in the active regions of a solar cell can reduce photo-collection through minority-carrier recombination. However, if we develop (p^+-n) GaAs solar cells with a thin emitter, a majority of the photo-current would be generated in the n-base region. If a group-VI dopant like Se or S is chosen to dope the n-GaAs base region, it has been suggested [4] that a favorable n/n^+ minority-carrier mirror could be naturally formed at the grain-boundaries due to preferential dopant segregation at these regions. These minority-carrier mirrors have been attributed to the relatively high short-circuit current density ($J_{sc} > 20$ mA/cm^2 under 1-sun) obtained in μm-grain-size poly-

GaAs solar cells [4]. While high J_{SC} values can be readily obtained in poly-GaAs solar cells, V_{OC} values have been low (~ 0.55 to 0.6 Volts) and the fill-factor has been low as well [4,5], probably due to the generation of dark currents at grain boundaries. We have proposed the use of a "thin" undoped spacer at the depletion layer of the p^+-n GaAs junction on poly-Ge and experimentally demonstrated that a significant reduction in dark current can be obtained, leading to an increase in V_{OC} and fill-factor [6]. In this paper, we briefly discuss further approaches to improve the J_{SC} and V_{OC} of GaAs cells on sub-mm grain-size poly-Ge substrates.

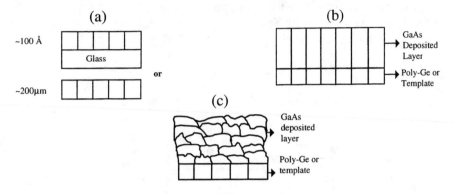

FIGURE 1. a) Schematic of a polycrystalline template or a substrate for cell growth; b) schematic of a layered, two-dimensional growth on poly-Ge or a polycrystalline template; c) schematic of a non-layered, three-dimensional, non-uniform growth on a polycrystalline template.

GaAs MATERIAL AND DEVICE OPTIMIZATION ON POLY-Ge

High-quality, uniform growth of GaAs-AlGaAs layers across the various crystalline orientations of a polycrystalline Ge substrate is critical to obtaining high-performance GaAs solar cells. The finish of poly-Ge substrate prior to the MOCVD growth of the GaAs cell is important in this regard; polishing defects such as dents or ledges or steps are detrimental to cell performance. Although the V_{OC} of the cells do improve with larger grain-sizes, these macro-polishing defects in the cell active-area tend to reduce the V_{OC} much more strongly than grain-boundaries. We have optimized the MOCVD growth process on large-grain and small-grain poly-Ge substrates, recognizing the need for maintaining a higher concentration of As-growth species on the growth-surface, to obtain uniform, two-dimensional layered growth. Optimization studies of the minority-carrier properties of GaAs layers on poly-Ge have suggested that lifetime-spread across the various grains can be reduced through the use of lower doping for the $Al_{0.8}Ga_{0.2}As$ layers [7].

Figure 2 is a schematic of the GaAs solar cell device structure that we have developed [6] for use on poly Ge substrates. The undoped spacer between the base and the emitter reduces the dark current and improves the cell V_{OC} and fill factor. Electron-beam-induced-current (EBIC) scans on p^+-n GaAs junctions on poly-Ge substrates, with a spacer of 0.1 to 0.3 μm, have indicated hole diffusion lengths ~1.2 μm in the vicinity of the depletion layer. We have also observed longer diffusion lengths away from the junction-region of the cell; this is consistent with the use of a spacer to reduce the doping near the depletion layer. We expect the regions away from the spacer to benefit from the formation of n/n^+ minority-carrier mirrors with Se doping for the n-base, leading to an enhancement in minority-carrier diffusion lengths.

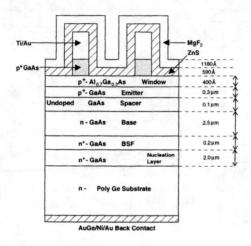

FIGURE 2. Schematic of a p^+n GaAs solar cell device structure on poly-Ge substrate.

The reduction in dark current with the use of the spacer is believed to be associated with the reduction of tunneling currents near the depletion-layer of the p^+-n junction in polycrystalline materials [8]. We have conducted an experimental study of dark currents in p^+-n GaAs junctions as a function of temperature (77K to 288K) to support this hypothesis. GaAs junctions without the spacer show a stronger reduction of dark current with temperature, with an apparent activation energy of ~0.07 eV, compared to junctions with the spacer. The band diagram of a p^+-n junction near the vicinity of a grain boundary, with n-type dopant segregation, is expected to resemble that of a p^+-n^+ junction. This would lead to electron tunneling (dark) current from the p^+-emitter to the n^+-base depending on the empty states below the Fermi-energy in the n^+-region. Thus the dark-current activation energy will be related to the amount of degeneracy in the

n^+-region. The measured activation energies in the range of 0.07 eV are typical for such degeneracy. We have seen similar behavior in p^+-n GaInP$_2$ junctions grown on poly-Ge. Thus, the reduction of tunneling dark-currents appears to be one of the keys to higher V_{oc} in III-V solar cells on polycrystalline substrates [8].

In order to further improve the efficiency of the p^+-n GaAs cells on poly-Ge, we have investigated various approaches to improve the blue-response of the cells. The use of a thinner $Al_xGa_{1-x}As$ window and a thicker p^+-emitter, an optimized AR-coat, and a reduction of the emitter grid-metallization coverage, have enabled us to obtain AM1.5 J_{sc} of as much as ~27 mA/cm^2 in GaAs cells on poly-Ge. We have optimized the spacer thickness towards the improvement of V_{oc} values. The effect of varying the spacer thickness on cell V_{oc}, for approximately similar grain-size in the GaAs cells, is shown in Table 1. Arguments based on the dark-current generation in the depletion-layer, in contrast to a diffusion-limited dark-current generation from the quasi-neutral base region, indicate that *as long as the spacer is thick enough to avoid tunneling of carriers between the heavily-doped emitter and the heavily-doped base regions formed near the grain-boundaries, a thinner spacer would lower the dark current and improve the V_{oc} of the solar cell.* Optimization of emitter thickness and the amount of emitter-grid metallization have improved the V_{oc} to 1.03 Volts. A significant component to dark-current generation appears to be from the emitter-contact region or from the $Al_xGa_{1-x}As$ window-GaAs cap hetero-interface in these poly-GaAs solar cells.

Table 1. Effect of spacer thickness on V_{oc} of GaAs cells on poly-Ge substrates, with all other cell structure parameters held constant. Cell area ~ 4 cm^2.

Cell #	Spacer Thickness (μm)	V_{oc} (V)
1-2766	0.22	0.90
1-2763	0.14	0.96
1-2782	0.14	0.96
1-2786	0.10	1.00

The above material and device optimization studies have enabled us to achieve significant improvement in large-area GaAs cell efficiencies on mm and sub-mm grain-size poly-Ge substrates. The typical grain-size and other polycrystalline features in the poly-Ge substrates, and replicated in the GaAs films, are as small as 400 μm. The data shown in Table 2 indicates that the recent progress in GaAs solar cell results are in spite of smaller grain-sizes in the starting poly-Ge substrates. We observe that large-area GaAs cell efficiencies in excess of 20% are achievable on sub-mm grain-size poly-Ge substrates. It is noteworthy that in the small-area, 21%-efficient cell, a V_{oc} of as much as 1.04 V and a J_{sc} ~

27 mA/cm^2 have been observed, thereby approaching some of the best single-junction GaAs cell result on single-crystal GaAs substrates [1].

Table 2. Progress of GaAs cell efficiency on poly-Ge substrates

Year	Grain-Size (cm)	Cell Area (cm^2)	AM1.5 Eff. (%)	Ref.
1993	0.5	1	15.8	[6]
1994	0.5	4	16.6	[7]
1995	0.5	4	18.2	[8]
1996	0.1	4	18.7	
Current	0.04-0.1	4	~19	
state-of-art	0.04-0.1	0.25	~21	

GaAs MATERIALS DEVELOPMENT ON GLASS AND MOLY

The high GaAs solar cell efficiencies on sub-mm poly-Ge substrates have motivated us to apply and extend the material and device optimization ideas to the development of high-performance GaAs solar cells on glass and on moly foils. Glass is attractive from the low substrate-cost and is being used by researchers involved in the development of thin-film solar cells in other materials. Thin moly foils are also low-cost substrates. Further, moly has good thermal expansion match with GaAs and may allow cell fabrication with minimal or no backside contact processing. Moly foils offer mechanical strength for manufacturing and possible use in flexible, light-weight solar arrays. In the following section we discuss our preliminary materials development with these two substrates.

Ge Templates on Glass

Ge films on glass were obtained using 99.999%-pure Ge source in an e-beam evaporator at typical pressures of 2E-6 Torr. The as-evaporated Ge films were amorphous, evidenced by the absence of any crystalline reflections in X-ray diffraction measurements as well as the extremely low mobility of carriers in Hall-effect experiments. Thermal annealing of the Ge films lead to polycrystallinity and improved quality of the Ge films as evidenced by a strong increase in carrier mobility and a significantly improved electrical conductivity (by over four orders of magnitude). However, we do observe no preferred crystalline orientation of the annealed Ge films. We have also investigated the use of a novel Ge/Sb/Ge sandwich structure; the n-type doping with Sb was found beneficial to obtaining an enhanced oriented film, similar to results with P-doped Ge films on glass [9]. The carrier mobilities and the minority-carrier lifetimes (measured by the RF-decay method) in the Ge/Sb/Ge sandwich structures appear to be higher compared to pure Ge layers [10].

We have investigated the growth of GaAs-$Al_{0.8}Ga_{0.2}As$ DH structures on Ge and Ge/Sb/Ge on glass substrates using MOCVD, with conditions similar to those used for the growth of ~20% efficient GaAs cells on poly-Ge. The DH structures were characterized by photoluminescence (PL) decay to determine the minority-carrier lifetimes. In general, the Se-doped DH structures indicated longer minority-carrier lifetimes than comparable Si-doped or Zn-doped DH structures. Higher growth temperatures were found beneficial for obtaining longer lifetimes in these GaAs thin-films. Since glass substrates cannot be exposed to temperatures above 650°C, we have been able to achieve only a maximum of ~60 psec for minority-carrier lifetimes in the GaAs DH structures.

The growth of GaAs on moly foils were carried out under similar conditions; the adhesion of GaAs film to moly was found to be excellent. The minority-carrier lifetimes in GaAs DH structures on moly were observed to improve at higher growth temperatures, similar to those on glass substrates. In this initial study, we have obtained the longer lifetimes at a growth temperature of ~775°C. Also, Se-doping was found beneficial for obtaining longer lifetimes. The data shown in Table 4 suggests the possible passivation of grain-boundaries with a group-VI dopant like Se, in obtaining longer minority-carrier lifetimes. The actual minority-carrier lifetimes in GaAs DH structures could be higher than those indicated in Table 4, considering that we have observed J_{sc} values of as much as 18 mA/cm^2 under 1-sun AM1.5 conditions in GaAs solar cells with these materials. Work is in progress to accurately estimate the minority-carrier lifetimes by PL-decay in these polycrystalline thin films, under conditions compatible with flat-plate solar illumination levels [11].

Table 3. Typical effect of dopants on measured minority-carrier lifetime in poly GaAs grown on moly for similar growth conditions

Substrate	Dopant	Minority-carrier Lifetime (psec)
Moly	Se	147 - 171
	Si	29
	Zn	42

SUMMARY

Approaches to high-performance GaAs solar cells, with AM1.5 1-sun efficiencies in the range of 20%, on sub-mm grain-size optical-grade poly-Ge substrates have been described; these are relevant to the development of GaAs solar cells on other low-cost substrates as well. The attainment of high-performance GaAs solar cells on sub-mm poly-Ge substrates have recently motivated us to apply and extend the material and device optimization ideas for the development of high-efficiency GaAs solar cells on glass and on thin moly foils. We have reported preliminary materials development with these two substrates. The minority-carrier lifetimes measured in GaAs-AlGaAs DH structures on Ge-on-glass templates are rather low, in the range of 60 psec. We have observed longer minority-carrier lifetimes (~170 psec) in GaAs DH structures on molybdenum. Initial results indicate that Se-doping of the GaAs thin-films on moly lead to longer lifetimes, consistent with our current understanding of p^+-n GaAs solar cells on poly-Ge substrates with Se-doping for the n-base region.

ACKNOWLEDGMENTS

This work has been performed under Subcontract No. YAL-3-1-3357-03 from NREL with Dr. Robert McConnell as the technical monitor.

REFERENCES

1) S.R. Kurtz, J.M. Olson, and A. Kibbler, Proc. of 21st IEEE Photovoltaic Specialists Conf. (IEEE Press, NY, 1990) p.138.

2) J.M. Olson, S.R. Kurtz, A.G. Kibbler, and P. Faine, Proc. of the 21st IEEE Photovoltaic Specialists Conf. (IEEE Press, NY, 1990) p.24.

3) M. Green and K. Emery, Progress in Photovoltaics: Research and Applications, Vol. 2, 231-234 (1994).

4) K.P. Pande, D. H. Reep, S.K. Shastry, A.S. Weiner, J.M. Borrego and S.K. Ghandhi, IEEE Trans. on Electron Devices, Vol. ED-27, 635 (1980).

5) S. Chu, T. Chu, and Y.T. Lee, IEEE Trans. on Elec. Dev., ED-27, 640 (1980).

6) R. Venkatasubramanian, M.L. Timmons, P.R. Sharps, and J.A. Hutchby, Proc. of the 23rd IEEE Photovoltaic Specialists Conf. (IEEE Press, NY, 1993) p.691.

7) R. Venkatasubramanian, B. O'Quinn, J. Hills, D. Malta, M.L. Timmons, J.A. Hutchby, R. Ahrenkiel and B.M. Keyes, in Proceedings of the 12th NREL Photovoltaic Program Review, Ed. by H.S. Ullal and C.E. Witt, AIP Conf. Proc. No. 353 (AIP, Denver, CO) 1995.

8) R. Venkatasubramanian, B. O'Quinn, J. Hills, P.R. Sharps, M.L. Timmons, J.A. Hutchby, H. Field, R. Ahrenkiel and B.M. Keyes, Proc. of the 25th IEEE Photovoltaic Specialists Conf. (IEEE Press, NY, 1995) p.31.

9) H. Atwater, Private Communication.

10) R. Ahrenkiel and R. Venkatasubramanian, To be published.

11) B. Keyes, Work in progress.

LIST OF PARTICIPANTS

Mansur Abdukhanov
Kompozit
4, Pionerskaya str.
Kozolev, Moscow Region
Kozolev 141070

Richard Ahrenkiel
National Renewable Energy Lab
1617 Cole Boulevard
Golden, CO 80401
Phone: (303)-384-6670
Fax: (303)-384-6604
Email: richard.ahrenkiel@nrel.gov

Dave Albin
National Renewable Energy Lab
1617 Cole Boulevard
Golden, CO 80402
Phone: (303)-384-6550
Fax: (303)-384-6430
Email: dalbin@nrel.nrel.gov

A. Paul Alivisatos
University of California, Berkeley
Dept. of Chemistry
B62 Hildebrand Hall #1460
Berkeley, CA 94720-1460
Phone: (510) 643-7371
Fax: (510) 642-6911
Email: alivis@uclink4.berkeley.edu

Wayne Anderson
SUNY at Buffalo
Dept. of ECE
217C Bonner Hill
Buffalo, NY 14260
Phone: (716) 645-2422 x1215
Fax: (716) 645-5964
Email: eleander@ubvms.cc.buffalo.edu

Hideo Aoyama
Photovoltaic Business Development
520-25 Omuro, Ami-machi
Inashiki-gun
Ibaraki-ken 300-03
Email: hiaoyama@mb.infoweb.or.jp

Harry Atwater
California Institute of Technology
1201 E. California Blvd.
Watson Labs 128-95
Pasadena, CA 91125
Phone: (818) 395-2197
Fax: (818) 795-7258

John Benner
National Renewable Energy Lab
1617 Cole Boulevard
SERF
Golden, CO 80401
Phone: (303)-384-6496
Fax: (303) 384-6481
Email: john_benner@nrel.gov

Andreas Bett
Fraunhofer Solar Energy Systems
Oltmannsstr 22
Freiburg D-79100

Bharat Bhargava
Ministry of Non-Con. Energy Sources
Block 14
CGO Complex, Lodhi Road
New Delhi 110003

Rambabu Bobba
Southern Unviersity
Baton Rouge, LA 70813
Phone: (504) 771-4130
Fax: (504) 771-2324
Email: rambabu@stark-phys.subv.edu

John Bockris
Texas A&M University
Dept. of Chemistry
College Station, TX 77843-3255
Phone: (409) 847-8861
Fax: (409) 845-4205
Email: bockris@chemvx.tamu.edu

Terry Brog
Golden Photon, Inc.
4545 McIntyre Street
Golden, CO 80403
Phone: (303) 271-7167
Fax: (303) 271-7410

Vladmir Bulovic
Princeton University
Dept. of Electrical Engineering
Princeton, NJ 08544
Phone: (609) 258-6688
Fax: (609) 258-1954

David Carlson
Solarex
826 NewtownYardley Road
Newtown, PA 18940
Phone: (215)-860-0902
Fax: (215) 860-2986
Email: dcarlson@solarex.com

Henry Chase
U.S. Flywheel Systems
1125 Business Center Circle
Newbury Park, CA 91320
Phone: (805) 375-8433
Fax: (805) 375-8432
Email: henryv.chase@prodigy.net

George Cody
Exxon Corporate Research
Route 22 East
Clinton Township
Annandale, NJ 08801
Phone: (908) 730-3022
Fax: (908) 930-3314
Email: gdcody@erenj.com

James Coleman
Monsanto Company
800 N. Lindbergh Blvd.
St. Louis, MO 63167
Phone: (314) 694-4585
Fax: (314) 694-3688
Email: jpcole@ccmail.monsanto.com

George Collins
Colorado State University
Dept. of Elec. Engineering
B117 Engr. Res. Center
Ft. Collins, CO 80523
Phone: (970) 491-8513
Fax: (970) 491-8316

Alexander Converse
Physica
P.O. Box 1349
Madison, WI 53701-1349
Phone: (608) 243-1780
Fax: (608) 243-1781
Email: HiPhysica@aol.com

Timothy Coutts
National Renewable Energy Lab
1617 Cole Boulevard
Golden, CO 80401
Phone: (303)-384-6561
Fax: (303) 384-6430
Email: tcoutts@nrel.nrel.gov

Richard Curry
PV Insider's Report
1011 W. Colorado Blvd.
Dallas, TX 75208
Phone: (214)-942-5248
Fax: (214)-942-5248
Email: pvir@nkn.net

Lawrence Curtin
Writer
215 Cranwood Drive
Key Biscayne, FL 33149
Phone: (305) 365-5825
Fax: (305) 365-0732
Email: lfcurtin@sprynet.com

420

Al Czanderna
National Renewable Energy Lab
1617 Cole Boulevard
Golden, CO 80401
Phone: (303)-384-6460
Fax: (303) 384-6530
Email: al_czanderna@nrel.gov

Greg Davis
Terrasun
4541 S. Butterfield Drive
Tucson, AZ 85714
Phone: 520)-512-1995
Fax: (520)-512-1997
Email: gdavis1608@aol.com

Ralph Dawson
Sandia National Labs
P.O. Box 5800
Dept. 1113, MS 0601
Albuquerque, NM 87185-0601
Phone: (505) 845-8920
Fax: (505) 844-3211
Email: rdawson@sandia.gov

Satyen Deb
National Renewable Energy Lab
1617 Cole Boulevard
Golden, CO 80401
Phone: (303)-384-6405
Fax: (303)-384-6481
Email: satyen_deb@nrel.gov

Richard DeBlasio
National Renewable Energy Lab
1617 Cole Boulevard
Golden, CO 80401
Phone: (303) 384-6452
Fax: (303) 384-6490
Email: deblasid@tcplink.nrel.gov

Lisa Dignard-Bailey
CANMET
1615 Boul. Lionel Boulet
Box 4800
Varennes, Quebec J3X 1S6
Phone: (514) 652-5161
Fax: (514) 652-5177
Email: lisa.dignard@nrcan.gc.ca

James Dunn
Ctr. for Technology Commercialization
1400 Computer Drive
Westborough, MA 01581-5043
Phone: (508) 870-0042
Fax: (508) 366-0101
Email: jdunn@ctc.org

Al Federici
U.S. Flywheel Systems
1125 Business Center Circle
Newbury Park, CA 91320
Phone: (805) 375-8433
Fax: (805) 375-8432

Suzanne Ferrere
National Renewable Energy Lab
1617 Cole Boulevard
Golden, CO 80401
Phone: (303) 384-6686
Fax: (303) 384-6655
Email: sferrere@nrel.nrel.gov

Halden Field
National Renewable Energy Lab
1617 Cole Boulevard
Golden, CO 80401
Phone: (303)-384-6685
Fax: (303) 384-6604
Email: halden_field@nrel.gov

Lewis Fraas
JX Crystals, Inc.
1105 12th Avenue, N.W.
Suite A2
Issaquah, WA 98027
Phone: (206)-392-5237
Fax: (206)-392-7303

Jonathan Francis
Swarthmore College
500 College Avenue
Dept. of Engineering
Swarthmore, PA 19081-1397
Phone: (610) 690-3566
Fax: (610) 328-8082
Email: jfranci1@swarthmore.edu

Arthur Frank
National Renewable Energy Lab
1617 Cole Blvd.
Golden, CO 80401
Phone: (303)-384-6262
Fax: (303)-384-6150
Email: afrank@nrel.gov

Daniel Friedman
National Renewable Energy Lab
1617 Cole Boulevard
Golden, CO 80401
Phone: (303)-384-6472
Fax: (303)-384-6531
Email: friedman@nrel.gov

Hiroshi Fujioka
University of Tokyo
7-3-1 Hongo
Bunkyo-ku
Tokyo 113
Email: fujioka@sr.t.u-tokyo.ac.jp

Charles Gay
Midwest Research Institute
425 Volker Boulevard
Kansas City, MO 64110-2299

James Gee
Sandia National Laboratories
MS-0752
Albuquerque, NM 87185
Phone: (505) 844-7812
Fax: (505) 844-6541
Email: jmgee@sandia.gov

David Ginley
National Renewable Energy Lab
1617 Cole Boulevard
Golden, CO 80401
Phone: (303)-384-6573
Fax: (303)-384-6430

Amit Goyal
Oak Ridge National Laboratory
P.O. Box 2008
Bldg. 4500-S, MS-6116
Oak Ridge, TN 37831-6116
Phone: (423)-574-1587
Fax: (423)-574-7659

Michael Grätzel
Swiss Fed. Inst. of Technology
EPFL, CH-1015 Lausanne
Lausanne CH-1015
Email: michael.graetzel@icp.dc.epfl.ch

Brian Gregg
National Renewable Energy Lab
1617 Cole Blvd.
Golden, CO 80401
Phone: (303)-384-6635
Fax: (303) 384-6655
Email: bgregg@nrel.nrel.gov

Robert H. Hammond
Stanford University
Hansen Physics Labs
Stanford, CA 94305
Phone: (415) 723-0169
Fax: (415) 725-2533
Email: hammond@cc.stanford.edu

Mark Hanna
National Renewable Energy Lab
1617 Cole Boulevard
Golden, CO 80401
Phone: (303) 384-6620
Fax: (303) 384-6655
Email: mhanna@nrel.nrel.gov

Jack Hanoka
Evergreen Solar, Inc.
211 Second Avenue
Waltham, MA 02154
Phone: (617)-890-7117
Fax: (617) 890-7141

Joshua Hill
Texas Southern University
School of Technology
3100 Cleburne Avenue
Houston, TX 77004
Phone: (713) 313-7007
Fax: (713) 313-1853
Email: tchajxhill@tsu.edu

James Hutchby
Research Triangle Institute
3040 Cornwallis Road
P.O. Box 12194
Research Triangle Pk, NC 27709
Phone: (919)-541-5931
Fax: (919)-541-6515

Osamu Ikki
Resources Total System Co.
2 Floor Kariya Bldg
2-7-11 Shinkawa Chuo-ku
Tokyo 104

Eric Jones
Sandia National Laboratories
Box 5800, MS-0601
Dept. 1113
Albuquerque, NM 87185-0601
Phone: (505) 844-8752
Fax: (505) 844-3211
Email: edjones@sandia.gov

Juris Kalejs
ASE Americas, Inc.
4 Suburban Park Drive
Billerica, MA 01821-3980
Phone: (508) 667-5900 x293
Fax: (508) 663-7555
Email: jpk.asepv@aol.com

Michael Kardauskas
ASE Americas, Inc.
4 Suburban Park Drive
Billerica, MA 01821-3980
Phone: (508)-667-5900 x239
Fax: (508)-663-2868

Victor Kaydanov
Golden Photon, Inc.
4545 McIntyre Street
Golden, CO 80403
Phone: (303) 271-7167
Fax: (303) 271-7410

Lawrence Kazmerski
National Renewable Energy Lab
1617 Cole Blvd.
Golden, CO 80401
Phone: (303) 384-6600
Fax: (303) 384-6604
Email: kaz@nrel.gov

Rahim Khoie
UNLV
4505 Maryland Parkway
Box 454026
Las Vegas, NV 89154-4026
Phone: (702) 895-3187
Fax: (702) 895-4075
Email: ark@ee.univ.edu

Lionel Kimerling
MIT
77 Massachusetts Avenue
Room 13-4118
Cambridge, MA 02139
Phone: (617) 253-5383
Fax: (617)-253-6782
Email: lekim@mit.edu

Richard King
Siemens Solar Industries
4650 Adohr Lane
PO Box 6032
Carmarillo, CA 93011
Phone: (805)-388-6263
Fax: (805)-388-6580

Sarah Kurtz
National Renewable Energy Lab
1617 Cole Boulevard
Golden, CO 80401
Phone: (303)-384-6475
Fax: (303)-384-6531
Email: sarahk@nrel.gov

Ron Larson
Solar Today
21547 Mountsfield Drive
Golden, CO 80401
Phone: (303)-526-9629
Fax: (303)-526-9629
Email: larcon@csn.net

423

Kyle Lefkoff
Boulder Ventures
1634 Walnut Street
Suite 301
Boulder, CO 80302

Xiaonan Li
National Renewable Energy Lab
1617 Cole Boulevard
Golden, CO 80401
Phone: (303)-384-6428
Fax: (303)-384-6430
Email: xli@nrel.gov

Guang Lin
Solarex
826 Newtown-Yardley Road
Newton, PA 18940
Phone: (215) 860-0902 x215
Fax: (215) 860-2986
Email: glin@solarex.com

Yunosuke Makita
Electrotechnical Laboratory
Umezono 1-1-4
Tsukuba-shi
Ibanagi-ken 305
Email: makita@etl.go.jp

Bill Marshall
National Renewable Energy Lab
1617 Cole Boulevard
Golden, CO 80401
Phone: (303) 275-3081
Fax: (303) 275-3097

Angelo Mascarenhas
National Renewable Energy Lab
1617 Cole Boulevard
Golden, CO 80401
Phone: (303) 384-6608
Fax: (303) 384-6481
Email: amascar@nrel.gov

Michael Mauk
Astro Power Inc
Solar Park
Newark, DE 19716-2000
Phone: (302)-366-0400
Fax: (302)-368-6474

Jeffrey Mazer
U.S. Department of Energy
1000 Independence Avenue, S.W.
EE-11
Washington, DC 20585
Phone: (202)-586-2455
Fax: (202)-586-5127
Email: jeffrey.mazwe@hq.doe.gov

Robert McConnell
National Renewable Energy Lab
1617 Cole Boulevard
Golden, CO 80401
Phone: (303)-384-6419
Fax: (303)-384-6481
Email: robert_mcconnell@nrel.gov

Andreas Meier
National Renewable Energy Lab
1617 Cole Boulevard
Golden, CO 80401

Gerald Meyer
Johns Hopkins University
Dept. of Chemistry, Remsen Hall
3400 Charles St.
Baltimore, MD 21218
Phone: (410)-516-7319
Fax: (410)-516-8420

Hartwig Modrow
Bonn University/LSU
CAMD
Baton Rouge, LA 70813
Phone: (504) 388-6565
Fax: (504) 771-2324
Email: hmodrow@camd.lsu.edu

Jeffrey Nelson
Sandia National Laboratories
P.O. Box 5800, MS-0601
Org. 1113
Albuquerque, NM 87185
Phone: (505) 844-4395
Fax: (505) 844-3211

Jenny Nelson
Imperial College
Dept. of Physics
Prince Consort Road
London SW7 2BZ
Email: jennyn@ic.ac.uk

David Norton
Oak Ridge National Laboratory
P.O. Box 2008
MS 6056
Oak Ridge, TN 37831-6056
Phone: (423) 574-5965
Fax: (423) 576-3676
Email: ntn@ornl.gov

Rommel Noufi
National Renewable Energy Lab
1617 Cole Boulevard
Golden, CO 80401
Phone: (303)-384-6510
Fax: (303)-384-6430
Email: noufi@nrel.gov

Arthur Nozik
National Renewable Energy Lab
1617 Cole Boulevard
Golden, CO 80401
Phone: (303)-384-6603
Fax: (303)-384-6655
Email: anozik@nrel.nrel.gov

Brian O'Regan
University of Washington
Chemical Engineering
P.O. Box 1750
Seattle, WA 98195-1750
Phone: (206) 685-2289
Fax: (206) 543-3778
Email: bcor@u.washington.edu

Joseph Olbermann
Unique Mobility, Inc.
Golden, CO
Phone: (303) 278-2002 x150

Bruce Parkinson
Colorado State University
Dept. of Chemistry
Ft. Collins, CO 80523
Phone: (970)-491-0504
Fax: (970)-491-1801
Email: parkinson@mail.chem.colostate.edu

Terry Peterson
EPRI
3412 Hillview Ave
Palo Alto, CA 94304
Phone: (415)-855-2594
Fax: (415)-855-8501
Email: tpeterso@epri.com

Tom Pieronek
TRW
One Space Park
MS:R4/2004
Redondo Beach, CA 90278
Phone: (310) 814-7468
Fax: (310) 814-4941
Email: tom.pieronek@trw.com

Thanu Pillai
Miss. Valley State University
Box 5004
Itta Bene, MS 38941
Phone: (601) 254-3385
Fax: (601) 254-3392
Email: pillai@netdoor.com

Hans-Joachim Queisser
Max Planck Festkorperforschung
Heisenbergstrasse 1
Stuttgart D-70569
Email: queisser@quasix.mpi-
stuttgart.mpg.de

David Reamer
Applied EPI
1290 Hammond Road
St. Paul, MN 55110
Phone: (612) 653-0488
Fax: (612) 653-0725
Email: daver@epimbe.com

Kitt Reinhardt
U.S. Air Force
920 Louisiana S.E., Apt. #35
Albuquerque, NM 87108
Phone: (505) 846-2637

Angus Rockett
University of Illinois
1-107 Engineering Sciences Bld
1101 W. Springfield Avenue
Urbana, IL 61801
Phone: (217)-333-0417
Fax: (217)-244-1630

Ajeet Rohatgi
Georgia Tech University
Electrical & Computer Eng.
Atlanta, GA 30332-0250
Phone: (404)-894-7692
Fax: (404) 894-5934

Fraser Russell
University of Delaware
IEC Building
Wyoming Road
Newark, DE 19716
Phone: (302) 831-6224
Fax: (302) 831-6226
Email: russell@che.udel.edu

Peter Searson
The Johns Hopkins University
102 Maryland Hall
Dept. of Materials Science &
Engineering
Baltimore, MD 21218
Phone: (410) 516-8774
Fax: (410) 516-5273
Email: searson@jhu.edu

Harvey Serreze
Spire Corporation
1 Patriots Park
Bedford, MA 01730
Phone: (617) 275-6000
Fax: (617) 275-7470
Email: spire.corp@channel1.com

Peter Sheldon
National Renewable Energy Lab
1617 Cole Boulevard
Golden, CO 80401
Phone: (303)-384-6533
Fax: (303)-384-6430
Email: peter_sheldon@nrel.gov

Eveny Sholomentsev
Kompozit
4, Pionerskaye str.
Korolev, Moscow Regioin
Korolev 141070

Jerry Smith
U.S. Department of Energy
Materials Sciences, ER-13/GTN
19901 Germantown Road
Germantown, MD 20874-1290
Phone: (301) 903-3426
Fax: (301) 903-9513
Email: jerry.smith@oer.doe.gov

Byron Stafford
National Renewable Energy Lab
1617 Cole Boulevard
Golden, CO 80401
Phone: (303) 384-6426
Fax: (303) 384-6490

Jack Stone
National Renewable Energy Lab
1617 Cole Boulevard
Golden, CO 80401
Phone: (303)-384-6470
Fax: (303)-384-6481
Email: jack_stone@nrel.gov

Thomas Surek
National Renewable Energy Lab
1617 Cole Boulevard
Golden, CO 80401
Phone: (303)-384-6471
Fax: (303)-384-6530
Email: surekt@tcplink.nrel.gov

Richard Swanson
SunPower Corporation
430 Indio Way
Sunnyvale, CA 94086
Phone: (408)-991-0908
Fax: (408)-739-7713
Email: rswanson1@aol.com

Hirose Takamitsu
Ministry of Posts & Telecommunications
1-3-2 Kasumigaseki
Chiyoda-ku
Tokyo 100-90
Email: t2-hitros@mpt.go.jp

Craig Taylor
University of Utah
Department of Physics
201 James Fletcher Bldg
Salt Lake City, UT 84112
Phone: (801)-581-3538
Fax: (801)-581-4246
Email: craig@mail.physics. utah.edu

John Thornton
National Renewable Energy Lab
1617 Cole Boulevard
Golden, CO 80401
Phone: (303)-384-6469
Fax: (303)-384-6490
Email: thorntoj@tcplink.nrel.gov

Lawrence Thorpe
Lawrence I. Thorpe & Co.
41 Woodstock Road
Roswell, GA 30075
Phone: (770) 993-1463
Fax: (770) 993-1484
Email: lit@mindspring.com

Kon Toshinori
Ministry Posts & Telecommunications
1-6-19, Azabu-dai
Minato-ku
Tokyo 106
Email: kon@iptp.go.jp

Gavin Tulloch
Sustainable Tech. Australia Ltd.
11 Aurora Avenue
P.O. Box 6212
Queanbeyan 2620
Email: sta@netinfo.com.au

John Tuttle
National Renewable Energy Lab
1617 Cole Boulevard
Golden, CO 80401
Phone: (303)-384-6534
Fax: (303) 384-6531

Harin Ullal
National Renewable Energy Lab
1617 Cole Boulevard
Golden, CO 80401
Phone: (303)-384-6486
Fax: (303)-384-6430
Email: ullal@nrel.gov

Rama Venkatasubramanian
Research Triangle Institute
3040 Cornwallis Road
Research Triangle Pk, NC 27709
Phone: (919) 541-6889
Fax: (919) 541-6515
Email: rama@es.rti.org

Bolko von Roedern
National Renewable Energy Lab
1617 Cole Boulevard
Golden, CO 80401
Phone: (303)-384-6480
Fax: (303) 384-6531
Email: vonroedb@tcplink.nrel.gov

Fritz Wald
ASE Americas, Inc.
4 Suburban Park Drive
Billerica, MA 01821-3980
Phone: (508) 358-4476 x286
Fax: (508) 358-4476
Email: fuwa@asn.com

Suhuai Wei
National Renewable Energy Lab
1617 Cole Boulevard
Golden, CO 80401
Phone: (303)-384-6666
Fax: (303)-275-4053

Ed Witt
National Renewable Energy Lab
1617 Cole Blvd.
Golden, CO 80401
Phone: (303)-384-6402
Fax: (303) 384-6481

John Wohlgemuth
Solarex
630 Solarex Ct.
Frederick, MD 21703
Phone: (301)-698-4375
Fax: (301)-698-4201
Email: jwohlgemuth@solarex.com

Eli Yablonovitch
UCLA Nano-Electronics Lab
Room 66-147K Engineering IV Bldg.
P.O. Box 951594
Los Angeles, CA 90095-1594
Phone: (310)-206-2240
Fax: (310) 206-4685
Email: eliy@ucla.edu

Takao Yonehara
Canon Inc.
Device Development Center
6770 Tamura
Hiratsuka 254

Henry Yoo
TECSTAR Inc.
15251 E. Don Julian RD.
City of Industry, CA 91745-1002
Phone: (818)-968-6581
Fax: (818)-336-8694

Gang Yu
UNIAX Corporation
6780 Cortona Drive
Santa Barbara, CA 93117
Phone: (805) 562-9293
Fax: (805) 562-9144
Email: gangyu@uniax.com

Arie Zaban
National Renewable Energy Lab
1617 Cole Boulevard
Golden, CO 80401
Phone: (303) 384-6190
Fax: (303) 384-6655
Email: zabana@tcplink.nrel.gov

Shengbai Zhang
National Renewable Energy Lab.
1617 Cole Boulevard
Golden, CO 80401
Phone: (303) 384-6622

Ken Zweibel
National Renewable Energy Lab
1617 Cole Boulevard
Golden, CO 80401
Phone: (303)-384-6441
Fax: (303)-384-6430
Email: zweibelk@tcplink.nrel.gov

428

AUTHOR AND SUBJECT INDEX

D

E

F

437

438

AIP Conference Proceedings

No. 324 Twelfth Symposium on Space Nuclear
Power and Propulsion
(Albuquerque, NM 1995)   94-73603   1-56396-427-9

No. 325 Conference on NASA Centers for
Commercial Development of Space
(Albuquerque, NM 1995)   94-73604   1-56396-431-7

No. 326 Accelerator Physics at the
Superconducting Super Collider
(Dallas, TX 1992-1993)   94-73609   1-56396-354-X

No. 327 Nuclei in the Cosmos III
Third International Symposium
on Nuclear Astrophysics
(Assergi, Italy 1994)   95-75492   1-56396-436-8

No. 328 Spectral Line Shapes, Volume 8
12th ICSLS
(Toronto, Canada 1994)   94-74309   1-56396-326-4

No. 329 Resonance Ionization Spectroscopy 1994
Seventh International Symposium
(Bernkastel-Kues, Germany 1994)   95-75077   1-56396-437-6

No. 330 E.C.C.C. 1 Computational Chemistry
F.E.C.S. Conference
(Nancy, France 1994)   95-75843   1-56396-457-0

No. 331 Non-Neutral Plasma Physics II
(Berkeley, CA 1994)   95-79630   1-56396-441-4

No. 332 X-Ray Lasers 1994
Fourth International Colloquium
(Williamsburg, VA 1994)   95-76067   1-56396-375-2

No. 333 Beam Instrumentation Workshop
(Vancouver, B. C., Canada 1994)   95-79635   1-56396-352-3

No. 334 Few-Body Problems in Physics
(Williamsburg, VA 1994)   95-76481   1-56396-325-6

No. 335 Advanced Accelerator Concepts
(Fontana, WI 1994)   95-78225   1-56396-476-7
                                                    (set)
                                                    1-56396-474-0
                                                    (Book)
                                                    1-56396-475-9
                                                    (CD-Rom)

No. 336 Dark Matter
(College Park, MD 1994)   95-76538   1-56396-438-4

No. 337 Pulsed RF Sources for Linear Colliders
(Montauk, NY 1994)   95-76814   1-56396-408-2

No. 338 Intersections Between Particle and
Nuclear Physics 5th Conference
(St. Petersburg, FL 1994)   95-77076   1-56396-335-3

	Title	L.C. Number	ISBN
No. 339	Polarization Phenomena in Nuclear Physics Eighth International Symposium (Bloomington, IN 1994)	95-77216	1-56396-482-1
No. 340	Strangeness in Hadronic Matter (Tucson, AZ 1995)	95-77477	1-56396-489-9
No. 341	Volatiles in the Earth and Solar System (Pasadena, CA 1994)	95-77911	1-56396-409-0
No. 342	CAM -94 Physics Meeting (Cacun, Mexico 1994)	95-77851	1-56396-491-0
No. 343	High Energy Spin Physics Eleventh International Symposium (Bloomington, IN 1994)	95-78431	1-56396-374-4
No. 344	Nonlinear Dynamics in Particle Accelerators: Theory and Experiments (Arcidosso, Italy 1994)	95-78135	1-56396-446-5
No. 345	International Conference on Plasma Physics ICPP 1994 (Foz do Iguaçu, Brazil 1994)	95-78438	1-56396-496-1
No. 346	International Conference on Accelerator-Driven Transmutation Technologies and Applications (Las Vegas, NV 1994)	95-78691	1-56396-505-4
No. 347	Atomic Collisions: A Symposium in Honor of Christopher Bottcher (1945-1993) (Oak Ridge, TN 1994)	95-78689	1-56396-322-1
No. 348	Unveiling the Cosmic Infrared Background (College Park, MD, 1995)	95-83477	1-56396-508-9
No. 349	Workshop on the Tau/Charm Factory (Argonne, IL, 1995)	95-81467	1-56396-523-2
No. 350	International Symposium on Vector Boson Self-Interactions (Los Angeles, CA 1995)	95-79865	1-56396-520-8
No. 351	The Physics of Beams Andrew Sessler Symposium (Los Angeles, CA 1993)	95-80479	1-56396-376-0
No. 352	Physics Potential and Development of $\mu^+\mu^-$ Colliders: Second Workshop (Sausalito, CA 1994)	95-81413	1-56396-506-2
No. 353	13th NREL Photovoltaic Program Review (Lakewood, CO 1995)	95-80662	1-56396-510-0
No. 354	Organic Coatings (Paris, France, 1995)	96-83019	1-56396-535-6
No. 355	Eleventh Topical Conference on Radio Frequency Power in Plasmas (Palm Springs, CA 1995)	95-80867	1-56396-536-4

	Title	L.C. Number	ISBN
No. 372	Beam Dynamics and Technology Issues for + - Colliders 9th Advanced ICFA Beam Dynamics Workshop (Montauk, NY, 1995)	96-84189	1-56396-554-2
No. 373	Stress-Induced Phenomena in Metallization (Palo Alto, CA 1995)	96-84949	1-56396-439-2
No. 374	High Energy Solar Physics (Greenbelt, MD 1995)	96-84513	1-56396-542-9
No. 375	Chaotic, Fractal, and Nonlinear Signal Processing (Mystic, CT 1995)	96-85356	1-56396-443-0
No. 376	Chaos and the Changing Nature of Science and Medicine: An Introduction (Mobile, AL 1995)	96-85220	1-56396-442-2
No. 377	Space Charge Dominated Beams and Applications of High Brightness Beams (Bloomington, IN 1995)	96-85165	1-56396-625-7
No. 378	Surfaces, Vacuum, and Their Applications (Cancun, Mexico 1994)	96-85594	1-56396-418-X
No. 379	Physical Origin of Homochirality in Life (Santa Monica, CA 1995)	96-86631	1-56396-507-0
No. 380	Production and Neutralization of Negative Ions and Beams / Production and Application of Light Negative Ions (Upton, NY 1995)	96-86435	1-56396-565-8
No. 381	Atomic Processes in Plasmas (San Francisco, CA 1996)	96-86304	1-56396-552-6
No. 382	Solar Wind Eight (Dana Point, CA 1995)	96-86447	1-56396-551-8
No. 383	Workshop on the Earth's Trapped Particle Environment (Taos, NM 1994)	96-86619	1-56396-540-2
No. 384	Gamma-Ray Bursts (Huntsville, AL 1995)	96-79458	1-56396-685-9
No. 385	Robotic Exploration Close to the Sun: Scientific Basis (Marlboro, MA 1996)	96-79560	1-56396-618-2
No. 386	Spectral Line Shapes, Volume 9 13th ICSLS (Firenze, Italy 1996)		1-56396-656-5
No. 387	Space Technology and Applications International Forum (Albuquerque, NM 1997)	96-80254	1-56396-679-4 (Case set) 1-56396-691-3 (Paper set)
No. 388	Resonance Ionization Spectroscopy 1996 Eighth International Symposium (State College, PA 1996)	96-80324	1-56396-611-5

	Title	L.C. Number	ISBN
No. 389	X-Ray and Inner-Shell Processes 17th International Conference (Hamburg, Germany 1996)	96-80388	1-56396-563-1
No. 390	Beam Instrumentation Proceedings of the Seventh Workshop (Argonne, IL 1996)	97-70568	1-56396-612-3
No. 391	Computational Accelerator Physics (Williamsburg, VA 1996)	97-70181	1-56396-671-9
No. 392	Applications of Accelerators in Research and Industry: Proceedings of the Fourteenth International Conference (Denton, TX 1996)	97-71846	1-56396-652-2
No. 393	Star Formation Near and Far Seventh Astrophysics Conference (College Park, MD 1996)	97-71978	1-56396-678-6
No. 394	NREL/SNL Photovoltaics Program Review Proceedings of the 14th Conference— A Joint Meeting (Lakewood, CO 1996)	97-72645	1-56396-687-5
No. 395	Nonlinear and Collective Phenomena in Beam Physics (Arcidosso, Italy 1996)	97-72970	1-56396-668-9
No. 396	New Modes of Particle Acceleration— Techniques and Sources (Santa Barbara, CA 1996)	97-72977	1-56396-728-6
No. 397	Future High Energy Colliders (Santa Barbara, CA 1997)	97-73333	1-56396-729-4
No. 398	Advanced Accelerator Colliders Seventh Workshop (Lake Tahoe, CA 1996)	97-72788	1-56396-697-2 (set) 1-56396-727-8 (cloth) 1-56396-726-X (CD-Rom)
No. 399	The Changing Role of Physics Departments (College Park, MD 1996)		1-56396-698-0
No. 400	High Energy Physics First Latin Symposium (Yucatan, México 1996)	97-73971	1-56396-686-7
No. 401	Thermophotovoltaic Generation of Electricity Third NREL Conference (Colorado Springs, CO 1997)	97-74374	1-56396-734-0
No. 403	Radio Frequency Power in Plasmas 12th Topical Conference (Savannah, GA 1997)	97-74472	1-56396-709-X
No. 404	Future Generations Photovoltaic Technologies First NREL Conference (Denver, CO 1997)	97-74386	1-56396-704-9